ANALYTICAL TECHNIQUES FOR MATERIAL CHARACTERIZATION

COVER
Field Ion Micrograph of Tungsten Metal ⟨110⟩ Orientation (Large Arrows) Shows a Grainboundary and {Small Arrows} Shows a Dislocation at the Centre. Resolution is ~ 2–3 A.
[Micrograph by: Chand Patel, Materials Evaluation Laboratory, Baton Rouge, Louisiana, 70810, U.S.A.]

WSPC—COSTED SERIES IN EMERGING TECHNOLOGY

ANALYTICAL TECHNIQUES FOR MATERIAL CHARACTERIZATION

Proceedings of the International Workshop
Baton Rouge, USA, 11-16 May 1987

Editors

W.E. Collins
Department of Physics
Southern University
Baton Rouge, LA, USA

B.V.R. Chowdari
Department of Physics
National University of Singapore
Singapore

S. Radhakrishna
Department of Physics
Indian Institute of Technology
Madras, India

COMMITTEE ON SCIENCE & TECHNOLOGY
IN DEVELOPING COUNTRIES

Published by

World Scientific Publishing Co. Pte. Ltd.
P.O. Box 128, Farrer Road, Singapore 9128

U. S. A. office: World Scientific Publishing Co., Inc.
687 Hartwell Street, Teaneck NJ 07666, USA

Library of Congress Cataloging-in-Publication data is available.

INTERNATIONAL WORKSHOP ON ANALYTICAL TECHNIQUES FOR MATERIAL CHARACTERIZATION

Copyright © 1987 by World Scientific Publishing Co Pte Ltd.

All rights reserved. This book, or parts thereof, may not be reproduced in any form or by any means, electronic or mechanical, including photocopying, recording or any information storage and retrieval system now known or to be invented, without written permission from the Publisher.

ISBN 9971-50-511-8
      9971-50-512-6 pbk

Printed in Singapore by Utopia Press.

WORKSHOP ORGANIZATION

ORGANIZERS
Department of Physics, Southern University, Baton Rouge, La, USA.
Committee on Science and Technology in Developing Countries, Madras, India.

INTERNATIONAL ADVISORY COMMITTEE
W. Eugene Collins	(U.S.A.) Chairman
James Barefield	(U.S.A.)
B.V.R. Chowdari	(Singapore)
Kuang Ding Bo	(China)
S. Radhakrishna	(India)
Chand Patel	(England)*

NATIONAL ADVISORY COMMITTEE
Tom Buck	(N.J.)
Sean P. McGlynn	(La.)
Steve McGuire	(Al.)
Ronald Mickens	(Ge.)
Norman Tolk	(Tn.)

LOCAL ORGANIZING COMMITTEE (DEPARTMENT OF PHYSICS)
Bobba Rambabu**	Ineatha Ruffin	K.H. Liu**	C.H. Yang**
Diola Bagayoko**	Z. Singh	Joseph Stewart**	Eloise Young
T. Wang**	Roosevetta Johnson	James Gist	R. Mohanty

WORKSHOP SECRETARY
Joyce P. Collins (Japco-Wecco Company)

WORKSHOP COORDINATOR
W. Eugene Collins

* Present Address : Baton Rouge, Louisiana
** Member of the Surface Physics and Material Science (SPAMS) Laboratory

ACKNOWLEDGEMENTS

In general, planning for an international workshop must start twelve to twenty-four months ahead of time. An opportunity to have this workshop was presented to me eleven months before the scheduled start date and a decision to host the workshop was not made until October, 1986 (seven months before the start date). Because of the lateness in planning, funding was a problem. Also, we found ourselves in the midst of a very busy academic-year schedule. Even so, we were able to do what was necessary to consumate a successful workshop. The success is due to the efforts of so many people and we thank them.

Once the decision had been made to host the workshop, two members of the international committee were extremely active in assisting me; they were B.V.R. Chowdari and S. Radhakrishna. Without our many conversations and correspondences, the workshop could not have been a success. Though much of the organizing work had been completed by March, 1987, we were very happy at that time to have Chand Patel to join our International Committee. He and the members of Materials Evaluation Laboratory played an important role in the success of the Workshop. Recognition must also be given to the other members of the national and international committee. They did well in meeting every request that I made of them; they worked from the very beginning. In fact, one member of the national committee, Tom Buck, did such an excellent job of contacting potential speakers who were nationally outstanding that I was unable to properly follow through with all of those contacts. It is good working with such excellent committee members.

Though the national and international committees played an important role in logistical planning, work must always be done by a local body. I therefore thank the department of physics for serving in that capacity. Special recognition must be given to five members who worked so tirelessly : K.H. Liu, C.H. Yang, I. Ruffin, J. Stewart, and B. Rambabu.

Due to the efforts of all those cited above, we had success. Yet there are others who must be recognized. Our conference secretary, Joyce P. Collins, was supplied by JAPCO-WECCO Company (SCIENTIFIC DIVISION) and performed in a superb manner. She did solicitations, planning, typing, record keeping, and report writing. She was and is doing a fantastic and indispensable job.

Of course, a conference of this type cannot be held without funding. For travel grants and some local support, we must thank our major sponsor, COSTED. For further significant financial support, we must recognize JAPCO-WECCO Scientific division, a scientific consulting and research firm. Thanks to Materials Evaluation Laboratory for sponsoring the reception on the first day of the meeting and Uniroyal Rubber Company for its helping support the banquet activities. These social functions provided an opportunity for enjoyment as well as informal and lively scientific discussions. Appreciation is given to many of the invited speakers and/or their institutions for providing their travel costs. Without such generosity, the workshop could not have materialized.

Thanks goes to all Southern University Staff who assisted us, especially to Dr. T. Bursh and the chemistry staff for their cooperation and use of facilities. We must also specifically mention the remarks given by Dean Martin and Vice Chancellor Spikes. You have our gratitude.

Finally, we must acknowledge the kind response to our solicitations from Pepsi Cola Bottling Company of Baton Rouge, Baton Rouge Little Theater and Mutual of Omaha Insurance Company.

With much gratitude,

W. Eugene Collins
Workshop Coordinator.

PREFACE

Preparation and characterization of materials of strategic importance are vitally significant for all countries. Materials are now the backbone of all applications and the range of materials required is staggering. Special high temperature materials, new superconducting materials which show superconductivity well above liquid nitrogen temperature (77K), optical fibers which are playing an increasingly important role in communications, and special high ionic conducting materials which are used in making solid state batteries for heart peacemakers are some of the new trends in material sciences. A need to develop manpower capabilities of understanding and utilizing these new materials is uppermost in the priorities of all developing countries. COSTED, the Committee on Science and Technology in Developing Countries of the International Council of Scientific Unions (ICSU) first formed in 1966 and organizes international workshops to promote exchange of ideas on latest trends in different scientific priority areas. This workshop on Analytical Techniques for Characterization of Materials was planned to present state-of-art reviews on different techniques used in characterizing important materials useful for a wide range of applications. Southern University, Baton Rouge, Louisiana, U.S.A. was chosen as the venue of the workshop as the University itself has been planning on initiating a major programme in materials science.

Over 60 participants from developing countries gathered at the Physics Department, Southern University, Baton Rouge, Louisiana during May 11-16, 1987, to have an intensive in depth discussion on the techniques used for characterizing materials used in various applications. The techniques used in characterizing materials are very varied and depend very strongly on the types of applications that are envisaged. Different techniques ranging from cheap, quick methods to sophisticated, costly and precise techniques are now well established. Invited review talks dealing with different techniques were delivered by several distinguished scientists . Several techniques for characterizing semiconductive materials, fast ion conducting glasses, optical fibres used in communications, nuclear materials etc were described. Besides the invited talks, several short contributions from the participants were also presented.

W.E. Collins
B.V.R. Chowdari
S. Radhakrishna

CONTENTS

Acknowledgments — vii
Preface — xi

I. INVITED LECTURES

Surface Science Probe Techniques and Their Application to Materials Characterization — 3
   Chand Patel

Fast Ion Conducting Glasses and Their Characterization — 15
   B. V. R. Chowdari & R. Gopalakrishnan

Characterization of Materials for Telecommunications — 35
   J. W. Mitchell

Applications of Rutherford Backscattering Studies — 43
   V. Lakshminarayana

Characterization of Nuclear Materials — 65
   C. K. Mathews

Compositional Characterization of Americium-Curium Mixtures — 79
   S. C. McGuire, D. E. Benker & J. E. Bigelow

Charge Densities : Comparison of Calculations and Experiment — 91
   Joseph Callaway

Pitfalls in the Numerical Integration of Differential Equations — 123
   Ronald E. Mickens

II. SHORT CONTRIBUTIONS

Structural Data from Solid-State Deuterium NMR Spectroscopy : A Karplus-Type Relationship for ^{2}H-C-C-X Torsion Angles — 147
   Leslie G. Butler

Characterization of Coating Materials and Coated Surfaces — 153
   Chand Patel & C. Cook

A Study of Laser Activated Gold-Silicon Interface Contact — 163
   Bill Dixon & Chand Patel

A Study of Gold-Strontium Flouride Interface Using
Surface Probe Techniques   185
    Chand Patel, R. Desbrandes & M. E. McConnaughhay

An XPS Study on the Structure of $Mg(PO_3)_2$-BaF_2-AlF_3
Glass System   197
    Chen Binjiang, Mi Qingzhou & Wang Shizhuo

An Investigation of Glass Structure on the $Mg(PO_3)_2$-BaF_2-AlF_3
System by Vibrational Spectroscopy   207
    Chen Binjiang, Mi Qingzhou & Wang Shizhuo

Additional Electron Beam Induced Voltage Contrast Due
to Skin Resistance   217
    Chen Boliang, Fang Xiaoming & Yu Jinbi

A Note on the Rutherford Back Scattering Technique   221
    W. Eugene Collins

Characterization of Gas-Nitrided Stainless Steel   225
    M. F. Chung & Y. K. Lim

X-Ray Structure Determination of $Fe_9Mn_6Al_5$ Precipitate
in a Fe-Mn-Al-Cr Alloy   233
    Tian-Huey Lu, Tseng-Fong Liu & Chi-Meen Wan

The Creation and Elimination of In Islands on Clean InP
Surfaces Studied by Electron Energy Loss Spectroscopy   243
    Xun Wang, Mingren Yu, Xiaoyuan Huo & Xiaofeng Jin

Performance Comparison of Capacitance Discharge Plasma (SPARK),
Inductively Coupled Plasma (ICP), and X-Ray Fluorescence (XRF)
Spectrometers in the Analysis of Austenitic Stainless Steels   261
    Claude R. Mount

Electrical and Dielectric Properties of $NaYF_4$ Thin Films   265
    Narasimha Reddy Katta

Determination of Traces of Sulphate Indirectly by Atomic
Absorption Spectrophotometer   271
    D. C. Parashar & A. K. Sarkar

Experimental Techniques for Characterization of Fast
Ion Conducting Materials   279
    P. Sathya Sainath Prasad & S. Radhakrishna

Material Characterization of $Ag_{1-x}Pb_xI$ Solid Solution
Ionic Conductor   317
    R. V. G. K. Sarma, P. Sathya Sainath Prasad & S. Radhakrishna

Advances in Fourier Transform Infrared Spectroscopy   335
    B. Rambabu & W. Eugene Collins

Characterization of Crystalline Phases Formed During Annealing
of Some Metallic Glasses Using X-Ray Diffraction   347
    S. B. Raju & G. Surya Prakasa Rao

Thermoluminescence Induced by TSFE — A New Analytical Tool   351
    S. Murali Dhara Rao, K. S. V. Nambi & M. P. Chougaonkar

Application of Fourier Transform Infrared (FT-IR) Spectroscopy
by Pyrolysis Technique to Polymeric Systems   363
    P. K. Seow, W. M. Yong & M. Mohinder Singh

Electrical Characterization of $Ag_4I_2WO_4$ Solid Electrolyte   373
    S. Austin Suthanthiraraj

Solid State Electrolytes and Intercalation Compounds of
High-Valence Ions   381
    Yu Wen-hai, Wang Da-zhi & Zhu Bin

Study of Phase Transition of Mordenite by IR and XRD   387
    Yuan Wang-zhi, Wang Da-zhi, Zhu Bin & Yu Wen-hai

Composite Pulse Magic Echo Sequence for the Excitation
of Three-Level Systems in Solid-State NMR   393
    Wang Dongsheng, Li Gengying & Wu Xuewen

List of Participants   401

I. INVITED LECTURES

SURFACE SCIENCE PROBE TECHNIQUES

AND

THEIR APPLICATION TO MATERIALS CHARACTERIZATION

CHAND PATEL

MATERIALS EVALUATION LABORATORY, INC.
17695 PERKINS ROAD
BATON ROUGE, LA 70810 U.S.A.
(504) 292-6070

ABSTRACT:

Although earlier "classical" methods for characterization of materials have been in existence for many years, modern techniques to analyze microstructural properties of materials have only been in practice since the late sixties. Presently, there are well over 300 techniques that have been used in some form or manner to characterize materials. In this review paper, techniques which are fairly common, as well as new ones that have made significant impact, will be described. A comparison of major advantages (capabilities), and disadvantages (limitations), will be discussed, including nondestructive testing, quantitative analysis, sensitivity, resolution, chemical state determination and speed with which data can be acquired and analyzed. Since no one technique alone can fully characterize a material, a new technique must offer a significant growth of the capabilities that exist already without setting undue limitations. (For example, high sensitivity with good spatial resolution.) This review will attempt to put a number of these new developments in perspective with regards to the more general techniques available for characterization of materials.

INTRODUCTION:

The world-wide explosion in information processing and in automation for industry is a direct result of the widespread availability of integrated circuits combined on small "chips" of silicon and recently Gallium Arsenide. In fields such as computers, communications and instrumentation, where performance is critical, speed means money. That is why the logical choice for future is characterization and processing of electronic material used in high speed devices. This demand will continue as we progress towards sub-micron level device technology.

In recent years, several new surface probes techniques for determining the elemental and chemical composition of solid surfaces and thin films have been developed. All of these techniques exhibit fundamental strengths and weaknesses, but some offer greater practical utility than others. X-ray photoelectron spectroscopy, also known as [XPS or ESCA], Auger electron spectroscopy, [AES], and Scanning Electron Microscopy [SEM], and numerous other techniques, have proven to be particularly valuable and are now being utilized to investigate a wide range of material problems. Since for total characterization, the needed information cannot be provided by any one technique alone, data from two or more techniques is required to solve many real-world material problems.

TECHNIQUES:

Recent research has exploited a number of new techniques to gather information on interface electronic and atomic structure, composition, reactivity and electrical properties. Surface spectroscopies such as AES [1], ultraviolet photoemission spectroscopy (UPS), ESCA (XPS), photon electron spectroscopy (PES), and field emission microscopy (FEM), measure the interface electronic structure and thereby give specific information about chemical bonding. When used in conjunction with depth profiling, composition and diffusion data are generated. Ion scattering and channeling methods give quantitative information on interface atomic structure, composition, and reactivity. SEM [5] is used to study surface and interface morphology, as well as to do channeling patterns. Many units are equipped with EDAX, which can be used to verify compositions. TEM yields information on phase and microstructure at atomic resolution [6]. Two of the newest techniques are surface EXAFS or SEXAFS (surface extended absorption fine structure measurements) and synchrotron radiation photoemission measurements. SEXAFS determines the nearest-neighbor environment of interfacial atoms. Synchrotron radiation photoemission measures band-bending and Schottky barrier heights at low metal coverages [7].

Clearly, all of these techniques are useful see Table 1. Another extremely valuable method which can be used in conjunction with many of the aforementioned research probe techniques is the in situ' formation of the silicide. Much of what is being studied — reactivity, composition, barrier height — is affected by the existence of oxygen or other impurities on the surface of the substrate. Metal deposition under ultra high vacuum will minimize this kind of complication.

This methodology is designed to provide a well-rounded set of data which can be integrated to provide an overall picture of material characterization. Some of the techniques which we are using are described below.

AUGER ELECTRON SPECTROSCOPY (AES):

This technique employs a beam of low-energy electrons to probe the sample surface. As a result of electronic rearrangements, Auger electrons are emitted whose energies are characteristic of the elements present. Only those Auger electrons emerging from the topmost two or three atomic layers contribute to the spectrum and this is what gives the technique its great surface sensitivity. The technique detects all elements except hydrogen and helium and its sensitivity is better than one atomic percent of a monolayer. AES is not only applicable to metals and semiconductors, but also to a wide range of insulators. Since the electron beam can be focused to a diameter of 0.5 um, the use of scanning electron microscope principles enables a magnified image of the sample surface to be obtained. This can then be used to pinpoint the precise area to be analyzed. In addition, the Auger signal from a single element can be detected as the beam is rastered over the surface. This enables the spatial distribution of the elements to be obtained and displayed either as a chemical map or as a line-scan. For further detail on AES see reference [8].Futher see schematic Figure 1.

X-RAY PHOTOELECTRON SPECTROSCOPY (XPS, ESCA):

In X-ray photoelectron spectroscopy or ESCA (electron spectroscopy for chemical analysis), the sample is illuminated with X-rays from a Mg K source. Photoelectrons are emitted from the surface as a result of X-ray excitation whose kinetic energy is characteristic of the elements present. Surface elemental analysis is determined from the positions of the peaks in the ESCA spectrum. Since the chemical state of the atom in the surface alters the electron binding energy, chemical shifts of up to 10 eV can occur in the kinetic energy of the photoelectric. This enables chemical bonding information to be derived from the precise positions and shapes of the ESCA peaks. In addition, ESCA is amenable to the widest range of surfaces since X-rays do not normally cause charging problems or beam damage. Hence the technique may be used for all materials and is particularly useful for delicate surfaces such as polymers, catalysts, and natural and synthetic fibers. At Materials Evaluation Laboratory, Inc., ESCA can be used in combination with AES to provide a most versatile system of analysis, with the advantages of both techniques see Figure 1.

COMPOSITION=DEPTH PROFILING:

Composition-depth profiling using sequential ion beam sputtering and surface analysis is a powerful method for the characterization of thin films and interfaces. Successive atomic layers are removed from the surface by sputtering so that changes in elemental concentration can be monitored as a function of depth. The technique can be used for depths in the range of 10 to 20 um. Tapering of the surface using external mechanical methods may be used over depths in the range 2 um to 100 um. A combination of point analyses and Auger line-scan techniques over the taper may then be used to obtain the depth profile. This technique is ideal for the investigation of surface coatings and diffusion profiles.

FIELD EMISSION MICROSCOPY:

E.W. Muller, who died in 1977, invented the Field Emission Microscope (FEM) in 1937, the Field Ion Microscope (FIM) in 1956, and the atom probe in 1968. This unique achievement is a fitting memorial to a man who not only invented three major techniques for surface study, but remained a principal contributor to surface physics throughout his life.

Field Emission and Field Ion Microscopy are in many ways complimentary techniques. While the FIM can depict surface structure in atomic detail (Figure 2), study of Field Electron Emission from the same specimen can yield information about the electronic structure of the surface layer, [11-13].

The Field Emission Microscope has been extensively used in the study of metal and semiconductor surfaces. The process of field emission is itself of great interest and a considerable amount of both theoretical and experimental work has been carried out in this field. The FEM also yields useful information of a more practical nature, such as the nature of bulk and surface impurity, diffusion, chemisorption, surface potential barrier, work function and activation energy determination.

A Field Emission Microscope is shown schematically in Figure 3. The field emitter in the form of a very sharp point is mounted on a filament, which may be heated resistively, facing a fluorescent screen coated onto the conducting inner surface of the glass vacuum tube. The applied field, F, at surface of an emitter is related to the applied voltage, V, to it by a simple expression F=V/kR, where R is the mean radius of curvature of the emitter surface and the constant k~5. A voltage of 1.0 to 3.5 kv must therefore be applied to an emitter of 1000 A radius to generate the fields of 3 to 6×10^7 v/cm commonly used in Field Emission Microscopy.

In experimental measurement, the emission current I (in amperes) is measured as a function of the electrostatic potential V between the screen and the tip. The current is related to applied potential by [11],

$$I = 1.54 \times 10^{-6} \frac{B V^2 A}{\sigma z^2 (y)} \exp\left(-6.83 \times 10^7 \frac{\sigma^{3/2}}{Bv} \cdot t(y)\right)$$

The curve is produced by plotting $\log(I/v^2)$ vs $10^4/v$ and is called the Fowler-Nordheim plot.

ADDITIONAL FACILITIES:

Other available facilities, including SEM, EDAX, TEM, and X-ray diffraction techniques, are well utilized in surface studies and so will not be described here. However, references [9] are provided for the reviewer. Two other powerful atomic resolution techniques also in operation in our laboratory are field-emission and field-ion (atom probe) microscopy [10-11]. Furthermore, we have numerous sample preparation techniques as well as vacuum (ultra-high) evaporators available for coating and growing thin films.See Table 2. Figures 4 and 5 shows a typical vacuum system and the different methods used for in deposition of thin films respectively.

REFERENCES:

1. P.S. Ho, G.W. Rubloff, J.E. Lewis, V.L. Moruzzi and A.R. Williams, Phys. Rev., B22 1781, (1980).

2. G.W. Rubloff in: Proc. 8th Intern. Vac. Congress, Vol. I, Thin Films, Cannes, 562 (1980).

3. N.W. Cheung, P.J. Grunthaner, F.J. Grunthaner, J.W. Mayer, and B.M. Ullrich, J. Vac. Sci. Technol., 18, 917 (1981).

4. (a) C. Patel, for FEM and FIM references see 4(b).
   (b) C. Patel and J.P. Jones, Surface Science, 80, 265 (1979).
   (c) C. Patel, FEM Study of Metal-Semiconductor Interfaces, ERO (UK), U.S. Army, 2nd Rept. 1982-83, Cont #BAJA 37-81-C005.

5. A.G. Gullis, S.M. Davidson and G.R. Brooker, eds., Proc. of Inst. Phys. Conf. on "Microscopy of Semiconducting Materials," Inst. of Phys., London (1983).

6. S.P. Murarka, D.B. Fraser, J. Appl. Phys., 51 (6), 1593, (1980).

7. G.W. Rubloff, Surface Science, 132 (1983) 268.

8. T.A. Carlson, Photoelectron and Auger Spectroscopy, Plenum Press, (1975).

9. M.T. Postek, K. Howard, A.K. Johnson and K.L. Michael, Scanning Electron Microscopy, A Student's Handbook, Ladd Research Ind. Inc., (1980).

10. E.W. Muller, T.T. Tsong, Field-Ion Microscopy, Principles and Applications, Elsevier Pub. Co., (1969).

11. C. Patel, Journal de Physique, Coll. C2, Suppl. au No. 3 Tome 47, 1986.

12. W.P. Dyke, W.W. Dolan, Adv. In Electron & Electron Physics, 8, 89, 1956,

13. R. Gomer, Field Emission and Field Ionization, Harvard University Press, Camb. Mass., 1961.

14. Materials Research Society, Symposia Proceedings, Volume 69, Materials Characterization, edited by Nathan W. Cheung and Marc-A. Nicolet.

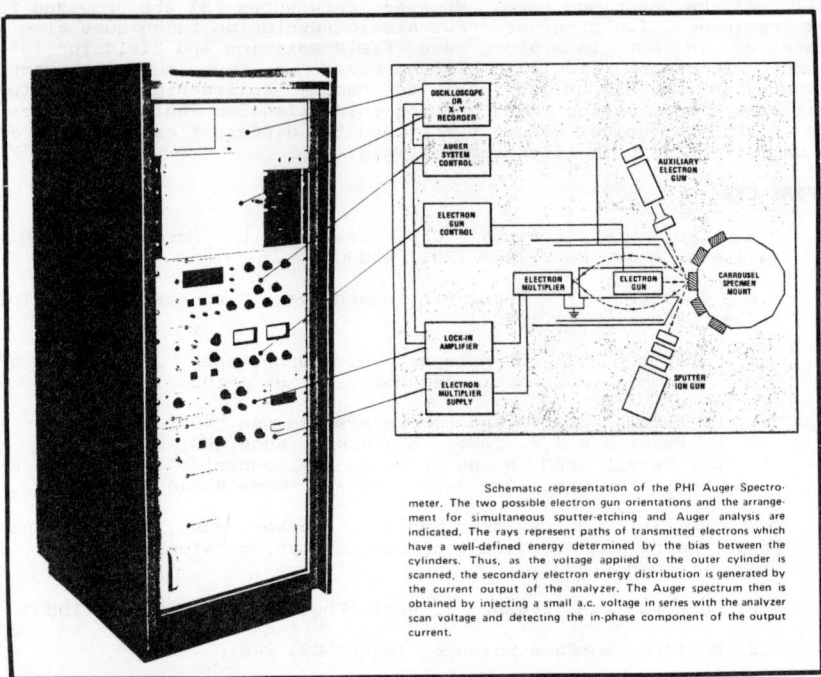

Courtesy of Physical Electronics and Perkin-Elmer

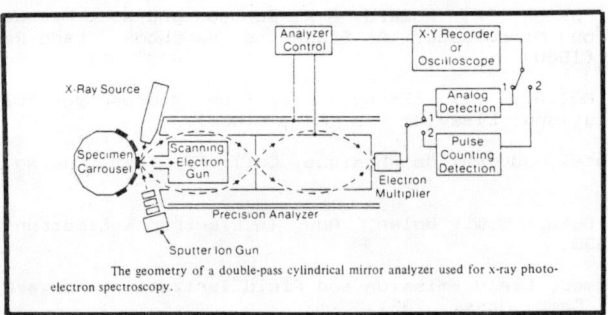

FIGURE 1

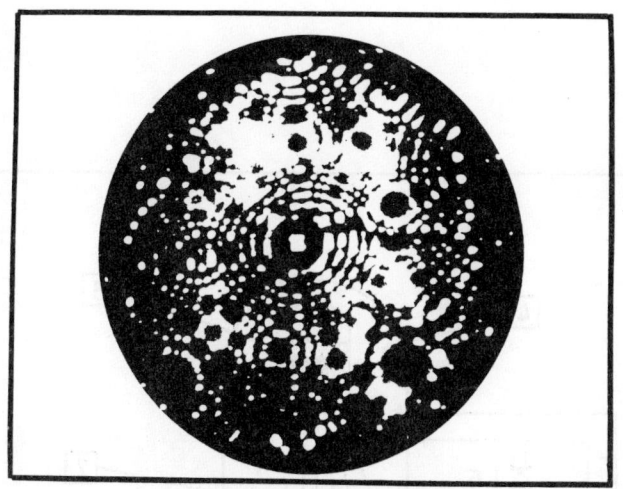

FIGURE 2: FIM Surface Structure in Atomic Detail
(Tungsten Surface)

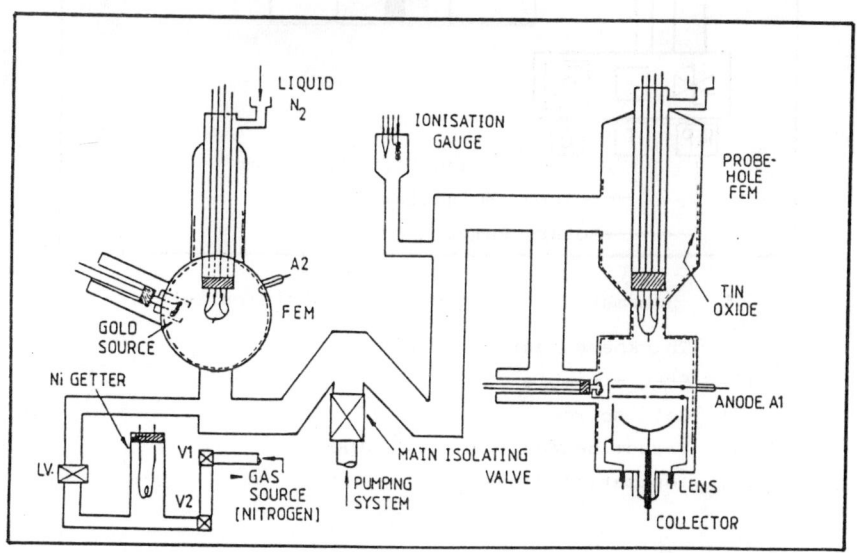

FIGURE 3: Schematic diagram of Field Emission and
Field Emission Probe Hole Microscope

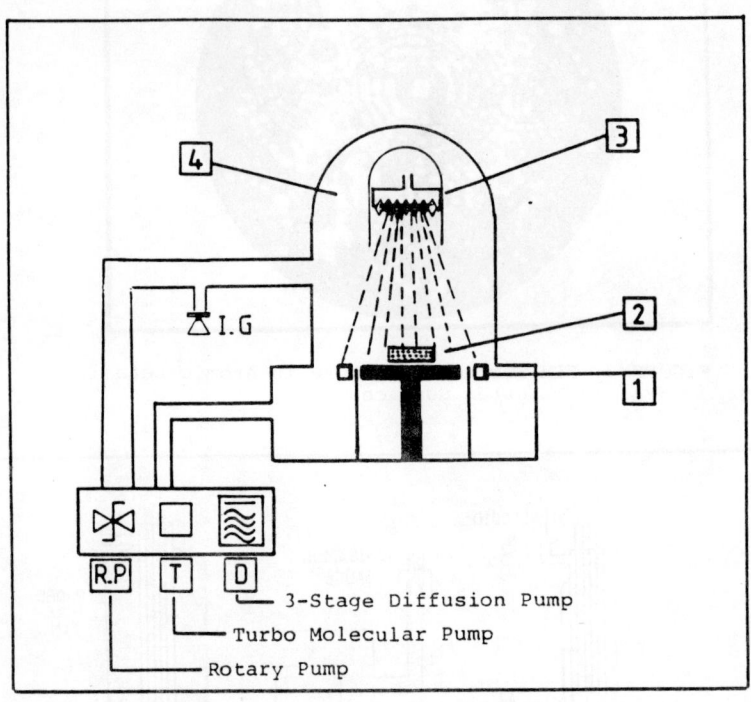

FIG. 4. Schematic Gold Thin Film Deposition Setup.

(1) Thickness monitor system
(2) SrF_2 - sample
(3) Gold source
(4) Vacuum better than 10^{-8} TORR
IG: Ionization Gauge

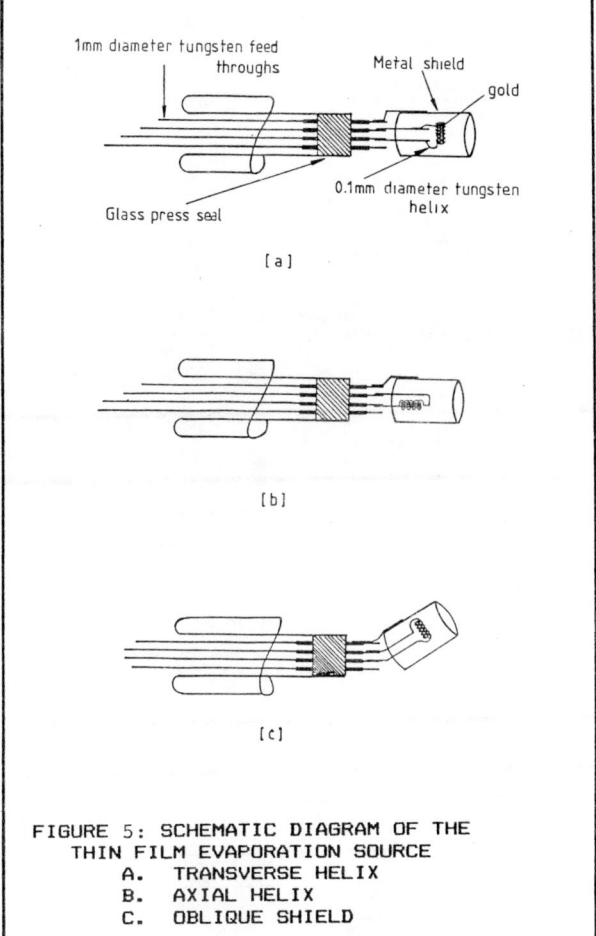

FIGURE 5: SCHEMATIC DIAGRAM OF THE
THIN FILM EVAPORATION SOURCE
A. TRANSVERSE HELIX
B. AXIAL HELIX
C. OBLIQUE SHIELD

TABLE 1

	Major Strengths	Major Weaknesses
RBS	• Quantitative Analysis • Nondestructive Testing	• Poor Lateral Resolution • Poor Sensitivity to Light Atoms in Heavy Substrates • Poor Depth Resolution Except for Very Thin Layers
XPS	• Chemical Phase Identification • Good Depth Resolution*	• Poor Lateral Resolution • Sensitivity Poor Compared to SIMS
AES	• Good Lateral Resolution • Simplicity of Operation • Good Depth Resolution • Chemical Phase Identification	• Sensitivity Poor Compared to SIMS
SIMS	• Good Sensitivity • Good Lateral Resolution now Available • Good Depth Resolution	• Poor Quantitative Ability** • Limited Chemical State Identification • Complexity of Operation
EDAX	• Good Lateral Resolution • Simplicity of Operation • Quantitative Analysis	• Poor Depth Resolution • Limited Chemical State Identification

* Small spot sizes necessary for profiling.
** Alleviated by laser ionization.

RBS	• More Flexible Commercial Systems
XPS	• Smaller Spot Sizes, Brighter Sources, More Photon Energies
AES	• Brighter Sources for Small Spot Sizes
SIMS	• Laser Ionization, Smaller Spot Sizes
EDAX	• Use of High Resolution Detectors, Sample Thinning for Better Lateral Resolution

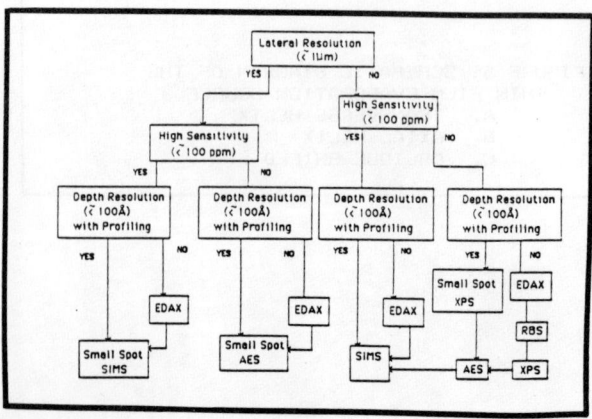

REFERENCE 14

TABLE 2

MATERIALS
EVALUATION
LABORATORY

IN-HOUSE EQUIPMENT AND SURFACE
PROBE TECHNIQUES CURRENTLY UTILIZED

1. AES - Auger Electron Spectroscopy
2. APM - Atom Probe Microanalysis (See FIM)
3. C-VP - C-V Plotting System
4. ESCA - Electron Spectroscopy for Chemical Analysis
5. FEEM - Field Electron Emission Microscopy
6. FEES - Field Electron Emission Spectroscopy
7. FIM - Field Ion Microscopy
8. ICP - Inductively Coupled Argon Plasma Spectroscopy
9. LEED - Low Energy Electron Diffraction
10. LES - Laser Ellipsometry
11. MS - Mass Spectroscopy
12. SEM-EDS - Scanning Electron Microscopy-Energy Dispersive Spectroscopy
13. SES - Spark Emission Spectroscopy
14. XRD - X-Ray Diffraction Spectroscopy
15. XRF - X-Ray Fluorescence Spectroscopy

FAST ION CONDUCTING GLASSES AND THEIR CHARACTERIZATION

B.V.R. Chowdari and R. Gopalakrishnan
Department of Physics
National University of Singapore
Kent Ridge,
Singapore 0511.

Abstract

The importance of fast ion conducting glasses, their applications and the role played by their constituents such as network former, modifier and dopant in determining the physical properties of these glasses are discussed. The applicability of differential thermal analysis (DTA) as a technique to characterize glasses is also discussed. Glass transition temperature (T_g) was found to decrease drastically with the addition of dopant in all the silver and lithium ion conducting glasses studied so far. The indispensability of complex impedance measurements for electrical characterization and the importance of the derived formalisms such as admittance, modulus and permittivity for the study of ionic transport are highlighted. In silver glasses the ionic conductivity reached a maximum value of 10^{-2} Scm^{-1} when systematically doped with AgI. dc conductivity, ion hopping rate and dielectric relaxation time were found to be Arrhenius in behaviour and the activation energies calculated from the respective physical processes were found to be favourably comparable. Parameters such as energy density, power density etc. of lithium and silver based electrochemical cells were compared and discussed. Results from our studies of AgI: Ag$_2$O:P$_2$O$_5$, AgI: Ag$_2$O:MoO$_3$, AgI:Ag$_2$O:MoO$_3$:P$_2$O$_5$ and LiCl:Li$_2$O:MoO$_3$:P$_2$O$_5$ glassy systems are used as typical data for the discussion.

1. Introduction

Characterization is an essential part of all investigations dealing with solid materials. Among the variety of experimental techniques available for characterizing the materials, a few are found to be more appropriate for a given category of materials. In this paper, the subject matter of fast ion conducting glasses and the experimental techniques appropriate for characterizing them are reviewed. Some of the results from our studies on silver and lithium ion conducting glasses are reported.

The electrical conduction in most of the normal crystalline ionic materials such as NaCl ($\sigma \simeq 10^{-12} Scm^{-1}$ at 25°C) is mainly due to the ionic defects such as vacancies and interstitials, the concentrations of which are usually low at ambient temperature. The abnormally high conductivity of some ionic materials like AgI ($\sigma \simeq 1 Scm^{-1}$ at 147°C) and $RbAg_4I_5$ ($\sigma \simeq 0.27 Scm^{-1}$ at 25°C) have been attributed to their abnormal crystal structure[1]. For a solid to be a good ionic conductor at moderate temperatures several conditions must be met. First, the potentially mobile species must be present as ions and not be trapped in strong covalent bonds. Second, a population of alternate sites that the ions can potentially occupy and that are not their principal crystallographic positions must also exist. Third, the energy to disorder the ions among the large population of alternate sites and the energy to move the ions among those sites must be low[2].

In addition to crystalline solids, glassy solids have also been considered as possible candidates for electrochemical applications. Since disorder happens to be one of the main criteria for a material to exhibit high ionic conductivity, consideration of glasses, where the atomic arrangement is random, as possible candidates appear to be appropriate.

2. Glass and it's structure

The term glass is commonly used to mean the fusion product of an inorganic material which has been supercooled so that the nucleation and growth of crystals have been suppressed[3]. The relation between crystal, glass and liquid is easily distinguishable by means of a volume-temperature diagram as shown in Fig. 1. When a liquid is

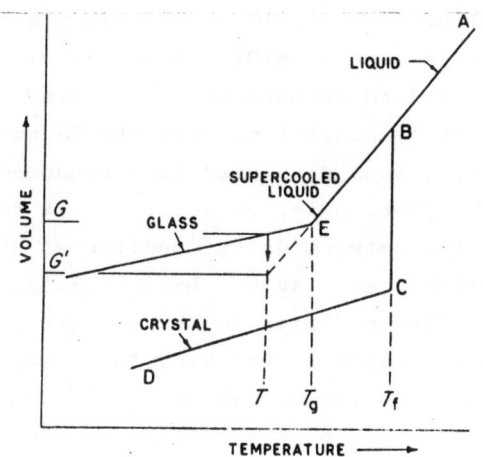

Fig. 1: Relationship between the glassy, liquid and solid states.

cooled from the initial state A, the volume decreases steadily along AB. The formation of nucleation begins at T_f called the freezing temperature. If the rate of cooling is slow, decrease in volume is seen along BC and thereafter the solid contracts with falling temperature along the line CD. On the other hand, if the rate of cooling is rapid, nucleation does not take place at T_f and the frozen liquid volume decreases along BE. At a characteristic temperature, T_g, the volume-temperature graph undergoes a significant change in slope (decline in fluidity) and follows the crystalline trend along EG'. T_g is called the glass transition temperature which distinguishes the glassy vitreous materials from amorphous materials. The stabilization of glass could be achieved if the temperature of glass is held constant at T, which is little below T_g and eventually the volume reaches the level G'.

It is known from theoretical predictions that any substance can be made into a glass if rapid quenching is employed (cooling rate 10^{6} °C), but in practice only a few glass formers such as SiO_2, B_2O_3, P_2O_5, GeO_2, GeS_2 etc...., are used to prepare glasses. Glass formation in oxides is generally consistent with the Zachariasen rules and in most cases they give rise to a random-corner linked tetrahedra[4]. Further, newer oxyanion matrix can be created by changing the network using 'modifiers' which include oxides or sulfides (eg. Ag_2O, Li_2O, Li_2S, Ag_2S etc). The purpose of adding modifier to the network former is to introduce ionic bonds by breaking the oxygen or sulphur bridge linking two former cations as shown in Fig. 2. modifier progressively breaks all the oxygen bonds, which results in the decrease of average length of a macromolecular chain.

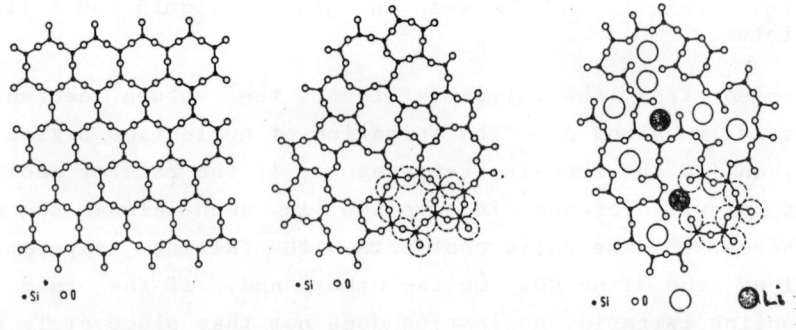

(a) (b) (c)
Fig. 2 : Difference between a) ordered, b) disordered, c) modified framework of SiO_2.

The typical acid-base reaction could be

$$-\underset{|}{\overset{|}{Si}} - O - \underset{|}{\overset{|}{Si}} - + Li_2O \rightarrow -\underset{|}{\overset{|}{Si}} - O^- \cdots Li^+ \quad Li^+ \cdots ^-O - \underset{|}{\overset{|}{Si}} -$$

bridging Oxygen   non-bridging Oxygen

Moreover, large vitreous domain could be obtained by adding alkali salts to the modified glasses. Alkali salts of large anions such as I^-, Cl^-, Br^-, F^-, SO_4^{--}, SCN^-, etc.......have been dissolved without affecting the glassy network to introduce notable changes in the physico-chemical properties[5]. All these modifications of the structure of the otherwise insulating glasses lead to the development of fast ion conducting glasses. The isotropic properties, ease of shaping into various forms, ease of thin film formation, wide selection of glass-forming systems and wide range of control of properties (like conductivity, refractive index) with changing chemical compositions have contributed to the enhanced interest in the study of fast ion conducting glasses.

3. Preparation of Glasses

Several silver and lithium ion conducting glasses of various compositions have been prepared by melting appropriate analar grade chemicals in required proportion in platinum crucible and by quenching the resultant melt to room temperature by pouring onto a stainless steel plate. A temperature drop of about 1000°K is accomplished in about a second. Composition of any glassy system is described by the following empirical parameters

$$x = \frac{(\text{dopant})}{(\text{dopant})+(\text{modifier})} \quad , \quad y = \frac{(\text{one former})}{(\text{sum of two formers})}$$

$$\text{and } n = \frac{(\text{sum of two formers})}{(\text{modifier})}$$

where the quantities in parentheses indicate the weight fraction of the various species. All the glasses are analysed by X-ray diffraction technique in order to confirm their amorphous nature. Typical results obtained for glasses such as $AgI:Ag_2O:P_2O_5$, $AgI:Ag_2O:MoO_3$ $AgI:Ag_2O:MoO_3:P_2O_5$ and $LiCl:Li_2O:MoO_3:P_2O_5$ are used for the discussion. The glassy nature of these materials and their ionic conductivity demand the usage of specialized experimental techniques for characterization.

4. Characterization of Glasses

4.1 Thermal Techniques

Thermal analysis is based upon the detection of changes in the heat content (enthalpy) or specific heat of materials as a function of temperature. The supply of thermal energy may induce a physical or chemical process in the sample e.g. melting or decomposition, accompanied by a change of enthalpy, the latent heat of fusion, heat of reaction etc. Hence thermal analysis can be conveniently used to study the thermal decomposition, phase transitions and phase diagrams of a wide range of materials. Several techniques such as thermogravimetric analysis (TGA), differential thermal analysis (DTA), and differential scanning calorimetry have been developed for thermal analysis of materials. In DSC measurements sample and the reference are maintained at the same temperature during the heating program and the extra heat input to the sample (or to the reference if the sample undergoes an exothermic change) required in order to maintain this balance is measured. The most significant use of DTA and DSC for glasses is to measure the characteristic temperature T_g (as discussed above) , since it represents the upper temperature limit at which the glass can be readily utilized. Fig. 3 represents the vitreous domain observed in the $Ag_2O:MoO_3:P_2O_5$ glassy system where P_2O_5 and MoO_3 together are network formers and Ag_2O is the modifier. The observation of T_g and its variation with

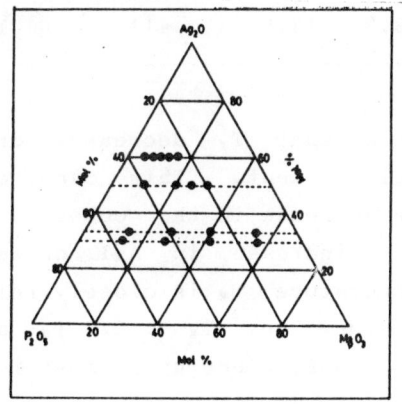

Fig. 3 : Phase diagram of $Ag_2O-MoO_3-P_2O_5$ system.

composition confirm both the glassy nature of this material and the involvement of each of the constitutents in glass formation. Since the specific heat, $C_p = (dQ/dT)_P$' measures the heat absorption from the temperature stimulus at the glass transition, C_p clearly appears as a 'step'. This measurement also can be used for the determination of T_g.

The effect of AgI doping in $Ag_2O:MoO_3:P_2O_5$ glassy system on the thermodynamic properties such as T_g and C_p is shown in Table 1.

Table 1 : Composition of $AgI:Ag_2O:MoO_3:P_2O_5$ glasses and their T_g and C_p.

Mol%				T_g	C_p at (T_g - 10K)
AgI	Ag_2O	MoO_3	P_2O_5	(K)	($J\ mol^{-1}K^{-1}$)
5.2	47.3	18.9	28.4	580	21.2
17.6	41.1	16.4	24.7	559	22.03
33.3	33.3	13.3	19.9	516	17.08
53.8	23.0	9.2	13.8	411	15.89

It can be seen that T_g decreases drastically with increasing AgI content. This can be accounted by considering the decrease in the cohesive energy of the glass with the increase in AgI content. The glass transition temperature T_g is closely related to T_m (the melting temperature) by $T_g = 2/3\ T_m$ and T_m in turn proportional to cohesive energy in most ionic solids[6].

4.2 Electrical Techniques

4.2.1 Temperature-frequency response and permittivity measurements

Measurements of accurate and meaningful conductivity values is the most useful characterization of fast ionic conductors, but often they are associated with interfacial problems[7]. For example when a voltage of about 100mV is applied across Na β-alumina crystal sandwiched between gold electrodes, Na^+ ions migrate preferentially towards the cathode and accumulate at the gold/ β-alumina interface. Hence the application of dc bias causes polarization due to the inability of the Na⁺ ions move across the electrode/electrolyte interface and ultimately the ionic current falls to zero. Hence the measurement of ohmic resistance using voltage and decaying current seems to be unreliable. The above problem can be overcome to some extent by using a four-probe technique or non-blocking (reversible) electrodes. But four-probe method is not always compatible with specimen geometry and suitable non-blocking, ion conducting electrodes for most ionic conductors also are uncommon (ex : for fluorine ion conductor). The problem of dielectric and electrode polarization can be successfully overcome by measuring the ac conductivity.

Generally, the conductivity of a material is measured at a fixed frequency say 1KHz. It is wellknown that several physical processes such as the migration of mobile ion through the bulk of the electrolyte, charge transfer across the interfacial zone, and condution via grainboundaries occur with the passage of electric current through a solid electrolyte. These individual processes could be easily resolved in the frequency domain since each of these have characteristic relaxation times. Hence the study of the a.c. response over a wide range of frequency and analysis of the data in complex plane are desirable[8].

Complex plane analysis is nothing but a mathematical technique which helps in the determination of individual components (R,C, and L) and involves the plotting of real and imaginary parts of complex electrical quantities. For example the total impedance of configuration ⎯R⎯⊢C⊢ is given by $Z = R - j\omega/C$ with $Z' = R$ and $Z'' = -1/j\omega C$. Similarly for configuration ⊣C⊢∥R 1/Z = 1/R + jωC = Y where Y is the complex admittance. On simplification

$$Z = R/(1+(\omega RC)^2) - j\omega R^2 C/(1+(\omega RC)^2) = Z' + jZ''\ldots(1).$$

A plot of Z" vs. Z' for each of these cases is shown in Fig. 4. Since any system under test can be represented in terms of an equivalent R,L,C network, the crux of the problem in analyzing the ac data of the fast ion conducting materials is to determine the appropriate equivalent circuit and to evaluate the various components in the equivalent circuit[9]. An ideal electrochemical cell with non-blocking electrodes can be represented by an equivalent circuit given in Fig.4 c.

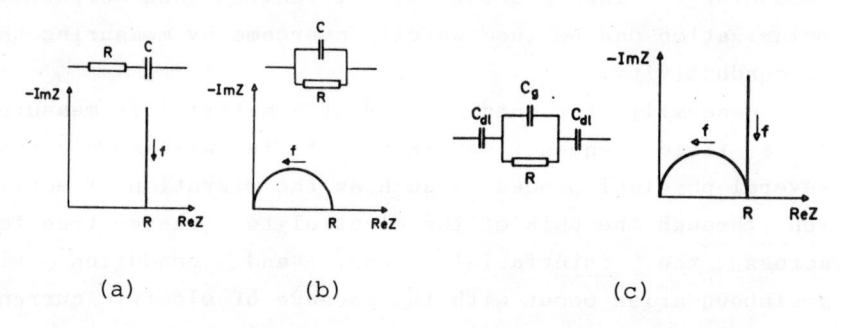

Fig. 4 : a), b) Complex impedance plots for some elementary circuits;

Fig. 4 : c) Equivalent circuit and complex impedance plot for a cell with blocking electrodes.

In addition to impedance and admittance formalisms, permittivity and modulus formalisms are also used for analyzing a.c. conductivity data. For a parallel RC network the complex permittivity $\varepsilon = \varepsilon' + j\varepsilon''$ where $\varepsilon' = C/C_0$ and $\varepsilon'' = -1/\omega C_0 R$ and the modulus, $M = M' + jM''$ where

$$M' = (\omega C_0 R)(\omega CR)/(1+(\omega CR)^2) \text{ and } M'' = (\omega C_0 R/1+(\omega CR)^2)\ldots(2)$$

Here C_0 is the vacuum capacitance.

In an impedance spectrum (Z" vs log f) information about the electrode/electrolyte double layer, intergranular effects and the bulk resistance are highlighted inview of the fact that the Z" is proportional to R of each simulated RC element. Secondly, information at high frequencies is supplemented by admittance and permittivity plots. Finally, the capacitance of the bulk assumes prominence in the modulus spectrum (M" vs log f), since its peak height is proportional to 1/C of each simulated RC element. Although the same experimental raw data is manipulated in different formalisms, it is advantageous to draw all four types of plots in order to obtain a valuable insight into the heterogeneous structure of solid electrolytes. A characteristic complex impedance diagram is shown in Fig. 5 for $AgI:Ag_2O:MoO_3$ glass for a wide range of temperature. It can be seen that a depressed

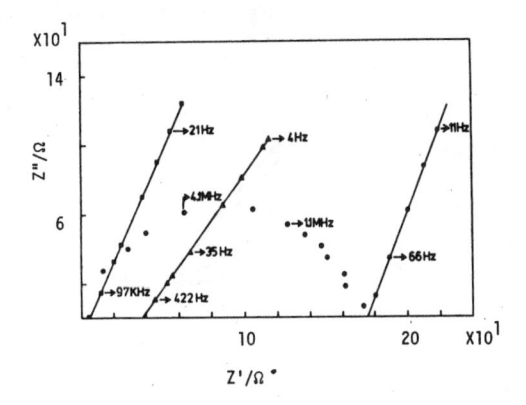

Fig. 5 : Complex impedance plots at various temperatures
1. (●) 223K, 2. (▲) 273K, 3. (■) 294K.

semicircle is followed by a spur at 223K. An increase in temperature causes the disappearance of semicircle. The centre of the semicircle lie on a straight line passing through the origin and making a very small angle with the real axis. Hence it may be concluded that the bulk relaxation is responsible to the arc in the high frequency region and electrode roughness is attributable to the deviation of inclined straight line from perpendicularity on the low-frequency region. Morever, the small departures from the semicircular arc in the high frequency region could be attributed to the deviation from single relaxation time behaviour. Its significance is highlighted in modulus spectrum analysis. The abscisa of the intersection with the real axis of the low-frequency extrapolation of the semicircular arc (equals ohmic resistance) is used to calculate dc conductivity. As discussed earlier, the impedance data is converted into the permittivity formalism and the dependence of relative permittivity ε' on frequency at various temperatures is shown in Fig 6. At low temperatures i.e. T<150°K the

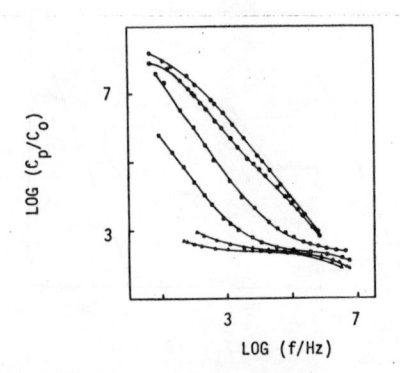

Fig. 6 : Capacitance vs. frequency at various temperatures.
1. (○) 294K, 2. (●) 273K, 3. (■) 223K, 4. (□) 173K.
5. (▲) 143K, 6. (△) 123K.

relative permittivity reaches a saturation values $C_\infty \simeq 100$, which is comparable to the value obtained for other silver glasses at high frequencies.

Number of research papers have been published recently on frequency dependent conductivity of single crystal β-alumina[10]. This crystalline material has been modeled by frequency dependent parameters A and n (from Jonscher's expression) and an expression for K'(mobile ion concentration factor) and ω_p (hopping rate) have been derived[11,12]. Extending these ideas and methods to AgI:Ag$_2$O: MoO$_3$ glassy system the following conclusions can be drawn from Fig. 7. The three different phenomena, high

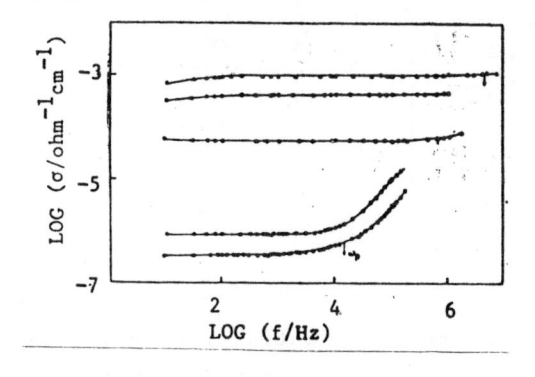

Fig. 7 : Conductivity vs. frequency at
1. (■) 164K,
2. (□) 174K,
3. (●) 223K,
4. (o) 273K,
5. (∆) 294K

frequency dispersion, low frequency dispersion and electrode polarization, are evident. The high frequency dispersion is dominant at low temperatures and the low frequency dispersion is dominant at high temperatures. The frequency dependent conductivity can be summarized by Jonsher's universal expression[11]

$$\sigma = \varepsilon_o \omega [(\omega/\omega_p)^{n_1-1} + (\omega/\omega_p)^{n_2-1} \ldots \ldots (3)$$

where ω_p is the frequency associated with the hopping rate and n_1, n_2 are the empirical parameters. Adopting the method proposed by Almond et al[12], the hopping rate ω_p has been calculated. Defining the, mobile ion concentration factor K' equals $\sigma T/\omega_p$, it is found that K' is invariant with temperature and ω_p is thermally activated. Moreover the factor K' is found to increase with AgI content in silveriodomolybdate glasses. Similar observation was also seen by making use of the data reported for Na-β-alumina[13].

The temperature dependence of d.c. conductivity obtained as discussed before is shown in Fig. 8.

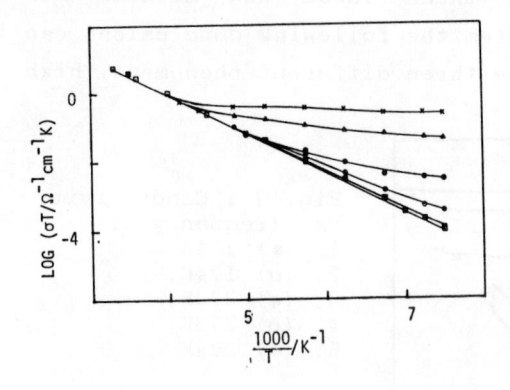

Fig. 8 : Plots of LOG (σT) vs. 1000/T for "dc" and fixed frequency values.
1. (□) 0 Hz,
2. (■) 1 KHz,
3. (o) 10 KHz,
4. (●) 100 KHz,
5. (▲) 1 MHz
6. (X) 10 MHz

Activation energy, E_{act}, and pre-exponential factor, σ_0, in the Arrhenius equation obtained for some silver and lithium ion glasses is given in Table 2.

Table 2 : Conductivity parameters for Ag and Li Glass

COMPOSITION (Mol%)	σ at r.t ($\Omega^{-1}cm^{-1}$)	σ_0	Activation energy (kcal/mole)
66.7 AgI: 25 Ag$_2$O: 8.3P$_2$O$_5$	1.6x10^{-2}	5.1	5.2
60 AgI: 20 Ag$_2$O: 20.MoO$_3$	1.2x10^{-2}	10	4.41
53.8 AgI: 23 Ag$_2$O: 11.5MoO$_3$: 11.5P$_2$O$_5$	1.24x10^{-2}		
53.8 AgI: 23 Ag$_2$O: 11.5MoO$_3$: 11.5V$_2$O$_5$	2.2x10^{-2}		4.61
25 LiCl: 37.4Li$_2$O: 7.4MoO$_3$: 29.9P$_2$O$_5$	3.5x10^{-5} at 425k		

In our constant effort to synthesize new materials with higher conductivity, it is found that AgI doped silver glasses reached a limiting value of 10^{-2} Scm^{-1} at room temperature[14] and for LiCl doped lithium glass σ maximize at 10^{-6} Scm^{-1}. Increase in conductivity is always associated with decrease in Tg and the realization of a material with high ionic conductivity and higher Tg is not far away from reality.

4.2.2 Real Dispersive Conductors and Modulus Analysis

An ideal ionic conductor is characterized by frequency independent conductivity (σ_0) and permittivity (ε'). This can be modeled by a single parallel RC element in which $R = k/\sigma_0$, $C = \varepsilon'\varepsilon_0/k$ where k is the cell constant and ε_0 is the permittivity of free space. The time constant $RC = \varepsilon_0\varepsilon'/\sigma_0$ is defined as conductivity relaxation time. Referring to the equations 2 and 3, it can be seen that the modulus and the impedance spectrum for the RC element have identical shapes and both the maxima peak at the same frequency as predicted by Debye theory of dielectric loss. Frequency dependence of real and imaginary parts of M' and M" is shown in Fig 9. But

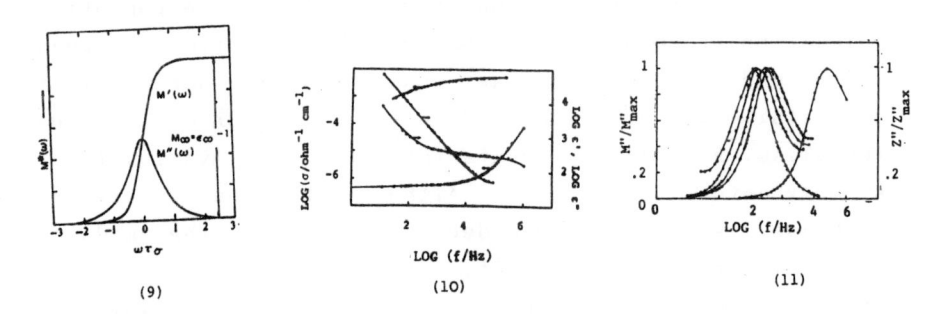

Fig. 9 : Frequency dependence of real and imaginary parts of (a)M' and (b) M"

Fig. 10 : Plots of σ, ε' and $\varepsilon"$ vs. Log f. 1. (□) σ at 294K, 2. (●) σ at 134K, 3. (□) ε' at 134K, 4. (●) $\varepsilon"$ at 134 K.

Fig. 11 : Normalized modulus spectrum at 1. ( ) 124K, 2. ( ) 129K, 3. (o) 134K, 4. (▲) 174K and impedance spectrum at (●) 124K

recently it has been shown that frequency dispersion is inevitable in ionic conductors and particularly in glassy solid electrolytes which change from normal conductors at low frequencies to leaky dielectrics at higher frequencies[15]. Fig. 10 represents the variation of conductivity and permittivity with frequency, measured at different temperatures. The deviations from the ideality could be accounted by introducing a third parameter in the admittance formalism defined by $Y(\omega) = A\omega^n + iB\omega^n$ where A, B, and n are emperical constants. Hence the real and imaginary part of the equivalent circuit

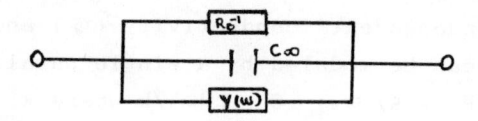

can be written as $Y^* = R_0^{-1} + A\omega^n + i(B\omega^n + \omega C_\infty)$ or $\sigma(\omega) = \sigma_0 + A\omega^n$ and $C(\omega) = C_\infty + B\omega^{n-1}$. The parameters A and B are related by $B = A \tan(n\pi/2)$. Characteristic parameters for 25 $LiCl:37.4$ $Li_2O:7.4$ $MoO_3:29.9$ P_2O_5 glasses are $R_0^{-1} = 2 \times 10^{-4}$ Scm^{-1}, $A = 1.73 \times 10^{-8}$ and $n = 0.49$. More recently the frequency dispersion of σ and ε' in fast ion conducting glasses is accounted by the distribution of relaxation time or by non-exponential relaxation process[16]. The absence of dielectric loss peak in the audio frequency range observed in the doped silver and lithium glassy system led to the attention of modulus formalism. Fig. 11 represents the plots of M" vs log f for a molybdate glass system. Although the characteristics such as width and non-debye nature of the modulus spectrum remains constant, the frequency at $M"_{max}$ increases with increase in temperature following the ionic glassy trend. Morever, the activation energies calculated from the Arrhenius plots of f at $M"_{max}$ are favourably comparable to the values obtained from the conductivity measurements (refer Table 1). Comparison of modulus and impedance spectra in Fig. 11 at T = 124K do indicate the non-debye nature of the practical solid electrolytes. The invariance of width of modulus spectra with temperature indicates that the distribution of relaxation time is independent of temperature.

The comparison of the width of modulus spectra of wide range of materials as shown in Fig. 12 has led us to gain further insight in explaining the distribution of relaxation time through a non-exponential decay formalism proposed by Moynihan et al[17]. They found the observed

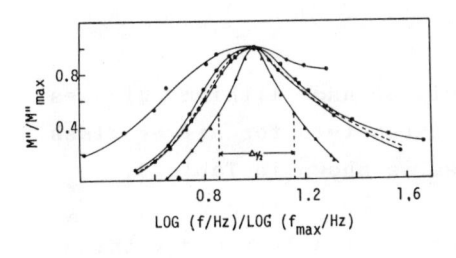

Fig. 12 : Comparison of electric modulus of ionically conducting materials.
1. (●) β-alumina at 113K,
2. (o) 100% Agβ-alumina at 113K,
3. (dashed curve)$K_2Si_3O_7$ glass at 298K,
4. (■) phosphate glass at 123K,
5. (▲) 27% Agβ-alumina at 208K,
6. (△) Debye type.

frequency dispersion can be empirically decribed in terms of the non-exponential decay of the electric field of the form

$$\phi(t) = \exp[-(t/\tau_p)^\beta] \quad , \quad 0<\beta<1 \ldots\ldots\ldots(4)$$

where τ_p is the characteristic relaxation time. When = 1, the decay is exponential with a single relaxation time. Extending the ideas proposed by Moynihan et al, it is observed that as the concentration of AgI or LiCl is reduced in the respective glasses, the full width at half maximum (FWHM) is reduced indicating the increase of parameter. Similar observations are seen in both insulating and conducting glasses[19]. Inspite of the different approaches a good theoretical explanation for the frequency dispersion in ionic conductors is yet to come.

5. Cell Characteristics

The cell characteristics of some of the silver and lithium glasses are discussed. The following cell configurations were used for the measurement of open circuit voltage and discharge characteristics.

a) $(Ag + GE_1)/GE_1/(I_2+C+GE_1)$
   (anode)   (cathode)

b) $Li/GE_2/(C+GE_2)$

where GE_1 and GE_2 refer to silver and lithium glasses respectively. Characteristic parameters for silver glass (from ref. 20) and lithium glass is shown in Table 3.

Table 3 : Characteristic parameters of silver and lithium cells.

Parameter	Silver cell	Lithium cell
Electrolyte		
thickness (cm)	0.11	0.1
diameter (cm)	1.3	1.1
conductivity($\Omega^{-1}cm^{-1}$) at 25°C	10^{-2}	10^{-7}
cell weight (g)	2.12	0.42
open circuit voltage (v)	0.687	3.00
current drain	0.2mA	0.044µA
power density (Wkg^{-1})	6.4×10^{-3}	0.314×10^{-3}
energy density (Wh kg^{-1})	2.1	0.421

It can be seen from the typical example of silver cell that these cells are limited to show cell voltages and low energy densities. However, it is desirable to research further utilizing the advantages of the glassy nature of the solid electrolyte. Although, lithium batteries (refer table) are characterized by lower weight and higher energy densities the problem of electrochemical compatibility either in primary or secondary batteries have to be overcome to realize an efficient solid state battery.

References

1. R.A. Huggins, Crystal Structures and Fast ionic Conduction in Solid Electrolytes, ed. by P. Hagenmuller and W. Van Gool, Academic press (1978).
2. S. Geller, Science 157, 3786 (1967), 30
3. Richard Zallen in 'The physics of amorphous solids' (John Wiley and Sons, 1983).
4. R.H. Doremus, in 'Glass Science' (John Wiley & Sons, 1973).
5. J. Wong and C.A. Angell, Glass: Structure by Spectroscopy, (Marcel Dekker, New York, 1976).
6. A.R. West in 'Solids State Chemistry' (John Wiley & Sons, 1984).
7. Suresh Chandra in 'Superionic Conductors' (North Holland, 1981) p. 145.
8. D. Ravine and J.L. Souquet, J. Chim. Phys. 71 (1974) 693.
9. I.M. Hodge, M.D. Ingram and A.R. West, J. Electroanal. chem. 74 (1976) 125.
10. D.P. Almond and A.R. West, Solid State Ionics 9/10 (1983) 277.
11. A.K. Jonscher, Nature 267 (1977) 673.
12. D.P. Almond, C.C. Hunter and A.R. West, J. Mater. Sci. 19 (1984) 3236.
13. C. Hunter, M.D. Ingram and A.R. West, Solid State Ionics 8 (1983) 55.
14. B.V.R. Chowdari and R. Gopalakrishnan, Solid State Ionics 18 & 19 (1986) 483
15. A.K. Jonscher in 'Dielectric relaxation in solids' (Chelsea Dielectrics Press, London, 1983)
16. F.S. Howell, R.A. Bose, P.B. Macedo and C.T. Moynihan, J. Phys. chem. 78 (1974) 639.
17. P.B. Macedo, C.T. Moynihan and R.A. Bose, phys. chem. Glasses 13 (1972) 171.
18. B.V.R. Chowdari and R. Gopalakrishnan, Solid State Ionics (In press).
19. C. Liu and C.A. Angell, J. non. Cryst. Solids 83 (1986) 162.
20. T. Minami, T. Katsuda and M. Tanaka, J. Electrochem Soc. 127 (1980) 1308.

Characterization of Materials for Telecommunications

J. W. Mitchell
AT&T Bell Laboratories
600 Mountain Avenue, Murray Hill, NJ 07974

ABSTRACT

Supply demand relationships for analytical chemistry within the telecommunications industry are discussed generally. Specific details are given to document needs for a broad range of expertise to respond adequately to characterization problems in manufacturing, development and research. The dual role of the analytical chemist as a problem solver and characterization scientist is described with examples.

Demands for Characterizations

The telecommunications industry, coupled intrinsically to semiconductor technology, has a critical dependency on materials with unusual properties resulting from extraordinary high purity, nearly perfect crystalline structure, precisely controlled composition, or unique processing. Bulk crystals, micron thin inorganic films, glasses, processing reagent chemicals, raw material gases, layer-structured devices, ceramics, metals, and polymers are required for fabricating network components and electronic/photonic products. Well known materials must be characterized to determine their conformance to increasingly stringent specifications for chemical composition and mechanical, physical or electrical properties. Newly discovered or evolved materials, must be characterized to correlate experimentally observed unusual properties with composition and structure.

To meet growing demands for complex materials analyses, expert characterization scientist are retained within the Analytical Chemistry Research Department, where opportunities abound for their immersion into a broad range of challenging work. Because materials characterizations are needed during most phases of research, development and manufacture, the work of analytical chemists spans the entire regime—service and support analyses, new chemical and instrumental development, and fundamental investigations of concepts and physical phenomena on which characterization methods depend.

The manufacturing plant has a strong dependence on analytical support. Prior to manufacture of a product, analytical characterizations are required for screening-out potential yield reducing poor quality raw materials and processing reagents. At intermediate stages of manufacture, chemical process components and partially completed products must be characterized on-line to be sure that specifications and performance criteria are on target. At the end of a product's life following accelerated laboratory testing or premature failure in the field, analyses are done to diagnose failure modes. Chemical and physical characterization methods are also essential for investigating problems in manufacturing, for monitoring the workplace environment, for characterizing by-product wastes, and for analyzing competitive products.

Materials characterization requirements in development and research continue to challenge the frontiers of analytical measurement science. Recent examples include the following. Expitaxially produced structures must be probed to identify ppb level of impurities in films, which are a few Anstromes thick. Distinguishing impurities in the top epilayer from those in the underlying substate of the same composition is being accomplished by photoluminescence techniques. Part per trillion levels of uranium in boron materials preclude their use in fabricating detectors for solar neutrinos. Chemical Preconcentrations and fission track counting techniques are being investigated for analyzing boric acid and trimyethlborate. Flouride glasses for long wavelength transmission optical fibers operating in the infrared region must be free of parts per trillion amounts of oxygen and lanthanide impurities. Measurements of these impurities at such low levels will require new advances in trace anlaysis. To respond to the full scope of materials characterization needs at a high-tech corporation, a broad range of analytical expertise is required.

Basic Characterization Methods

The central analytical chemistry research organization executes the functional responsibility of maintaining and sustaining a diverse group of experts in the areas traditionally important to the telecommunications industry. Expertise is maintained in the key areas of trace and major element analysis, organic analysis, surface and thin film analysis, and microanalysis. Polymer, defect mechanical, and electrical characterization methods are maintained also. Within each of these areas, several techniques and methods are used to provide the broad critical coverage needed. For trace analyses the expertise listed in Table 1 permits responses to a broad range of inorganic characterization problems. Similarly, organic analyses are vital in the telecommunications industry and the methods given in Table II provide broad coverage for organic and polymeric materials. Major element and microanalyses are critical needs as well. The methods listed in Table III are maintained within the central analytical department. Surface and thin film characterizations have obvious importance in the semiconductor technology. In this regime of analysis, secondary ion mass spectrometry with resonance ionization, ellipsometry, X-ray fluorescence, and photoluminescence techniques are used conveniently.

The central resources and expertise provided by members of the Analytical Chemistry Research Department are augmented by the simultaneous existence of other analytical group within AT&T. The combined work of the much larger overall organization depicted structurally in Fig 1, provides the entire corporate entity with the full spectrum of needs in analytical chemistry from service analysis to materials characterization science research. As shown in Fig 1, other independent analytical groups co-exist within the corporation. Collaborative interactions between analytical groups are initiated; completely autonomous works are executed; and where appropriate, outside analytical laboratories are utilized. AT&T chemists and engineers in need of analytical expertise converse directly with either members of the cent~~l analytical department, with staff of other dedicated corporate analytical groups, or with outside chemists of service analytical laboratories.

Solving Materials Characterization Problems

The role of the modern industrial analytical chemist involves solving challenging materials characterization problems while performing analytical chemistry research. Examples of these dual responsibilities will be given during a description of investigations of the analytical chemistry of silicon nitride. Results of investigations using scanning Auger microscopy (SAM), electron for spectroscopy (ESCA), scanning electron microscopy with X-ray fluorescence (SEM-XRF), 14 MeV neutron activation (NAA), alpha track image analysis, Rutherford backscattering spectrometry (RBS), and chemiluminescence will be discussed.

Table I. Trace Analysis Techniques

Atomic Absorption Spectrometry	(AAS)
Induction Coupled Plasma - Atomic Emission Spectrometry	(ICP-AES)
Neutron Activation Analysis	(NAA)
Spark Source Mass Spectrometry	(SSMS)
X-ray Fluorescence Spectrometry	(XRF)
Fourier Transform Infrared Spectromerty	(FT-IR)
Derivative IR diode Laser Absorption	
Chemiluminescence Analysis	
Ion Chromatography	

Table II. Characterization Methods for Organic and Polymeric Materials

Gas Chromatography	G-C
Mass Spectrometry	MS
GC - MS	GC-MS
Fourier Transform MS	FT-MS
Fourier Transform Infrared Spectrometry	FT-IR
Liquid Chromatography	LC
Nuclear Magnetic Resonance	NMR
Gel Permeation Chromatography	GPC
Spectrophotofluorometry	

Table III. Major Element and Microanalysis Methods

Major Element Techniques

Traditional wet Chemistry

Fast Neutron Activitation Analysis

Coulometry

Microanalysis Methods

Laser Microprobe Emission Spectrometry

Electron Microbe Analysis

Nuclear Image Analysis

Micro Infrared Spectroscopy

Laser Induced Photoluminescence

Scanning Electron Microscopy

42

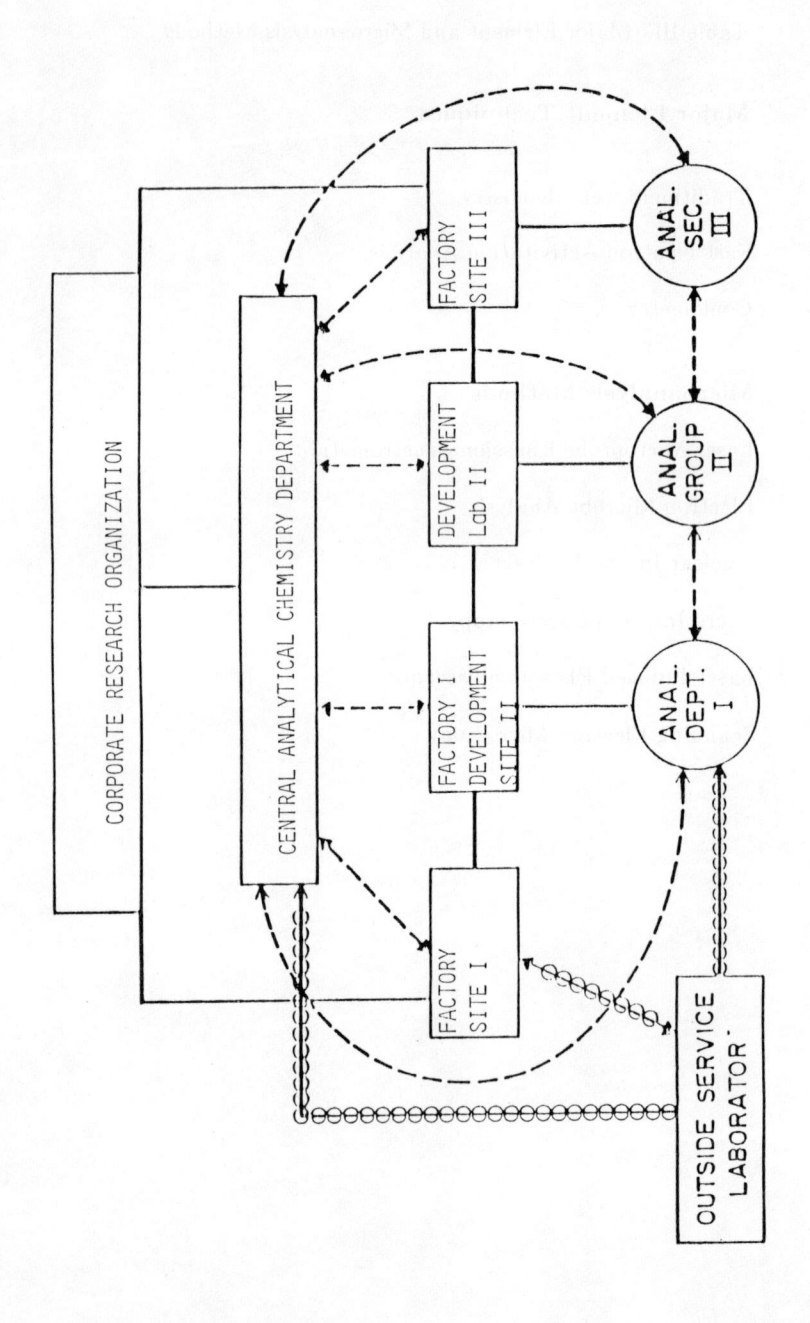

APPLICATIONS OF RUTHERFORD BACKSCATTERING STUDIES
V. LAKSHMINARAYANA
Nuclear Physics Department, Andhra University,
Visakhapatnam - 530 003, INDIA

Rutherford Backscattering techniques[1-3] are being used routinely for 'material characterisation'. These studies involve identification, determination of relative amounts and depth distributions of the scattering atoms contained in a specimen. Since low-energy ions (upto a few MeV) are employed in these studies, only a few microns of the material at the surface can be studied. The principal advantage of the technique is its ability to obtain quantitative information without using standards, quickly and non-destructively.

A typical experimental arrangement is shown in Fig.1. Ions of mass m and energy E_o strike a specimen of thickness t. An atom of mass M at the surface scatters the ion though an angle θ and modifies its energy to KE_o where K is called the "Kinetic Factor", which depends on the mass of the scattering atom M. The same kind of atom situated at a depth 't' inside the specimen, scatters the incident ion of energy E_1 (which is less than E_o) and the energy of the ion after scattering is KE_1. When it emerges out of the specimen in the same direction as the previous ion (scattering angle θ) and detected by the detector D, the energy is decreased to E. The energy difference ΔE($=KE_o$ - E) is a measure of the thickness.

It can be shown from the usual conservation laws that

$$K = \frac{E_\theta}{E_0} = \left\{ \frac{m\cos\theta + \sqrt{M^2 - m^2\sin^2\theta}}{M+m} \right\} \quad \ldots (1)$$

For a given system (m and θ) K is thus a function of M only indicating that E_θ, the ion energy at a scattering angle θ carries "mass-specific" information through the kinematic factor K. The values of K for the different atomic species (M) can be estimated from the relation (1). Thus from the experimentally determined energy, E_θ, in a given situation it is possible to identify the different atomic species contained in a specimen.

The percentage loss P of the projectile energy on scattering, as a function of the scattering angle is shown in Fig.2. for different projectiles. It can be seen that the "mass-resolution" is the best at a scattering angle of π. It can be shown from relation (1) that the mass resolution $\frac{M}{\Delta M}$ is given by

$$\frac{M}{\Delta M} = \frac{E_0}{\Delta E} \cdot \frac{2\mu \left[(1-\mu^2 \sin^2\theta)^{\frac{1}{2}} + \mu \cos\theta\right]}{\left[(1+\mu)^3 (1-\mu^2 \sin^2\theta)^{\frac{1}{2}}\right]} \times \left[1+\mu - \cos\theta \left\{(1-\mu^2 \sin^2\theta)^{\frac{1}{2}} + \mu \cos^2\theta\right\}\right] \quad \ldots (2)$$

where $\mu = (m/M)$ and $E_0/\Delta E$ is the energy resolution. Best mass resolution conditions are usually achieved for $90° < \theta < 180°$ and $0.2 < \mu < 0.4$. For $E/\Delta E = 500$, $\theta = 150°$, $M = 70°$, $M/\Delta M$ increases from 88 ($\Delta M = 0.8$) to 195 ($\Delta M = 0.36$) when going from $^4He^+$ to ^{20}Ne projectile beams.

SCATTERING YIELD

The scattering yield Y per unit time is given by

$$Y = Q N \sigma(\theta) \Omega \quad \cdots \quad (3)$$

where Q is the projectile current, N is the number of scattering centres per unit area, $\sigma(\theta)$ is the coulomb scattering cross-section for the scattering angle θ, and Ω is the solid angle subtended by the detector at the target. Neglecting the recoil effects, the differential coulomb scattering cross-section $\frac{d\sigma}{d\Omega}$ is given by

$$\frac{d\sigma}{d\Omega} = D^2 / \sin^4 \theta/2 \quad \cdots \quad (4)$$

where $D = \dfrac{Z_1 Z_2 e^2}{(4\pi \varepsilon_0) E_0}$ is the Collision diameter.

Z_1 and Z_2 are the atomic numbers of the projectile and the scattering atom. The scattering yields can therefore be used to determine the amounts of the scattering atomic species in the specimen. The yield varies as Z_1^2 and Z_2^2. The technique is therefore ideally suited for the study of heavy element traces in a light matrix.

DEPTH SCALE

The experimental arrangement for the study of depth profiles is illustrated in Fig.3. The respective energies may be estimated from the energy loss factors (dE/dx) in the target.

$$E_1 = E_0 - \int_0^{t/\cos\theta_1} \left(\frac{dE}{dx}\right) dx \quad \cdots \cdot (5)$$

$$E = K E_1 - \int_0^{t/\cos\theta_2} \left(\frac{dE}{dx}\right) dx \quad \cdots \cdot (6)$$

The energy difference ΔE ($=KE_0 - E$) of the two ions is given by

$$\Delta E = K \int_0^{t/\cos\theta_1} \left(\frac{dE}{dx}\right) dx + \int_0^{t/\cos\theta_2} \left(\frac{dE}{dx}\right) dx \quad \cdots (7)$$

ΔE can be experimentally measured. With a knowledge of (dE/dx) which varies as a function of energy, t can be estimated. If the thickness is small, the best way of evaluating the integrals is to adopt the 'mean-value approximation'. The final expression is given by

$$E = (S)\, t \quad (8)$$

where the back-scattering energy loss-factor (S) is given by

$$(S) = \frac{K}{\cos\theta_1} \left(\frac{dE}{dx}\right)_{\overline{E}in} + \frac{1}{\cos\theta_2} \left(\frac{dE}{dx}\right)_{\overline{E}out} \quad (9)$$

For normal incidence $\theta_1 = 0$ and $\theta_2 = \pi - \theta$ where θ is the scattering angle. For this case,

$$(S) = K\left(\frac{dE}{dx}\right)_{\overline{E}_{in}} + \frac{1}{|\cos\theta|}\left(\frac{dE}{dx}\right)_{\overline{E}_{out}} \quad \cdots (10)$$

Usually $\overline{E}_{in} = (E_0 - \frac{\Delta E}{4})$ and $\overline{E}_{out} = (E + \frac{\Delta E}{4})$. Equation (8) furnishes a method for the determination of thickness, using the RBS technique. This relation translates "energy scale" in RBS spectra to "dpeth scale", with the values of (S) computed from $(\frac{dE}{dx})$ - tables for any scattering species.

The 'depth resolution' is essentially limited by the "energy resolution". For a given detector system, the depth resolution depends on θ_1 and θ_2. It can be shown that the depth resolution can be increased by nearly 'one order' for θ_1 and $\theta_2 \approx 80 - 85°$. Such an experimental arrangement is called "glancing angle RBS arrangement". An alternative approach to improve the depth resolution[4] is to use an electrostatic analyser for the determination of the energies of the scattered ions.

EXPERIMENTAL ARRANGEMENT

The experimental arrangement used at the National University of Singapore is shown in Fig.4. A vertical crosssection of the 10" Scattering chamber is shown in the figure. The 2 MeV projectile beam (^{4}He$^+$) accelerated in the HVEC 2.5 MeV Vande Graaff Accelerator, collimoted to a spot of diameter 1mm. with Ta discs, passes through an annular detector mounted at one end of the scattering chamber and strikes the specimen at 12 cm. from the detector. The vertical position of the specimen on the target holder could be adjusted from outside the chamber. Six specimens could be mounted on the target holder and any one of them could be brought into projectile beam. An Si surface barrier detector, placed at $145°$, also was used to detect the scattered particles. The target current could be measured with a calibrated digitiser unit and automatic programming for a specified total charge could be carried out. A Canberra model 35 Multichannel Analyser with a router and two external ADC units was used, by interfacing it to an Apple II Microcomputer system for recording the two detector-spectra simultaneously. The system is being used routinely for a variety of investigations.

APPLICATIONS

A few examples are given below by way of illustrations on the utility of the technique.

A) Surface Impurities

A carbon substrate which has been contaminated with many trace impurities on its surface was analysed with 2.3 MeV $^4He^+$ ions and the resultant spectrum is shown in Fig.5. The relative concentration could be determined from the respective areas of the peaks of elements Bi, Ba, Pd, Y, Ni, Fe, Ti, K, S, Si, Mg, F, O and N, which could be fully resolved.

As an example of the sensitivity, a monolayer of Au($N = 10^{15}$ atoms/cm^2) on Si, annalysed with 500 na of 1.6 MeV. (dσ/dΩ (170°) = 12.8 barns) He ions for 40 Secs (1.25 x 10^{14}. He particles; 20 μc of charge) generates a signal (peak area) in the detector of 6400 counts above the background, when the detector subtends a solid angle of 4 mSr. Only 1 out of 2 x 10^{10} alpha particles are scattered into the detector. With a background signal of 1 c/s, the detection limit is about 10^{11} Au atoms/cm^2.

B) <u>Ion Implantation</u>

A silicon target was implanted with 200 KeV As ions (2 x 10^{15} ions/cm^2) and the sample was analysed with 2 MeV ^{4}He$^+$ ions. A typical spectrum is shown in Fig.6. The position of As Peak can be used for the estimation of range of As ions in Si. The FWHM of the AS peak gives information on the range straggling. The area under the peak yields the dose of implanted ions.

Implantation is a widely used technique in "material modification". The physical understanding of this phenomenon was made possible through the use of the RBS technique. As an example the report of Madakson[5] may be mentioned on the role of impurity mobility in the material modification through implantation. An Al specimen (chemical composition by wt % 98 Al, 0.7 Fe, 0.5 Si, 0.1 Cu, 0.1 Zn and 0.1 Mn) was implanted with 340 KeV Sn^+ ions (10^{15} ions/cm^2). RBS spectrum was recorded immediately after implantation and again after annealing at 300°C for 5 hours. The intensities of the resulting impurity peaks in each case were helpful in providing an understanding of the modifications of the surface properties (friction, wear, hardness, corrosion, oxidation and fatigue). The spectrum immediately recorded after implantation showed increased concentration per unit depth of the impurities Fe, O and Cu. Annealing, further increased these concentrations. Similar studies were also carried out with implantations of other ions (Pb, N, Cu).

C) <u>Thin Film Processes and Reactions</u>

The RBS technique has been extensively used in the study of layered structures. Typical examples of multiple layers are shown in Fig.7. The upper part of the figure shows resolved structure, while the lower part shows the type of distribution that can result from an incomplete resolution.

The thicknesses in both cases can be estimated. In the heavy atom region the problem is more severe, as the individual contributions are hard to resolve. In such situations the RBS spectra and Ion-Induced X-rays are recorded simultaneously to aid the identification of the individual constituents.

When two metal films or a metal film and a semi conductor film are placed in contact they will inter-diffusse and form compounds at the interfaces. One of the major subjects of study using the RBS technique is the silicide formation. The physics of the process as well as the dynamics of the process are of great interest as a variety of silicides are involved in semiconductor technology.

One interesting technique in understanding the process is the use of implanted rare-gas ions as "diffusion markers". An example of the use of Xe ions is illustrated in Fig.8. After implanting Xe ions in a Si substrate a thin layer of Ni was evaporated. The RBS spectrum (open circles) shows clearly the Ni film distribution, the Xe distribution peak and the matrix distribution. On a heat-treatment at 325°C for 20 minutes Ni_2Si was formed at the interface. The RBS spectrum shows three effects.

1. The Xe peak moves closer to the expected Xe surface energy.
2. Nickel peak shows two contributions. The peak at higher energy corresponds to surface Ni while the inner one corresponds to the Ni in Ni_2Si.
3. Si distribution shows a step. The higher energy step corresponds to Si in Ni_2Si while the lower-energy step is due to Si in the matrix.

The spectra show that Ni moves inside during the heat treatment and forms the silicide.

Some times inhibition of silicide formation is of interest. In the development of VLSI devices, it is of interest to obtain reliable shallow contacts which minimise the consumption of substrate Si during the thermal annealing process. Ho and Nicolet[6] studied the role of nitrogen implantation in the Co-Si system. They evaporated a Co film of 181.4 nm on a Si wafer. ^{15}N ions of energy 200 KeV were implanted as N^+_2 with an expected range of 96 nm of Co. Implantation dose was 10^{16} ions/cm^2. Silicide growth was monitored by 2 MeV $^4He^+$ ions. They showed that nitrogen implanted in Co on Si diffused rapidly to the Co-Si interface at the silicide formation temperature of 450°C. A nitrogen diffusion barrier of maximum density 8.5 x 10^{15} ions/cm^2 forms at the interface inhibiting CO-Si reaction upto 550°C. for lower doses of implantation no inhibitation was noticed. For higher doses, the excess nitrogen escapes from the surface.

Possible mechanisms of diffusion and different types of other processes that occur at interfaces are now-a-days routinely studied using RBS techniques, in addition to a variety of surface analytical technique.

As an example the recent study of Sood[7] et al. on ion-beam induced demixing in the Ni-Pb system. They studied different layer structures of Ni-Pb-C on SiO_2 after bombarding with 200 KeV Kr ions to different doses (0.5 to 3 x 10^{16}) ions/Sq.Cm.), using RBS and SEM techniques. They observe that Ni migrates to the surface through the overlay film of Pb and large sputtering loss of Pb which is minimised by C cap. An island structure of Ni on SiO_2 was observed whenever Kr ions reach the interface of SiO_2/Ni. Possible mechanisms are radiation induced seggregation and radiation enhanced diffusion.

A comparison of the different techniques used in depth microscopy is given below:

Table 1

Comparison of the different Analytical Techniques

Technique	Primary Range (μm)	Signal escape depth (μm)	Sensitivity mass fraction	Depth resolution Å	Accuracy without standards
RBS	0.5 - 2	< 2	10^{-1} to 10^{-5}	20 - 500	2%
AES	0.05 - 0.5	$10^{-3} - 10^{-2}$	10^{-3}	20 - 100	50%
SIMS	$10^{-3} - 10^{-2}$	5×10^{-4}	$10^{-6} - 10^{-8}$	20 - 100	Factor of 3
XRF	0.5 - 5	0.5 - 5	$10^{-6} - 10^{-7}$	none	3 - 10%

In most of the work on RBS so far carried out conventional particle accelerators were used. The requirements of RBS however are modest. A few nano amperes of (10 to 100) 2 MeV He ions are needed in most of the applications. The current trend is to develop dedicated systems providing one type of ion of fixed energy for the commercial production of RBS systems to enable a large scale use of the technique by non specialists.

REFERENCES

1. New Uses of Ion Accelerators
   Ed by J.F. Zieglar
   Plenum Press (1975).

2. Back-Scattering Spectrometry
   W.K. Chu, J.W. Mayer and M.A. Nicolet
   Academic Press (1978).

3. Beam Interactions with Materials and Atoms
   Nucl. Instr. and Meth. B 19/20 (1987)

4. M. Hage - Ali and P. Siffert
   Nucl. Instr. and Meth. 166 (1979). 411 - 418.

5. P.B. Madakson
   Nucl. Instr. and Meth. 218 (1983) 537 - 541.

6. K.T. Ho and M.A. Nicolet
   Thin Solid Films 127 (1985) 313 - 322.

7. D.K. Sood et. al.
   Nucl. Instr. and Meth. B20 (1987) 632 - 637.

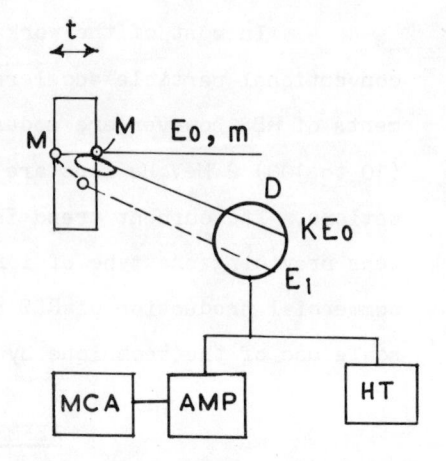

FIG.1. TYPICAL EXPERIMENTAL ARRANGEMENT OF RBS TECHNIQUE

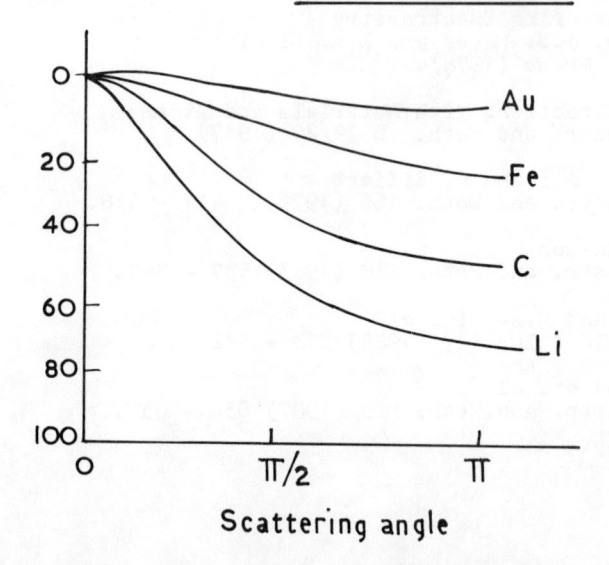

FIG.2. A PLOT OF THE PERCENTAGE LOSS OF THE PROJECTILE ENERGY ON SCATTERING AS A FUNCTION OF SCATTERING ANGLE θ

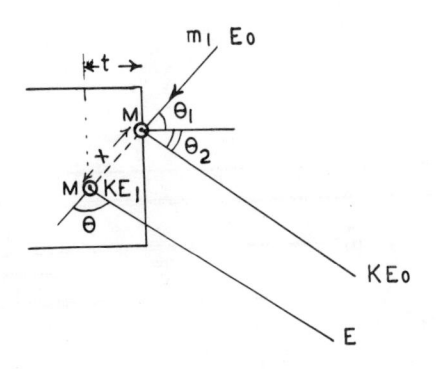

FIG.3. DEPTH SCALE IN BACK-SCATTERING

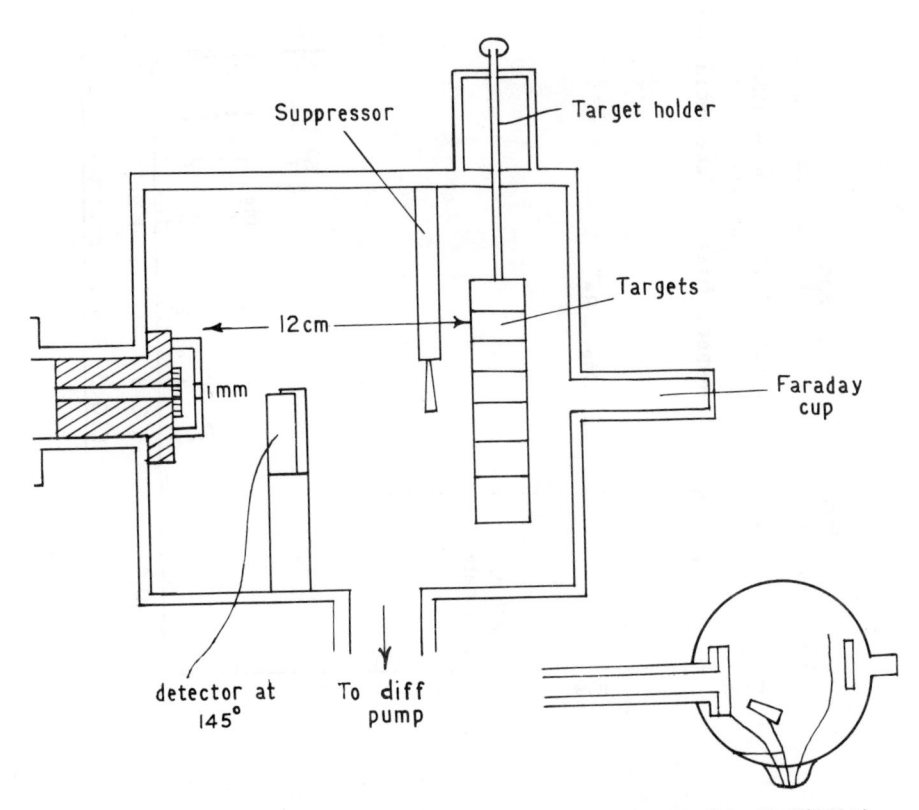

FIG.4. SCHEMATIC OF THE SCATTERING CHAMBER NATIONAL UNIVERSITY OF SINGAPORE.

FIG.5. R.B.S Spectrum of surface impurities.

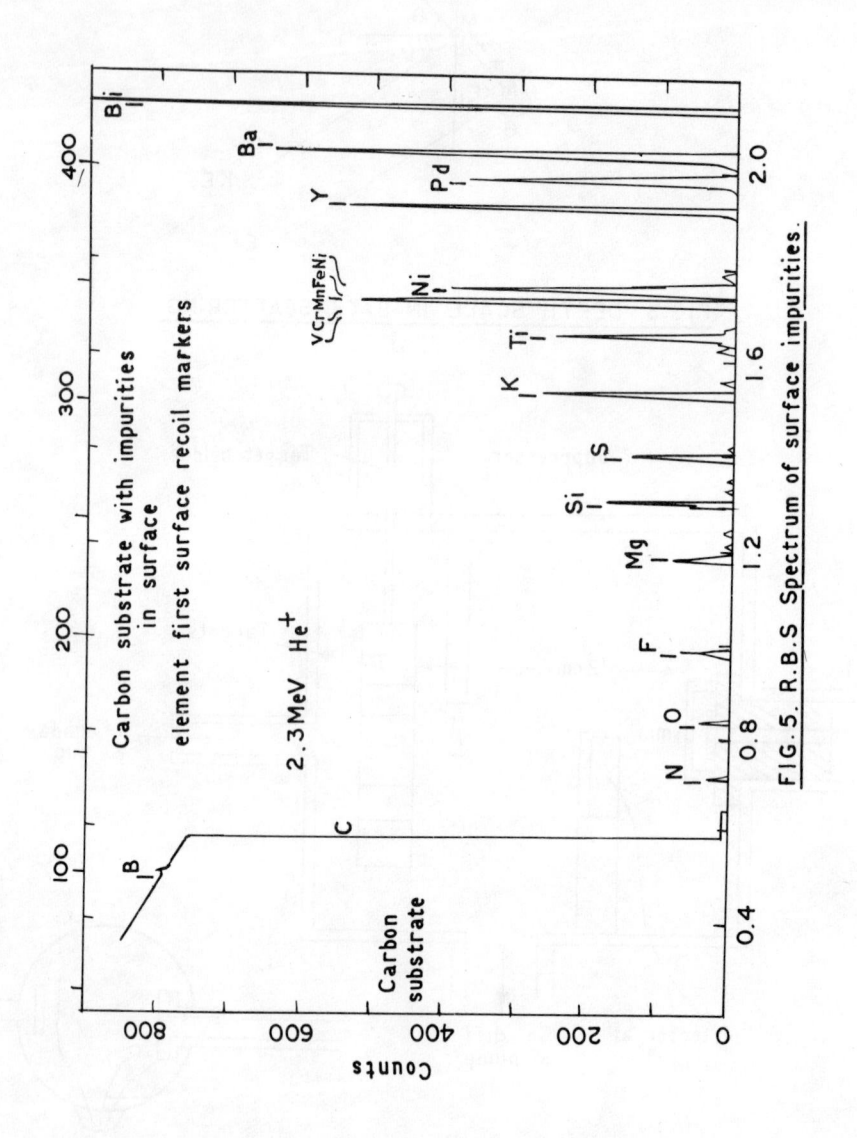

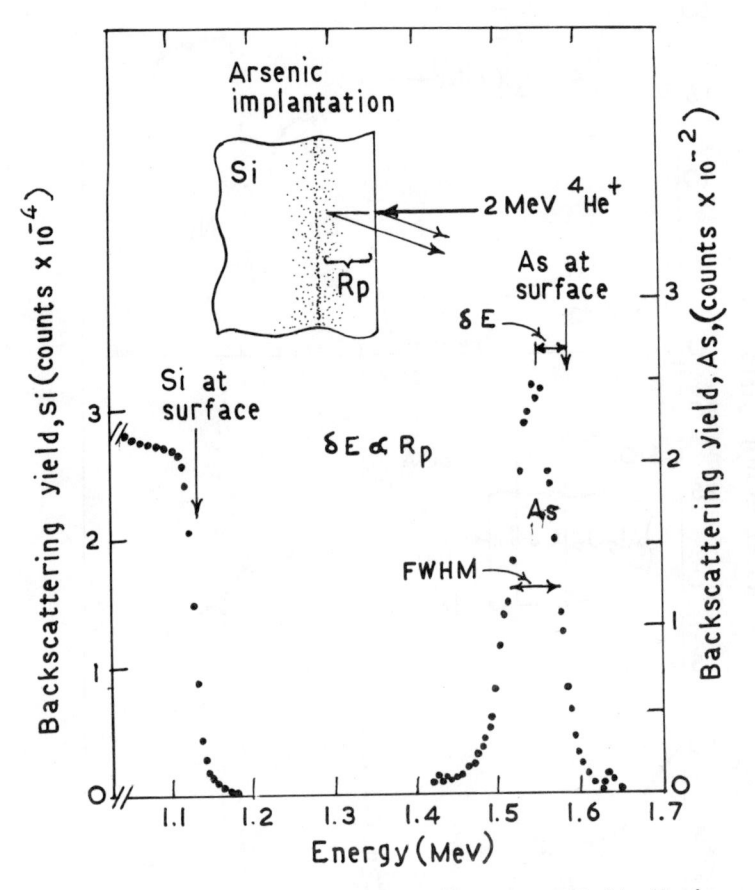

Fig. 6. Backscattering spectrum of 2 MeV He ions on an As implanted Si sample

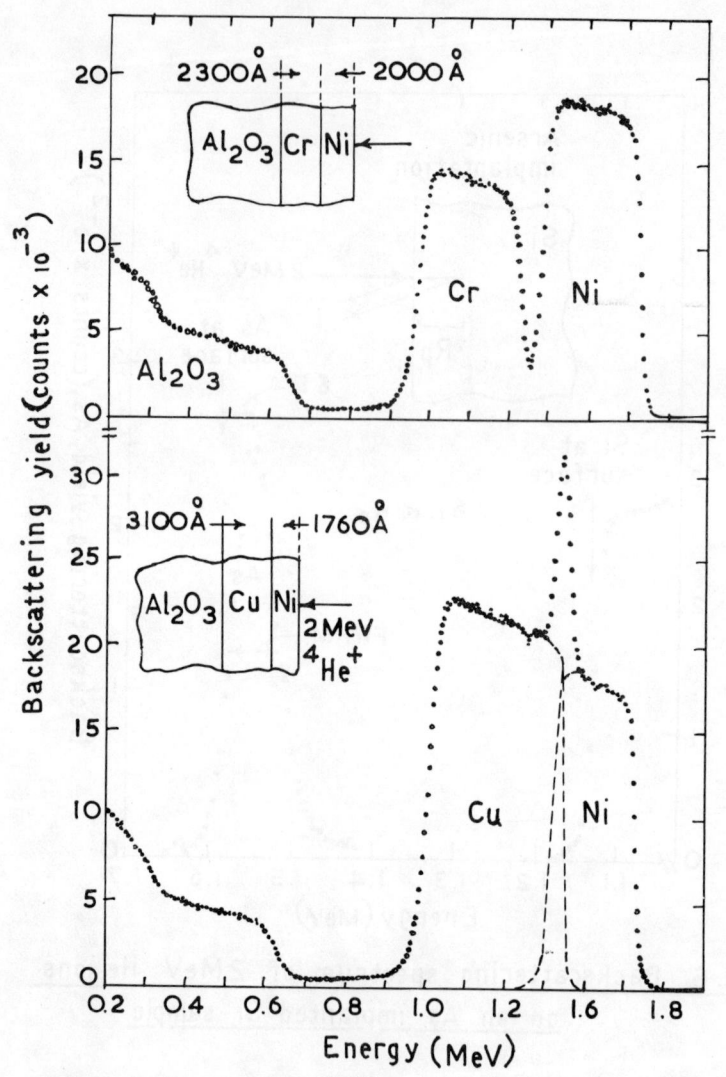

Fig. 7. Spectra from metal layers on Al_2O_3 substrates.

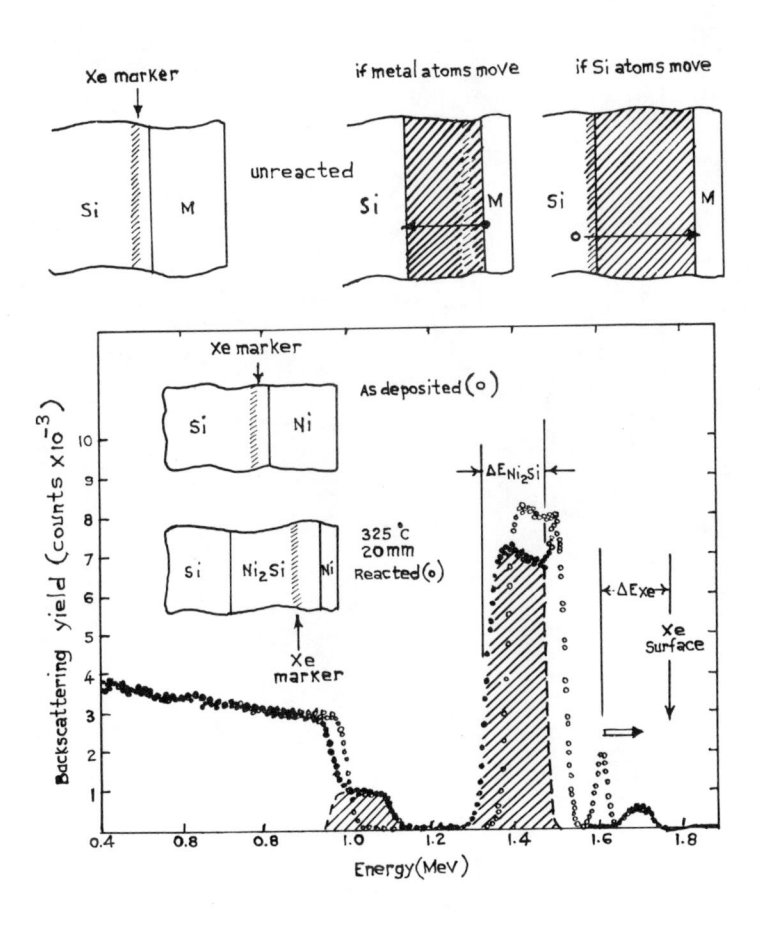

Fig. 8. The concept (above) and application of backscattering to diffusion marker studies

CHARACTERISATION OF NUCLEAR MATERIALS

C.K. MATHEWS
Radiochemistry Programme
Indira Gandhi Centre for Atomic Researach
Kalpakkam 603 102

1. INTRODUCTION

A nuclear reactor mainly contains three materials - fuel, coolant and structural materials. In the reactor fuel heat is produced through the fission process. This heat is transported by the coolant for the production of steam. Structural materials contain these materials in appropriate geometric shapes. Thermal neutron reactors contain a moderator in addition, in order to slow down fission neutrons to thermal energies. Fast breeder reactors [1] do not employ a moderator. The fuel here is usually a mixed oxide of uranium and plutonium. Mixed carbides and nitrides are also under development as potential fuels. Liquid sodium is the coolant in fast reactors, because it transports heat from the high density source in the core with high efficiency. Austenitic steels are generally employed as structural materials. Characterisation of nuclear materials will be discussed here in taking fast reactor materials as examples.

2. ANALYTICAL CHARACTERISATION OF SODIUM

The specifications of nuclear grade sodium are given in Table 1. Various techniques were developed/standardised in our laboratory for the chemical quality control of reactor grade sodium. Sodium being a very reactive metal, special techniques are necessary to handle it. The inert atmosphere boxes set up in our laboratory for this purpose and the sampling techniques which ensure negligible pick up of impurities, especially oxygen and moisture, are described elsewhere [2].

In the determination of non-volatile metallic impurities as well as non-metallic impurities like oxygen and carbon the impurities are first separated by distilling off sodium using a vacuum distillation procedure. The sample of sodium is taken in an appropriate crucible [material choice dictated by the impurity to be analysed] and placed in the crucible holder inside the chimney under high purity argon, the whole operation being carried out in a sodium box. The distillation apparatus is then connected to a vacuum system and evacuated to 10^{-3} torr pressure. An induction coil placed around the chimney at the sample location is used to distill off the sodium which gets coated on the inner surface of the chimney. The crucible is later taken out and the residue dissolved in an appropriate reagent for analysis. For the analysis of oxygen, a nickel crucible is used and the residue dissolved in water for the determination of Na (remaining as Na_2O after distillation) by flame emission. For carbon analysis, the sample is taken in an alumina crucible and the residue is subjected to combustion followed by CO_2 analysis. Trace metals are dissolved in a suitable acid (crucible material: Ta) and determined by atomic absorption spectrophotometry (AAS). Volatile metals like Cd, Pb and Zn are analysed by anodic stripping voltammetry or by a direct atmoisation method using graphite furnace atomisation.

3. ATOMIC ABSORPTION SPECTROPHOTOMETRY (AAS)

Atomic absorption spectroscopic analysis depends on the conversion of a part of the sample into an atomic vapour whose absorbance is then measured at a specific wavelength. Unknown concentrations are determined by comparison with absorbance measurements made on standards of known composition. The relationships between absorption and concentration have been derived in several papers and in standard reference texts [3-5]. Thus the most important part of an AAS is the atom cell where neutral atoms of the analyte element are produced in the ground state. Generally laminar flow combustion flames such as air/C_2H_2 ($2300^{O}C$) and N_2O/C_2H_2 ($2800^{O}C$) are very useful atom cells. Electrically heated graphite furnaces are also used when very high sensitivities ($10^{-11} - 10^{-12}g$) are required.

Hollow cathode lamps or electrodeless discharge lamps are generally used as light sources. The construction and principle of working of these lamps are described in text books

[3-5]. The important property of these lamps is that these could emit lines which are sharper than the absorption lines in the flame. Hence the peak absorption measurements could be made which is proportional to concentration [6]. The sample solution after nebulisation into a fine spray using either a pneumatic nebuliser or an ultrasonic nebuliser is introduced into the flame. A monochromator serves to isolate the analyte line of interest and the light intensity measurements are done using the detector/read out system.

The advantages of AAS are:
1. The technique uses the sharp characteristic resonance lines of the elements. Hence it is very specific and accurate. Spectral interferences are very few.
2. Very good precision of the order of 1-2% rsd are routinely attainable and the method is very rapid.
Its disadvantage is that it is a single element technique and that there are possibilities of spectral, chemical and ionization interference which must be taken into account.

Graphite Furnace Atomiser

The atomiser takes the form of a small graphite tube which is electrically heated (in an inert atmosphere) through three different stages of controlled heating, viz., dry, char and atomization. In the first stage (dry), the solvent is removed slowly at a low temperature (80-100°C). In the second stage (char), a suitable intermediate temperature is used (400°C - 900°C) to remove as much as possible of the bulk of the matrix, without sacrificing the analyte. In the third stage (atomisation), the graphite tube is heated to a high temperature (2000-2800°C) at a very rapid rate (800°C/sec). The peak height of the transient absorbance pulse is measured.

The high sensitivity obtained with graphite furnace atomiser arises from the small volume of the atomiser (6 mm dia, 30 mm long) the higher residence time of the atoms (~500ms - 1 sec) as compared to 10^{-3} - 10^{-4} sec. in the case of flames, the absence of dilution by expanding flame gases, and the highly reducing environment inside the incandescent graphite tube.

Characterisation of Sodium in AAS

The following methods have been standardised in our laboratory for the chemical characterisation of sodium.

1. Vacuum distillation cum flame AAS:

In this method, the residue obtained after vacuum distillation [see Section 2] is dissolved in 1:1 aqua regia, made up to volume and the impurities determined using flame AAS. The results of the analysis of nuclear grade sodium and the detection limits obtained are presented in Table 2.

2. Direct method using graphite furnace aas:

This method has been developed to determine volatile impurities such as cadmium which co-distil with sodium and hence cannot be determined by the above method. In this method, sodium is carefully dissolved in water in argon atmosphere and the resulting solution is converted to sodium nitrate. 25 ul of 0.2% (Na W/V) solution is loaded into the graphite furnace and analysed. This method [7] has been standardised for the following elements viz., Cd, Pb, Ni, Co and Cr. The detection limits obtained are presented in Table 2.

4. ON-LINE MONITORS

For the on-line monitoring of the purity of liquid sodium several electrochemical meters have been developed [8-11]. They are based on the principle of concentration cells which are represented as follows:

Oxygen Meter : In, In_2O_3 / YDT / $[O]_{Na}$

Hydrogen Meter : Li, LiH / $CaCl_2$-CaH_2 / $[H]_{Na}$

Carbon Meter : $[C]_{Ni}$ / Na_2CO_3-Li_2CO_3 / $[C]_{Na}$

In all these cases sodium containing the impurity being measured is the sample electrode which is separated from a reference electrode by an electrolyte. Yttria doped thoria (YDT) is used as the solid electrolyte in the oxygen probe while $CaCl_2$-CaH_2 serves this function in hydrogen meter. A liquid electrolyte of molten carbonate is used in the carbon meter. The EMF measured between the reference and sample electrodes gives a measure of the impurity concentration in sodium. Sub-ppm sensitivities are available with all these meters.

5. NUCLEAR FUEL CHARACTERISATION

A mixed oxide of uranium and plutonium is the fuel commonly used in fast breeder reactors. Mixed carbide and nitride are under development as advanced fuels. For ensuring good performance of the nuclear fuel in the reactor, the composition of the fuel as well as the impurity levels have to be carefully controlled. A typical specification for $(U,Pu)O_2$ fuel is given in Table 3. It can be seen that many impurities have to be maintained at levels of a few tens of ppm. The characterisation of the nuclear fuel material at various stages during preparation and fabrication is therefore a very important requirement. In general, the characterisataion of the nuclear fuel involves (i) determination of chemical composition and (ii) measurement of physical parameters such as dimension, density, pore size, grain size etc. Apart from this, uniform distribution of Pu in the fuel is to be ascertained. The chemical characterisation of mixed oxide or carbide fuel consists of determination of the metallic constituents (U & Pu), the non-metallic constituents (O or C) and impurities-metallic, non-metallic and gaseous. Apart from this, the isotopic composition of U & Pu also needs to be determined as this is important from the neutronics point of view.

A number of methods exist for the determination of U and Pu in the fuel materials (12). Some of these have the attraction of sequential determination of the two elements in one sample. The most important method for uranium determination is the Davies & Gray method (13). The determination of U & Pu by the electrochemical methods described above yields results accurate & precise to +0.1% or better. Another interesting aspect of these titrimetric methods is that the titration is carried out on a weight basis rather than volume basis. This helps in obtaining precise results with U or Pu amounts in aliquots as low as 2 mg.

The oxygen-to-metal ratio is another important parameter that affects the physico-chemical behaviour of the oxide which in turn has a decisive influence on its behaviour during irradiation.

Of the methods available for the determination of O/M, the most popular is the thermogravimetric method (14) which involves equilibration of the sample at 800°C with flowing Ar/H_2 mixture saturated with water at 0°C. This equilibration converts the oxide, whether hypostoichiometric or hyper, to O/M = 2.000. The weight gain or loss as a result of the equilibration can then be used to calculate the O/M. The precision of O/M determination by this method is typically +0.002 O/M units.

Another important method for O/M determination is based on the measurement of the EMF developed in the cell

$$Pt, MO_{2+x} // YDT // Ni, NiO$$

due to the difference in oxygen potentials between the oxide electrodes (15).

At IGC, the characterisation of U & Pu carbides has a special significance in view of the use of mixed (Pu,U) carbide fuel in our fast reactor. In the characterisation of carbides, apart from U & Pu, carbon must also be determined. The characterisation of oxygen and nitrogen in carbides is also important as these impurities affect the carbon potential of the fuel.

6. OPTICAL EMISSION SPECTROGRAPHY (OES)

The principle of the technique, which is described in textbooks [16] may be briefly summarised as follows. An aliquot of the sample to be analysed is volatilised and atomised; a fraction of the free atoms is raised to an excited state, usually in an electrical discharge. Optical radiation from the discharge is gathered, dispersed and either photographically or photoelectrically recorded. Elements are identified from the position of the spectral lines with respect to each other. The line intensities, determined by a suitable calibration procedure, serve as a basis for quantitative analysis.

Optical Emission Spectrograph

The instrument used in our laboratory is a grating spectrograph (Jarrel Ash 3.4 M Ebert).

D.C. arc is most often used as the excitation source. Here an arc discharge is established between two electrodes. The analytical samples are mounted in the anode, where sample volatilisation into the arc column is more efficient. The efficient thermal vaporisation process of the arc makes it the most sensitive discharge available for the determination of trace impurities. This technique has been used to determine trace impurities in uranium oxide using the carrier distillation technique [17]. Silver chloride (5%) has been used as the carrier. The boiling point of AgCl is less than that of uranium oxide, but more than the impurity elements of interest. Hence AgCl serves to volatilise only the impurity elements into the arc stream. Uranium, which has a very complex spectrum, is effectively contained and is not allowed to come into the arc.

Analytical Procedure

Synthetic standards were prepared by thoroughly mixing the spec-pure oxides of analytes (Johnson-Mathey) with nuclear grade uranium oxide. The master standard was successively diluted to get the working standards. The samples and standards were mixed with AgCl (5%). 100 mg of the charge was loaded on to the carrier distillation electrodes and excited by a 10 amp d.c. arc. The spectra were recorded on Kodak SA (1) photographic plate and the line transmittances were read by microphotometer. The transmittance values were converted to intensity values and background corrected using the emulsion calibration graph drawn using a seven step filter. The final calibration was drawn between $\log [I^{analyte}/I^{int.\ std.}]$ vs concentration of the analyte.

The results of the analysis of uranium oxide sample (sample no. SRM 64) received from IAEA for inter-laboratory comparison purposes, are presented in Table 4.

7. X-RAY FLUORESCENCE SPECTROMETRY (XRF)

X-ray fluorescence spectrometry is a convenient and, in favourable cases, non-destructive method of analysis. In this technique, x-ray excitation is caused by irradiating the sample with x-rays or gamma rays. The emitted (secondary) x-ray spectrum contains the characteristic lines of the elements of interest in the sample. By analysing this with a crystal spectrometer (wavelength dispersion) or with a Si(Li) detector coupled to a multichannel analyser (energy dispersion), the intensities of the characteristic lines can be measured and related to elemental concentration. The area under the diffraction peak gives a measure of the concentration of the element in the sample. XRF is a well known method of steel analysis [18] and it is extensively used in our Laboratory [19].

The general procedure consists of the following steps:
1. Specimen preparation and presentation.
2. Excitation of charactersitic x-ray spectra.
3. Wavelength/energy dispersion.
4. Detection.
5. Correlating intensity to concentration.

The advantages of XRF are that it (i) is a non-destructive technique, (ii) has a wide composition range (ppm-100%) and (iii) has a wide elemental range (Z=9-94). Its disadvantages are that (i) it is essentially a surface technique limited by the depth from which the x-rays can emerge from the sample; (ii) measurements at low z are difficult (require special arrangements); (iii) there are strong matrix effects. There are several techniques for overcoming matrix effects.

XRF as a trace analysis technique :

The technique has good precision and accuracy (2-5% at ppm level).and high sample versatality (solid, liquid, powder, gases absorbed in suitable liquids, particulate matter on filters and so on). In nuclear technology, XRF can be employed [20] for the determination of nuclear fuel materials (Th, U, Pu) even at trace levels. XRF can also be used for the quality control of nuclear fuels [21], e.g., the determination of metallic impurities like Ni, Fe, Cr, W in UO_2 & $(U,Pu)O_2$. We have standardised a method for the determination of Fe and Ni in UO_2 [22].

8. INDUCTIVELY COUPLED MASS SPECTROMETRY (ICP-MS)

ICP-MS is a new method for elemental and isotopic analysis which is attracting a great deal of interest. This technique allows the direct, rapid and convenient recording of the mass spectra of trace elements in solutions. Potentially attractive features of this technique are low detection limits (sub ng/ml) simple spectra and the availability of isotope ratio information (as against ICP-OES).

ICP as an ion source:

ICP has already been well established as the most efficient excitation source in optical emission spectrometry. An overview of ICPs by Fassel [23] and deatailed review by Barnes [24] can be found in literature. Samples are sprayed into the plasma with a nebuliser. The ICP heats the sample to approximately 7000 K where the dissolved solids are vaporised, atomised and ionised. Since the ICP efficiently converts the elements of the sample into ions, it is a very useful ion source.

ICP-MS

The working principle of ICP-MS can be understood with reference to the fig. 1 which shows a schematic of the ICP-MS system. ICP is used in an end on configuration. A fraction of the ions is sampled through a differentially pumped region at one torr and enters the ion optics. This region consisting of the sampler (1.125 mm dia) and skimmer (0.875 nm) constitutes the crucial ICP-MS interface which was developed by Gray, Date and Douglas [25,26] and still is being improved to bring down the molecular ions and doubly charged ions. The ion optics incorporates a central stop to reduce photon noise. The ions then enter the quadrupole mass analyser and then exit to the channel electron multiplier. The detector is mounted off axis to further reduce photon noise from the plasma. As the elemental ions formed are essentially singly charged, unit mass resolution attainable throughout the mass range (3-300 amu) with the quadrupole mass analyser has been found to be quite adequate for mass analysis. Rapid scanning of the entire or selected mass range (3-300 amu) is possible. Alternatively peak hopping to the elemental peaks of interest is conveniently done at high speed (100 vs setting time).

The advantages of both the ICP and MS are combined nicely in this technique. Thus there is:

1. Freedom from matrix interference and ionisation interference.
2. Very high sensitivity and low detection limits (sub ng/ml).
3. Wide applicability [It has been shown [27] that in ICP some 54 elements are ionized with an efficiency of 90% or more].
4. Excellent linearity and dynamic range (5-6 orders of magnitude).
5. Simple spectra; and
6. Possibility of accurate analysis through isotope dilution.

This technique will be particularly very useful in the trace charactorisation of uranium for both the common impurities and rare earths. In optical emission spectroscopy, the technique generally used for this purpose, uranium gives a very complex spectrum and hence it is necessary to use the carrier distillation technique for common impurities and the laborious solvent extraction procedure for the determination of rare earths. On the other hand, in ICP-MS uranium just gives two lines (masses 238, 235) which are far away from the impurity lines. Hence the concentrations of the impurities are readily determined by directly aspirating the uranium nitrate solution into the ICP.

References

1. Status of liquid metal cooled Fast Breeder Reactors, Tech. Report No. 246, IAEA, Vienna (1985).
2. C.K. Mathews, Pure & Appl. Chem. 54, 807 (1982).
3. A.C.G. Mitchell and M.W. Zemansky, Resonance Radiation and Excited Atoms, Cambridge Press, 1934.
4. G.F. Kirkbright and M. Sargent, `Atomic Absorption and Fluorescence Spectroscopy´, Academic Press, London, 1970.
5. B.V.L´Vov, Atomic Absorption Spectrochemical Analysis, Adam Hilger, London, 1970.
6. A. Walsh, Spectrochim Acta, 7, 108 (1955) p. 252.
7. T.R. Mahalingam, R. Geetha, A. Thiruvengadasami and C.K. Mathews, Analytica Chimica Acta, 142 (1982) 189-195.
8. M.R. Hobdell and C.A. Smith, J. Nucl. Mater. 110 (1982) 125-139.
9. P. Roy and B.E. Bugbel, Nucl. Technol. 39, 216-218 (1978).
10. E.C. Subba Rao and H.S. Maiti, solid State Ionics 11 (1984) 317-338.
11. S. Rajendran Pillai and C.K. Mathews, J. Nucl. Mater. 137 (1986) 107.
12. M.V. Ramaniah, P.R. Natarajan and P. Venkataraman, Radiochim. Acta, 22 (1975) 199.
13. W. Davies & W. Gray, Talanta, 11 (1964) 1203.
14. C.E. McKeilly & T.D. Chikalla, J. Nucl. Mater., 39 (1971) 39.
15. T.L. Markin and R.J. Bones, AERE-R-4178.
16. T. Torok, J. Mika and E. Gegus, `Emission Spectrochemical Analysis´, Adam Highear, Briston, 1978.
17. B.F. Scribner and H.R> Mullin, J. Res. Bur. Stand. 37 (1946) 979.
18. A.P. Vikolsky, C.R.C. Critical Reviews in Anal. Chem. 13 (1982) 373.
19. K.V.G. Kutty, S. Rajagopalan, R. Asuvatha Raman and C.K. Mathews [To be published as a Technical Report].
20. A.H.E. von Baeckmann, D. Ertel and J. Neuber, Adv. X-ray Analysis, 18 (1974) 62.
21. B.L. Taylor, G. Phillips and G.W.C. Milner, Analytical Methods in Nuclear Fuel Cycle, IAEA (1972) 237.
22. K.V.G. Kutty, S. Rajagopalan, S.K. Ananthakrishnan and C.K. Mathews, IGC-89 (1986).
23. V.A. Fassel, science, 202 (1978) 183.
24. R.M. Barnes, CRC Crit-Rev. Anal. Chem. 7 (1978) 203.
25. A.R. Date and A.L. Gray, Spectrochim Acta, 38B (1983) 29.
26. D.J. Douglas, E.S.K. Quan and R.G. Smith, Spectrochim Acta, 38B (1983) 39.
27. R.S. Houk, Mass Spectrometry of Inductively Coupled Plasmas, Anal. Chem. Vol. 58, NI, Jan. 1986.

Table 1. Specifications of reactor grade sodium

Impurity	Specified maximum (ppm)	Impurity	Specified maximum (ppm)
Al	5	In	30
Ag	20	K	200
B	5	Li	10
Ba	10	Mg	10
C	30	Mn	10
Ca	10	O	10
Cd	10	P	20
Co	10	Pb	10
Cr	5	Rb	50
Cs	6	S	10
Fe	25	Si	10
H	10	Sn	10
Halogens	15	U	0.2

The sodium metal content shall be 99.95% or better

Table 2. Determination of trace metals in sodium

A. Flame AAS

SL.NO.	ELEMENT	WAVELENGTH nm	CONCENTRATION IN NUCLEAR GRADE SODIUM, ppm	DETECTION LIMIT (BASED ON 5g SAMPLE WEIGHT) ppm
1	Al	309.3	<5.0	5.0
2	Ba	553.5	1.0	1.0
3	Bi	223.1	<2.0	2.0
4	Ca	422.7	<0.5	0.5
5	Cr	357.9	<0.5	0.5
6	Cu	324.7	0.2	0.2
7	Co	240.7	<0.3	0.3
8	Fe	248.3	1.3	0.4
9	In	303.9	<2.0	2.0
10	Li	670.8	<0.2	0.2
11	Mg	285.2	0.05	0.02
12	Mn	279.5	<0.2	0.2
13	Mo	313.3	<2.0	2.0
14	Ni	232.0	3.5	0.3
15	Pb	283.3	<1.0	1.0
16	Sn	235.5	<7.5	7.5

B. Direct Method

1.	Cd	228.8	0.03	0.02
2.	Pb	283.3	<0.5	0.5
3.	Ni	232.0	0.8	0.5
4.	Cr	357.9	0.3	0.2
5.	Co	240.8	0.2	0.1

Table 3

Specification of $(U,Pu)O_2$

QUANTITY	SPECIFIED VALUE
TOTAL Pu CONTENT	+ 1 % RELATIVE OF NOMINAL VALUE
TOTAL U CONTENT	DO
Am CONTENT, WT % OF Pu	0.25
CARBON (ug/g)	150
FLUORINE (ug/g)	10
NITRIDE NITROGEN (ug/g)	200
BORON (ug/g)	20
BERYLLIUM (ug/g)	20
LITHIUM (ug/g)	10
SUM OF Dy, Gd, Eu, Sm (ug/g)	100
GAS (STP, cc/g)	0.09
WATER (ug/g)	30
O/M RATIO	1.95 TO 1.99

Table 4

Determination of impurities in uranium (IAEA intercomparison exercise)

ELEMENT	CONCENTRATION RANGE	CONCENTRATION IN IAEA SAMPLE
B	0.15-2.5 PPM	0.68 PPM
Be	0.10-2.5 "	<0.1 "
Cd	0.25-10 "	<0.1 "
Co	1-10 "	6.2 "
Cr	5-100 "	3.1* "
Cu	5-100 "	2.5* "
Fe	25-500 "	77.5 "
In	0.5-25 "	<0.5 "
Mg	5-100 "	<5 "
Mn	2.5-100 "	18.63 "
Ni	3-100 "	15.8 "
Pb	2-50 "	1.68* "
Sm	1-50 "	<1 "
Sb	2-100 "	<2 "
Ti	10-100 "	<10 "
V	1-25 "	<1 "
W	100-2500 "	<4* "
Zn	25-500 "	<25 "

* BY SOLVENT EXTRACTION TECHNIQUE

FIGURE 1

Schematic diagram of an I C P - M S system

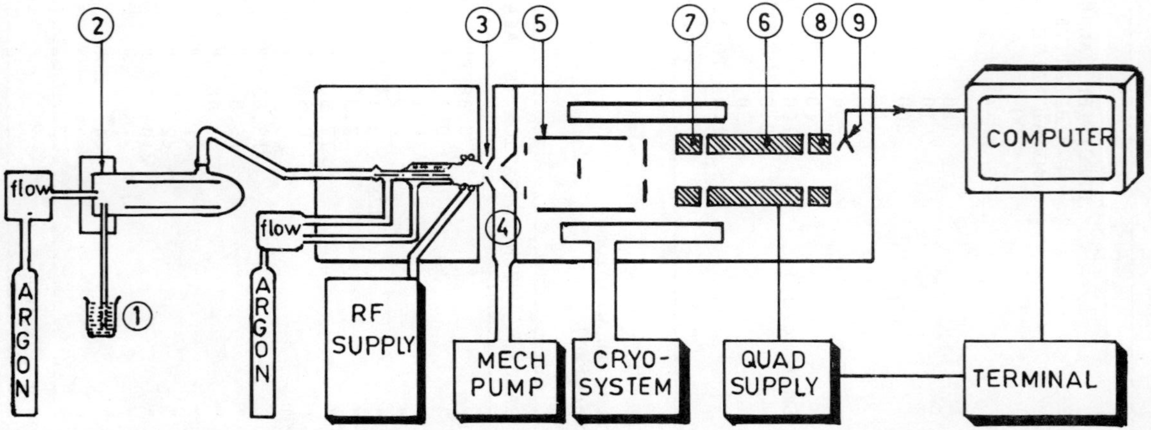

①Sample solution ②Nebulizer ③Plasma tail flame ④Differentially pumped interface ⑤Ion optics incorporating a central stop ⑥Quadrupole rod set ⑦Entrance R F only rod set ⑧Exit RF only rod set ⑨Electron multiplier detector

COMPOSITIONAL CHARACTERIZATION OF AMERICIUM-CURIUM MIXTURES

S. C. McGuire
Department of Physics
Alabama A&M University
Huntsville, Alabama   35762

D. E. Benker and J. E. Bigelow
Chemical Technology Division
Oak Ridge National Laboratory
Oak Ridge, Tennessee   37831

ABSTRACT

A method is described for the rapid characterization of purified curium obtained via high pressure cation exchange operations at the Oak Ridge National Laboratory's(ORNL) Transuranium Processing Plant(TRU). The major actinide activities in the feed to the ion exchange runs were due to 244Cm, 243Am and 241Am. Prompt photon counting is used to identify and determine quantitatively the prominent nuclides, 244Cm and 243Am ,in the product fractions. The method is further used to assess the decontamination factors(DFs) acheived during the separation process.

INTRODUCTION

This paper describes a method for the rapid characterization of purified curium obtained from high pressure cation exchange operations at the Oak Ridge National Laboratory's(ORNL) Transuranium Processing Plant(TRU). The availability of chemically separated curium is important to the study of its fundamental chemical and physical properties[1,2]. Curium is used as the target material from which higher transuranics(Bk,Cf,Es and Fm) are generated via neutron capture in the central core region of the High Flux Isotope Reactor[3]. In addition, ^{248}Cm which is produced from the radioactive decay of ^{252}Cf, has been used as a target material in heavy ion studies aimed at producing superheavy elements[4]. The primary objective of the present work has been to develop a radioanalytical procedure that permits the identification and quantitative determination of ^{243}Am in the presence of relatively large amounts of ^{244}Cm where dilute evaporated samples are used. When the samples are observed on an alpha particle counting system employing a room temperature solid state surface barrier detector, difficulties arise due to the limited resolution of the system. Specifically, the events belonging to the less intense ^{243}Am activity can be obscured by the low energy tail of the main ^{244}Cm lines. An example of an alpha particle spectrum taken on a system which was in operation at the time of this work is shown in Figure 1. This difficulty can be avoided by using instead an intrinsic germanium detector to observe the prompt photons at 42.84 and 74.67 keV which are emitted in the decay of ^{244}Cm and ^{243}Am, respectively. As indicated by the decay data[6] listed in Table 1., the absolute intensity of the 74.67 keV line favors its observation in the gamma spectrum and its energy is well removed from the 42.84 keV line of ^{244}Cm. Prompt photon detection has been used successfully in performing radiochemical analyses of these actinides in addition to ^{239}Np, ^{154}Eu, ^{144}Ce and ^{106}Ru for the control of pressurized ion exchange processes[7]. We find that this approach has immediate application in the characterization of curium product fractions produced by means of high pressure cation exchange operations at TRU.

EXPERIMENTAL METHODS

The pressurized ion exchange system used to separate the transplutonium elements is contained inside one of the hot cells at TRU. A flow diagram of the system is shown in figure 5. The equipment components are maintained and operated through the use of master-slave manipulators that penetrate the 1.4 m thick shield wall of the cell. In a typical separation run the feed material is first loaded onto a relatively short column via a vacuum-pressure pot of about 1 liter capacity. This column is subsequently connected to the longer separation column and the eluent solutions are forced through both by means of a positive displacement pump. Silicon surface barrier detectors are used as neutron and alpha particle probes to monitor the progress of the elution. A major advantage of using the two column system is that most of the radiation damage to the resins can be confined to the relatively small volume of the short column. A detailed discussion of the equipment specifications for the system has been given by Benker et al.[5].

The counting samples in this work were made up to have a total alpha activity of approximately 0.23 MBq. This amount of activity was given a volume of 100 μl in 5 $\underline{M}$ HNO_3 solution which was then dried onto a 24mm-diameter watch glass sample dish using infrared heat. The sample dish was seal-mounted onto an 80mm x 64mm sample card having a 25mm-diameter hole in its center. The sample holder was made of lucite to minimize photon scattering into the detector. A cross sectional view of the counting sample is shown in Figure 2.

Photon spectra were taken with a 200mm^2 x 10mm thick intrinsic germanium detector fitted with a thin Beryllium entrance window. The detector was interfaced to a Nuclear Data 6600 computer system equipped with a dual hard disk drive CRT display, line printer and X-Y plotter.

The block diagram for the data acquisition system is shown in Figure 3. Gamma ray standards of ^{226}Ra and ^{57}Co were used to calibrate the system with respect to energy and photopeak detection efficiency.

Data acquisition times of 3.6 ks proved sufficient to obtain reliable estimates for the decontamination factors(DFs). Acquisition times up to 36 ks were used in the case of some samples to improve counting statistics. A low energy photon spectrum showing the photopeaks corresponding to the prominent actinides is provided in Figure 4. Gaussian peak fitting routines were used to determine the energy and intensity of the individual lines in a given spectrum. Under the conditions of the measurements, the DFs could be derived from a comparison of the ratios of the 42.82 and 74.67 keV photopeak count rates in the feed and the product samples. The count rate for the ith isotope in the sample is given by

$$CR_i = N_i \lambda_i I_i E_i \quad , \qquad (1)$$

where N_i is the number of atoms of the ith isotope present, λ_i is its intrinsic decay constant, I_i is the intensity(photons/decay) of the emitted photon and E_i is the absolute efficiency with which the photon is detected in the system. We form the ratio,

$$\frac{CR_3}{CR_4} = \frac{N_3 \lambda_3 I_3 E_3}{N_4 \lambda_4 I_4 E_4} \quad , \qquad (2)$$

where the subscript 3 refers to ^{243}Am and 4 to ^{244}Cm. The decontamination factor can then be defined as

$$DF = \frac{\left[CR_3/CR_4\right]_o}{\left[CR_3/CR_4\right]_k} \quad , \qquad (3)$$

where the subscripts o and k refer to the feed and kth product samples, respectively.

RESULTS AND DISCUSSION

Strictly speaking, in order to determine the decontamination factor at the time of separation one must take into account the time period between the taking of the sample and its counting. This delay however would result in less than 0.01% error in the decontamination factors reported here because of the relatively long half lives of ^{243}Am($T_{1/2}$ = 7.37x10^3 λy) and ^{244}Cm($T_{1/2}$ = 18.10 y). It can be taken into account via the inclusion of an additional factor in equation (3) yielding

$$DF = \exp[-(\lambda_4 - \lambda_3)(\Delta t_k - \Delta t_o)] \times \frac{\left[CR_3/CR_4\right]_o}{\left[CR_3/CR_4\right]_k} , \quad (4)$$

where t is the time that elapses between taking a sample and counting it.

Another experimental consideration is the buildup and subsequent decay of ^{239}Np. Gamma rays with energies of 278 and 228 keV accompany the decay of ^{239}Np which grows into secular equilibrium[8] with the ^{243}Am activity via the decay chain,

$$^{243}\text{Am} \xrightarrow[7,370y]{\alpha} {}^{239}\text{Np} \xrightarrow[2.35d]{\beta^-} {}^{239}\text{Pu} \xrightarrow[24,110y]{\alpha} \cdots .$$

These relatively high energy gamma rays contribute to a platform of Compton scattering[9] events that interfere with the observation of the lower energy 42.84 and 74.67 keV lines in the spectrum. This interference can be reduced considerably by counting the product fraction sample with a minimum delay after it is taken and by using a larger volume detector of comparable resolution.

Typical values for the DFs achieved in this work are given in Table 2. The somewhat large statistical uncertainties in the reported values were the result of the subtraction of a 74.8 keV interference from the K α_2 x-ray of bismuth, a decay daughter of ^{222}Rn, which was present in the background. The lower limits given for three of the samples reflect a detection sensitivity for the apparatus of one part ^{243}Am per 10 million parts ^{244}Cm. Also included is a comparison of the ^{244}Cm content in the product fraction as found via the present procedure and a formerly used technique involving a combination of alpha particle and gamma ray counting. In general the agreement between the ^{244}Cm measurements is seen to be good.

SUMMARY AND CONCLUSIONS

We find that the principal advantage of prompt photon counting is that it provides a rapid and relatively straightforward means of determining the effectiveness of the separation process. Former procedures relied upon a combination of alpha spectra measurements together with delayed counting of photons. The alpha spectra were used solely to account for the curium in the samples and photon counting was performed on a duplicate set of samples to determine the amount of ^{243}Am present. This method required ageing the duplicate samples for two weeks under appropriate storage conditions to allow for the ingrowth of ^{239}Pu from ^{243}Am. In the present technique, both ^{244}Cm and ^{243}Am are determined quantitatively at the same time. Reliable estimates of the DFs are obtained at the time of separation thus eliminating the need for duplicate sets of samples for additional measurements. It is expected that using a larger volume photon detector with a multicomposition(eg. Pb-Cd-Cu) background shield will result in an improvement in the sensitivity of the method.

ACKNOWLEDGEMENTS

The authors express their appreciation to J. Cooper, F. F. Dyer and J. F. Emery for assistance in the early stages of this work. This research was sponsored by the Office of Basic Energy Sciences, U. S. Department of Energy, under contract W-7405 eng-26 with the Union Carbide Corporation.

REFERENCES

1. E. F. Worden, J. G. Conway and J. Blaise, "Electronic Structure of Neutral and Singly Ionized Curium," *Americium and Curium Chemistry and Technology*, Eds. N. Edelstein, J. D. Navratil and W. W. Schultz, D. Reidel Pub. Co., 1985, p. 123.

2. F. Weigel and R. Kohl, "Preparation and Properties of Some New Curium Compounds," *Americium and Curium Chemistry and Technology*, Eds. N. Edelstein, J. D. Navratil, and W. W. Schultz, D. Reidel Pub. Co., 1985, p. 159.

3. J. E. Bigelow, B. L. Corbett, L. J. King, S. C. McGuire and T. M. Sims, "Production of Transplutonium Elements in the High Flux Isotope Reactor(HFIR),"*Transplutonium Elements - Production and Recovery*,Eds. J. D. Navratil and W. W. Schultz.(ACS Symposium Series, 161) 1981, p. 3.

4. D. C. Hoffman et al., *Phys. Rev. C, Vol. 31*, No. 5, May, 1985, p. 1763; S. Bjornholm and W. J. Swiatecki, *Nucl. Phys. A391*, p. 471(1982).

5. D. E. Benker, F. R. Chattin, E. D. Collins, J. B. Knauer, P. B. Orr, R. G. Ross, and J. T. Wiggins, "Chromatographic Cation Exchange Separation of Decigram Quantities of Californium and Other Transcurium Elements," *Transplutonium Elements - Production and Recovery*, Eds. J. D. Navratil and W. W. Schultz, (ACS Symposium Series 161) 1981, p. 161.

6. C. M. Lederer and V. S. Shirley, editors, *Table of the Isotopes*, 7th ed., Wiley, New York, 1978.

7. M. A. Wakat, and S. F. Peterson, "On-Line Radiochemical Analysis for Controlling Rapid Ion Exchange Recovery of Transplutonium Elements," *Nuclear Technology*, Vol. 17, January, 1973, p. 49.

8. R. D. Evans, *The Atomic Nucleus*, McGraw-Hill Co.,1955, p.484.

9. W. J. Price, *Nuclear Radiation Detection*, McGraw-Hill Co., 1958, pp. 20-26.

Table 1. Decay Characteristics of Principal Isotopes.

Nuclide	TBq/g	α%	E_α (MeV)	I_α (%)	E_γ (keV)	I_γ (%)
^{243}Am	7.3x10^{-3}	100	5.275 5.234	87.9 10.6	74.67	66
^{244}Cm	2.99	100	5.805 5.763	76.4 23.6	42.82	0.0255

Table 2. Decontamination factors and ^{244}Cm content comparison.

Sample Number	δT Counting time, h	Am DF	^{244}Cm(mg) Spectrometer x-ray	α-particle	δ% = (x/α -1)x100
CX-CM-71	10	2591±30%	53.3	51.9	+2.70
CX-CM 96	10	1071±39%	64.2	64.8	-0.13
CX-CM 72	10	502±23%	33.6	35.9	-6.41
CX-CM-75	10	390±53%	10.0	9.5	+5.26
CX-CM-78	10	413±54%	88.6	88.3	+0.34
CX-CM-83	1	≥200	63.4	65.5	-3.21
CX-CM-84	1	≥150	96.0	87.9	+9.22
CM-CX-97	1	≥100	36.4	38.9	-6.43

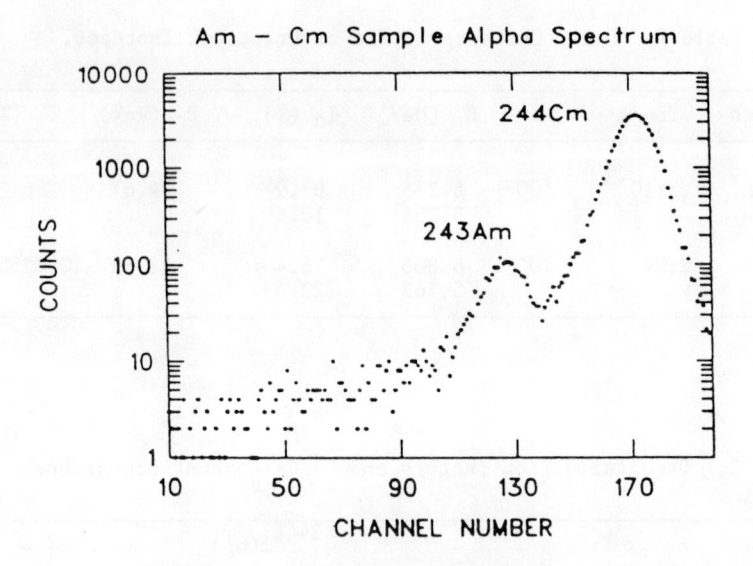

Figure 1. Sample alpha particle spectrum showing principal peaks from the americium and curium.

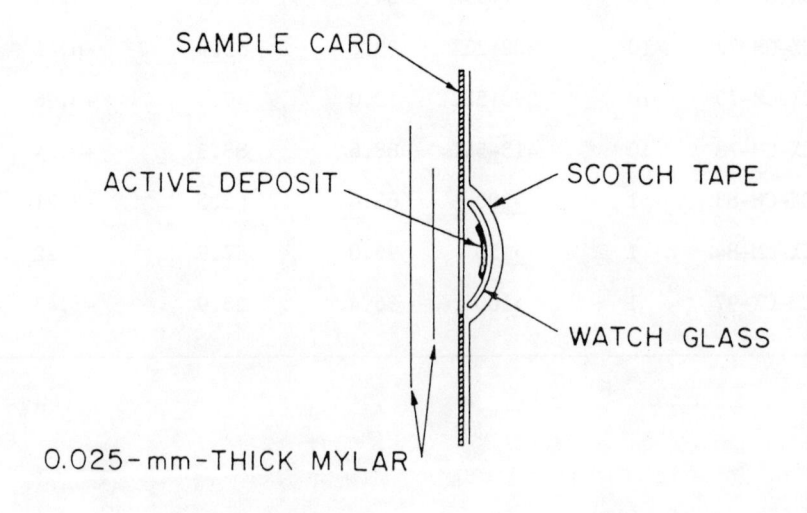

Figure 2. Side view of the counting samples used in this work.

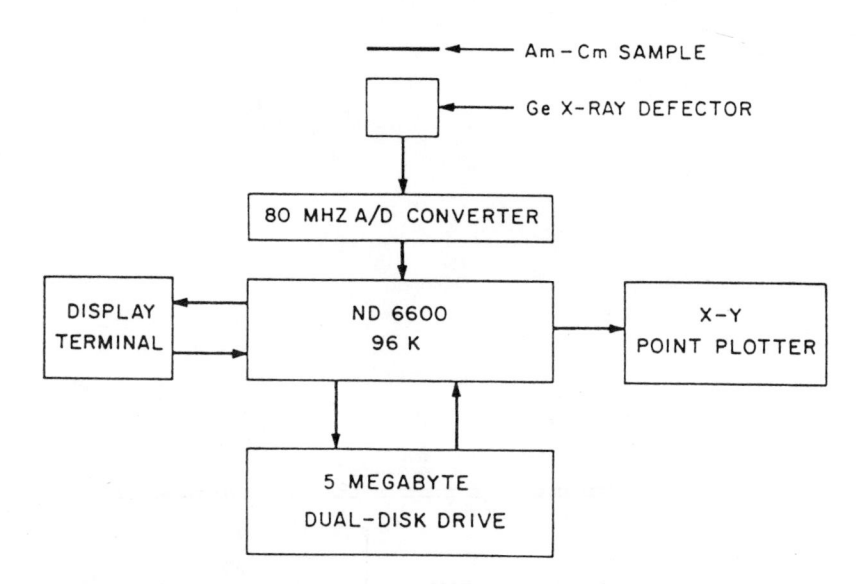

Figure 3. Block diagram of the data acquisition system.

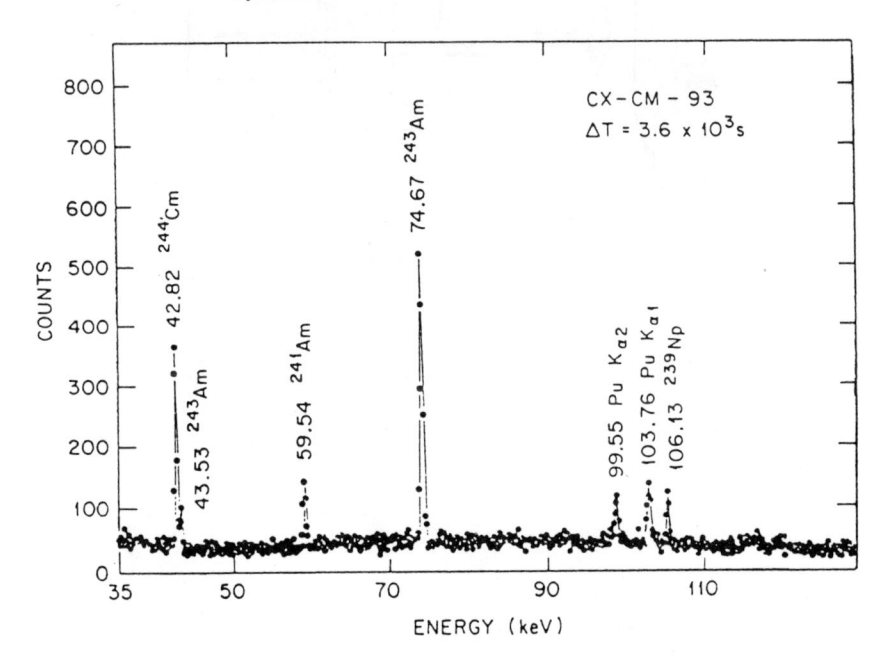

Figure 4. Low energy spectrum of photons emitted from a sample containing an Am-Cm mixture.

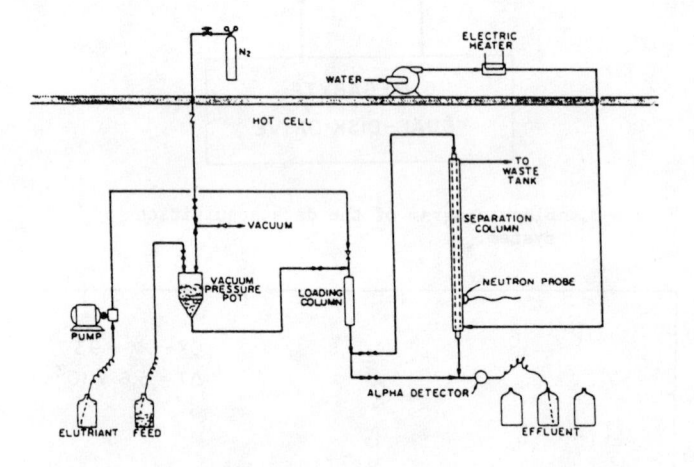

Figure 5. Diagram of the pressurized ion exchange system used to separate transplutonium elements.

Charge Densities: Comparison of Calculations and Experiment

Joseph Callaway
Department of Physics and Astronomy
Louisiana State University
Baton Rouge, Louisiana 70803-4001
U.S.A.

Abstract

The role of the charge density as a basic quantity in quantum mechanics is discussed on the basis of the Hohenberg-Kohn theorem. One method of calculating this quantity for solids is described briefly. Theoretical and experimental results for the charge density form factor are compared in some test cases, specifically beryllium, vanadium, chromium, iron, copper, silicon, germanium, and gallium arsenide.

I. The Charge Density as a Fundamental Quantum Mechanical Quantity.

The electron density in solids has been a quantity of interest to both experimentalists and theorists ever since there have been techniques to calculate and to measure it. However, the fundamental nature of the electron density acquired additional emphasis with the development of density functional theory [1-3], based on the Hohenberg-Kohn (HK) [4] theorem. This theorem pertains to consider a system of electrons interacting with each other, and with a local external potential V_{xt}. [The restriction to local external potential is essential]. Then for such a system, the Hamiltonian is completely specified in principle, up to an additive constant, if the ground state electron density is known.

Actually, as the theorem is stated in [4], V_{xt} is shown to be a universal functional of ρ, apart from an additive constant. This implies the form of the theorem given above.

It is therefore implied that the energies, wave functions, etc. of all states are determined in principle if the ground state density is known, i.e; are functionals of the density. However in general, we have little idea as to how to make such a computation.

In the case of the ground state of the system it is, however, possible to proceed further because there is a minimum principle. This principle is just a straightforward application of the general variational principle of quantum mechanics. In the present context, the second part of the HK theorem asserts that if the ground state energy is computed from the density, it has a minimum with respect to variation of the density (subject to the condition of correct normalization) when the density is correct:

$$\frac{\delta E_g(\rho)}{\delta \rho} = 0 \text{ for } \rho = \rho_g , \qquad (1)$$

provided

$$\int \rho_g(r) \, d^3r = N . \qquad (2)$$

Actually, this is somewhat stronger than what was actually proved by HK. They showed that (1) applies only to those charge densities which are "v-representable", that is, derivable by solution of the Schrodinger equation containing some external potential V_{xt}. The content of this class of charge densities is not clear. This difficulty (and another restriction involving the assumption that the ground state should not be degenerate) have been removed by Levy [5], who shows that it is permissible to consider a much wider class of ρ: those which can be obtained from some antisymmetric N body wave function.

An important extension of the H.K. theorem was obtained by Mermin [6]. He showed that the grand potential, for a system at temperature T, is a functional of the (electron) density in the system at that temperature. A minimum theorem also applies in this case.

The calculational problem now focuses on the determination of the functional $E_g[\rho]$ (we shall restrict our discussion here to T=0). Only certain parts of this functional are known exactly. We can separate the functional as follows:

$$E_g[\rho] = T[\rho] + \int \rho(r) \, V_{xt}(r) d^3r + \frac{e^2}{2} \int \frac{\rho(r)\rho(r')}{|r-r'|} d^3r' \, d^3r$$

$$+ E_{txc}[\rho] . \qquad (3)$$

The two terms written explicitly are the interaction with the external potential and the average electrostatic interaction of the electrons. The first term $T[\rho]$ is intended to represent the kinetic energy. However, an exact expression for $T[\rho]$ is not known -- if it were, one could proceed through a simple generalization of Thomas-Fermi theory. Instead, one must follow the procedure of Kohn and Sham [7], and approximate $T[\rho]$ by a form which is correct for non-interacting Fermions in an external potential. This will be described below. Finally, $E_{txc}[\rho]$ contains what is left over: the exchange and correlation portions of the electron interaction, plus that portion of the kinetic energy not included in the approximate $T[\rho]$.

In order to carry out the variation, the charge density is expressed as the sum of the squares of N orthonormal single particle functions $u_i(\vec{r})$

$$\rho(\vec{r}) = \sum_i^N |u_i(\vec{r})|^2 \quad . \tag{4}$$

Such a representation is always possible, and examples can be given. Eq. (4) is exact, in principle. The expression for the kinetic energy is

$$T[\rho] = \int \sum_i u_i^*(\vec{r}) \left(\frac{-\hbar^2}{2m} \nabla^2\right) u_i(\vec{r}) d^3r \quad , \tag{5}$$

and this is not exact. However, it is evidently a good approximation, and its relative goodness accounts for the superiority of density functional theory in the present form to Thomas Fermi theory. We may proceed to vary the charge density by varying the functions $u_i(\vec{r})$, subject to maintaining the normalization. The result is a set of effective Schrodinger-like equations for the $u_i(\vec{r})$

$$\{-\nabla^2 + V_{xt}(\vec{r}) + e^2 \int \frac{\rho(\vec{r}')}{|\vec{r}-\vec{r}'|} d^3r' + V_{txc}[\rho(\vec{r})]\} u_i(\vec{r}) = \varepsilon_i u_i(\vec{r}) \quad , \tag{6}$$

in which
$$V_{txc}[\rho(\vec{r})] = \frac{\delta E_{txc}[\rho]}{\delta \rho} . \qquad (7)$$

The quantities ε_i, which are introduced as Lagrange multipliers in the variational process, can be shown to have the significance of one electron energies in regard to the application of Fermi statistics [3,8]. This means that one may reconstruct the charge density for an infinite system in accord with Eq. (4) from the N functions $u_i(\vec{r})$ which have the lowest values of the energy parameter ε_i. In the case of a finite system (an individual atom) it would be better to choose the N functions u_i for which $E(\rho)$ is the smallest.

If we know the exact E_{txc}, we could derive according to (7) a (non-local) V_{txc} such that presumably the exact charge density would be obtained. Of course we don't know E_{txc} very well. An exact formal expression for this quantity exists [9],

$$E_{txc}(\rho) = \frac{e^2}{2} \int d^3r \, d^3r' \, \frac{\rho(\vec{r}) \rho(\vec{r}')}{|\vec{r}-\vec{r}'|} \int_0^1 d\lambda \, (g_\lambda(\vec{r},\vec{r}') - 1) , \qquad (8)$$

in which $g_\lambda(\vec{r},\vec{r}')$ is the pair correlation function for electrons in a situation in which the electron-electron interaction is λe^2 ($0 \le \lambda \le 1$). Of course, we do not know this quantity accurately in general. In practice, we know E_{txc} with reasonable accuracy only for the highly idealized system of a free electron gas, for which the density is constant. The most accurate of these calculations is probably that of Ceperley and Alder [10], from which a correlation potential has been deduced by Painter [11].

Suppose then we know E_{txc} as an ordinary function of the (spatially constant) ρ for a free electron gas. The standard sequence of approximations is to replace this constant ρ by the spatially varying $\rho(r)$ for a real system. We can then replace the variational derivative in (7) by an ordinary derivative; the result is a local exchange-correlation potential $V_{txc}(\vec{r})$ which depends on r through its dependence on the local density. This is the so-called local density approximation. This procedure is justified only in a situation in which the electron system has nearly constant density; a situation which obviously does not apply to real solids in which the charge density actually varies rapidly and dramatically with position near any nucleus. However, in spite of several attempts to improve matters which I will not discuss here we really don't know how to proceed beyond the local density approximation in the case of a solid.

There is one extension of the density functional formalism that needs to be mentioned here. This is the spin density functional procedure: a spin polarized extension of the ordinary density functional approach. In this method, we consider the density of electrons of spin σ, $\rho_\sigma (\sigma = \uparrow\downarrow)$ as the basic variable; and calculate it by the obvious generalization of Eqs. (4) and (6) in which now appears a spin dependent exchange correlation potential V_{xc_σ} [12]. This procedure has a firm theoretical foundation in the relativistic generalization of the Hohenberg-Kohn theorem [13]. Unfortunately, it appears that there is no complete extension of the Hohenberg-Kohn theory to momentum space, e.g; to calculate the electron density in momentum space. The difficulty is that the Coulomb potential of interaction between electrons and fixed nuclei becomes a non-local operator in momentum space. However, the use of the Fourier transforms of the one particle functions $u_i(\vec{r})$ to calculate Compton profiles has been given a firm foundation by Lam and Platzman [14] and Bauer [15]. There is a correction term involving the difference between momentum distribution functions in interacting and non-interacting electron gases.

II. Methods of Calculation

Now I want to consider briefly the techniques for making calculations of the charge density. It will be assumed temporarily that $V_{txc\sigma}[\rho]$ is known to an adequate degree of accuracy. Of course, it is just this assumption which is ultimately to be tested by comparison of theory and experiment. We are concerned with methods of solution of Eq. (6).

It first must be voted that Eq. (6) is nonlinear, and $V_{txc\sigma}(\rho)$ may be a fairly complicated expression. In the simplest case, in which only the exchange contribution to $E_{txc}[\rho]$ is considered,

$$V_{txc\sigma} \approx \frac{\partial E_x}{\partial \rho_\sigma}[\rho] = -2e^2 \left(\frac{3\rho_\sigma}{4\pi}\right)^{1/3} , \qquad (9)$$

the potential depends on the cube root of the charge density. Eq. (6) must be solved numerically and self-consistently.

I shall consider here only the case of calculations for solids, which leads me to a brief discussion of the energy band problem. There are indeed many methods of solving Eq. (6) for crystals: including the Green's function method, the augmented plane wave method, the orthogonalized plane wave method, the pseudopotential method, and the tight binding method [15]. Each of these methods has spawned variants, which may be in some cases, more powerful than the original method. It is obviously impossible to review techniques of energy band calculation here. All methods are in fact capable of doing what is claimed: solving the single particle wave equation with a given potential. There are now well documented comparisons between methods which demonstrate that this assertion is correct.

However, this does not mean that all methods are equally useful for self consistent calculations of the charge density and related properties. The complexity of the boundary conditions imposed on the single particle wave function by the requirement that it should satisfy Bloch's theorem forces a common element into many otherwise different procedures: it is desirable to expand the unknown function in a set of function which obey Bloch' theorem. The expansion coefficients can then be determined by a variational procedure. Methods differ primarily in the choice of basis functions for the expansion, and some methods are easier to apply than others for the calculation of charge, spin, and momentum densities.

I claim that the LCGO (Linear Combination of Gaussian Orbitals) method [1], which is a variant and extension of tight binding, is particularly useful for the calculation of densities. In this method, the one electron functions are expanded in a set of Gaussian orbitals placed on the lattice sites of the crystal

$$u_n(\vec{k},\vec{r}) = \sum_j c_{nj}(\vec{k}) \phi_j(\vec{k},\vec{r}), \tag{10}$$

the c_{nj} are eigenvectors of the band problem,

$$\phi_j(\vec{k},\vec{r}) = \frac{1}{\sqrt{N}} \sum_\mu e^{i\vec{k}\cdot\vec{R}_\mu} g_j(\vec{r}-\vec{R}_\mu), \tag{11}$$

g_j is a Gaussian orbital; e.g.

$$g_j(r) = N_j \, r^{\ell(j)} e^{-\alpha_j r^2} K_j(\theta,\phi). \tag{12}$$

Here N_j is a normalization constant, $\ell(j)$ is the angular momentum state considered, α_j is an orbital exponent, and K_j is an angular function (normalized Kubic harmonic for cubic crystals). I shall omit other details [see Ref. 17].

The charge density can be calculated either in position space or in momentum space. For comparison with experiments which involve x-ray scattering, it is useful to have the dimensionless form factors.

$$\rho(\vec{K}) = \int \rho(\vec{r}) \, e^{i\vec{K}\cdot\vec{r}} \, d^3r \,, \tag{13}$$

in which the integration is carried out over the entire crystal, but ρ is normalized so that

$$\rho(\vec{K}=0) = \text{number of electrons in a unit cell}.$$

The reason for the usefulness of the LCGO method is that the $\rho(\vec{K})$ are obtained quite directly in terms of quantities called generalized overlap integrals which have to be calculated in the program.

$$\rho(\vec{K}) = \sum_{\substack{\vec{n}\,\vec{k} \\ i\,j}} c^*_{ni}(\vec{k}) \, c_{nj}(\vec{k}) \, S_{ij}(\vec{k},\vec{K}) \tag{14}$$

the c's are the eigenvectors referred to in Eq. (10); the sum includes all occupied bands and wave vectors, and

$$S_{ij} = \int \phi^*_i(\vec{k},\vec{r}) \, e^{i\vec{K}\cdot\vec{r}} \, \phi_j(\vec{k},\vec{r}) \, d^3r \,. \tag{15}$$

The S_{ij} are the generalized overlap integrals (the ϕ's are given in Eq. (11)). So we can obtain the ρ at the end of the band calculation quite directly, and without additional labor. Ultimately, the usefulness of the method results from the fact that the Fourier transform of the product of two Gaussians as in Eq. (12) can be evaluated analytically.

If the charge density is constructed in real space, it is possible to draw interesting contour diagrams illustrating, for example, the increase of charge density along the convalent bond in semiconductors such as Si. However, I don't know of any way in which these diagrams can be compared with experiment directly. One would have, in effect, to sum the entire Fourier series whose components are $\rho(\vec{K})$, using the experimental form factors. This does not seem to be a reasonable procedure at the present time, and so I shell consider primarily the comparison of theoretical and experimental $\rho(\vec{K})$.

There is one significant exception in which we can compare computed and measured charge densities in real space. Occasionally we can obtain values for the charge density at the nuclear site from isomer shift or internal conversion measurements [18]. The case of internal conversion is particularly valuable since one can distinguish between electronic states at least to the extent of determining, say, the charge density of band electrons at the nuclear site separately from the core electrons. Other experiments, which measure hyperfine fields determine the net spin density of the nuclear site: $\rho\uparrow(0)-\rho\downarrow(0)$. Unfortunately, calculations of charge densities at the nuclear sites are more difficult to get right because there are corrections which need to be for relativistic effects and even for the finite size of the nucleus. It may be suspected that a Gaussian orbital basis causes problems here by giving a less accurate representation of wave functions near a nucleus then would be obtained with a Slater orbital basis, however, numerical tests that I made for free transition metal atoms show that the difference between bases is not large; ~ 4% or less.

III. Results

I will now consider the nature of the results which have been obtained, and discuss the comparison of theory and experiment. Our group has made calculations for lithium [19], sodium [20], aluminum [21], calcium [22], vanadium [23], chromium [24], iron [25], nickel [26] and copper [27]. Calculations by the same method have been for a number of common semiconductors [28]. Unfortunately, the form factors have not been published in all of these calculations. I must also mention extensive calculations of charge densities in real (position) space for many metals by the IBM group (Moruzzi, Janak, and Williams [29]), who use the Green's function method. However, I am not aware that they have published charge density form factors from these calculations. There are calculations of charge densities in semiconductors based on a pseudo-potential formalism [30], and some Hartree-Fock calculations as well [31].

I am shall first compare theory and experiment for selected metals in order of increasing atomic number, and then consider convalent semiconductors. I shall emphasize the results of ab-initio calculations, and have very little to say about simple models (e.g. one OPW or renormalized free atom [32] models). However it is important to note at the outset that in the present state of knowledge in which there are serious disagreements between the results of different experimental groups in regard to x-ray form factors which are only slowly being cleared up, that simple models may do as well as complicated first principles calculations in comparison with experiment.

Also, it has been known for a long time that form factors in general are dominated by large contribution from core electrons. It is believed these contributions should be essentially unchanged in a solid compared to the free atom. We are interested in the outer, or valence electrons. In most cases, we are concerned therefore with the values of the charge form factor for only the first few reciprocal lattice vectors. Even for these, a high degree of precision and reliability of experimental results (we need in general 1% accuracy) is required to distinguish between the results of different theoretical methods. When the experimental situation ultimately settles down, I would expect that there will then be meaningful and informative discrepancies between theory and experiment which will serve to develop our understanding of the electronic structure solids.

Beryllium

Elements of low atomic number ought to be particularly interesting because the effect of core electron on the form factors are less overwhelming. Beryllium is a very important case in this respect. There are large deviations from free electron behavior. We would like to understand the importance of p type functions in the electronic structure.

Unfortunately, beryllium has the hexagonal close packed structure. This means that theoretical calculations, at least by some methods, are significantly more difficult than for cubic metals. There is a (probably) rather accurate calculation for Be using density functional methods [33], but charge form factors have not been published. In addition, a pseudo-potential calculation has been reported which contains form factors [34].

There is a self consistent Hartree Fock calculation [35] in which form factors are given. However, not only are Hartree-Fock calculations in solids rather difficult, but they suffer from pathologies related to the band structure (for example, band widths and band gaps are typically much too large perhaps a factor of 2). It is probable that charge densities are given with reasonable accuracy. This is indeed the case as shown by the comparison of form factors from the Hartree Fock calculation cited above, and measurements of Hansen, Schneider, and Larsen [36] (Table I). In this case, which will not be followed in the data for other elements that I will present later, the theoretical values have been corrected (rather than the experimental ones) for thermal motion of the atoms. This correction is not small (~ 10%). It is seen that there is general agreement, with discrepancies at the 3% level. I do not know how significant these discrepancies are. There are much greater discrepancies with regard to some earlier experimental results [37], which now appear to be erroneous. Moreover, a simple renormalized free atom model calculation fits the data about as well as the elaborate Hartree Fock calculation.

The results of pseudopotential calculation of Chou, Lam, and Cohen [34] are also presented in Table I. The agreement with experiment is on the whole slightly better than obtained in the Hartree Fock studies.

Vanadium

Table II shows calculated charge density form factors for vanadium. Results of two recent calculations are shown. The computation of Laurent, Wang, and Callaway [23] used a Gaussian orbital basis and the Kohn-Sham potential. The other calculation, that of Wakoh and Kubo [38] employs the augmented plane wave method with a potential which contained an adjustable parameter which alters the relative position of t_{2g} and ε_g states. However, it will be seen that the agreement between the two quite different calculations is rather good, with an exception discussed subsequently.

Experimental results of Mazzone and Diana [39], of Linkoaho [40], and of Terasaki, Uchida, and Watanabe [41] are presented. Again, I conclude that the agreement is reasonably good, although in quite a few cases, the theoretical results are outside of the error has given by Linkoaho. It simply is not clear, in view of the tendency in many x-ray experiments for the authors to be overly optimistic about their possible errors, whether the discrepancy is a meaningful indication of inadequate theory. Certainly the degree of overall agreement does not enable one to discriminate between the different calculations.

However there is one sort of disagreement between theory and experiment here which may be significant. The ratio of the form factors for wave vectors $\vec{K}$ of the same magnitude but different orientation, such as (330) and (411) or (431) and (510) is a measure of the angular anisotropy of the charge distribution. These ratios have been measured by Ohba, Sato, and Saito [42]. The measured ratios indicate the presence of more angular anisotropy in the charge distribution than is predicted by the calculations using a local exchange potential [23]. On the other hand, Wakoh and Kubo are able to adjust the relation between potentials for the ε_g and t_{2g} states to give greater anisotropy.

The discrepancies between the density functional type theory and experiment in regard to these ratios are small, and I would tend to ignore the difficulty if it occurred in this case only. However, something of this nature is present in other experiments, including some not involving x-rays at all. I consider the accumulation of results such as this to be an indication that density functional theory as presently applied in the local density approximation leads to an underestimate of angular anisotropy. But no definite conclusion can be stated at this point. We shall see that in the case of chromium to be discussed next, that the anisotropies given from density functional theory are in reasonable agreement with experiment for that material.

Chromium

In the case of chromium we have calculations by Laurent et al [22] who used the density functional approach with an exchange correlation potential of the von-Barth Hedin type, and a calculation by Ohba, Saito, and Wakoh [43] who employ an adjusted potential somewhat similarly to the previously mentioned calculation of Wakoh and Kubo for vanadium [38]. The results are compared with each other and with some experiments in Table III. The experiments are those of Ohba et al [43], Diana and Mazzone [44], and Terasaki et al [41].

In spite of the differences of potentials, the agreement between the theoretical calculations is rather good. The only significant differences are in the anisotropies. The potential used by Ohba et al. gives somewhat larger anisotropies than the straightforward local density calculation; however in contrast to the case of vanadium the agreement with experiment is not better. The experimental form factors of Diana and Mazzone are likely to be too small in magnitude. The disagreement with theory in this respect is much outside the probable theoretical error. On the other hand, the agreement with the measurements of Terasaki et al [41] is rather good.

Iron

There have been a substantial number of calculations and of experiments concerning charge density form factors. In order to avoid overwhelming the reader with too many numbers, I present in Table IV results only for three reciprocal lattice vectors and for one ratio connected with the anisotropy. These numbers are sufficient to indicate the essential features of the situation.

Let us first consider the results of calculations by our group using the same calculational method (LCGO) but two different exchange correlation potentials [25]. The notation VBH indicates that the computation was based on the exchange-correlation potential of von-Barth and Hedin [45]; KS indicates use of the Kohn Sham exchange only potential [46]. The inclusion of an approximate representation of correlation causes the VBH potential to be somewhat more attractive than the KS potential, therefore the form factors from VBH are slightly larger than those from KS. However, this difference is quite small, about 0.1%.

Results from three other, earlier, and somewhat less technically accomplished calculations are presented. Those of Wakoh and Yamashita [47] and DeCicco and Kitz [48] are in reasonable agreement with our work; however those of Duff and Das [49] are quite different. The differences between the results of Duff and Das and the other calculation are at a level such that experiment should be able to indicate which is more nearly correct.

If we now consider the experimental results, we see that the measurements of Terasaki et al [41], Paakkari and Suortti [50], Radchenko and Tsvetkov [51], and Hosoya [52] support the first group of calculations, while the measurements of Batterman et al [53] and Diana and Mazzone [54] agree more closely with the calculations of Duff and Das. This was a serious concern. Now there are many indications that there are serious problems with the calculation of Duff and Das, and that their results are most probably badly wrong. However, we really can't distinguish between the other calculations on the basis of the form factor evidence. We do see in the work of DeMarco and Weiss [55] and Diana and Mazzone [54] an indication of the same kind of problem we noticed for vanadium: the density functional calculations seem to give a smaller anisotropy to the charge density.

It is interesting to consider, in the case of iron, the question of charge and spin densities at a nuclear site. Results from our Gaussian orbital density functional calculations are presented in Table V. I have broken the results down so that the contribution of individual electron shells can be seen. These are experimental measurements, from internal conversion, of the charge density of the outer (band) electrons at the nucleus [56]. These seem to agree fairly well with the calculations, although the errors are not small. These are also rather precise measurements (based on the Mossbauer effect) of the spin density [57]. The negative sign indicates that minority spin electrons predominate at the nuclear site. While the calculations agree with the sign of the measurements, the numerical value is somewhat low. I am inclined to regard this discrepancy as a manifestation of relativistic effects ignored in the calculations as discussed earlier. Even in a material with a nuclear charge as low as that of iron [26] the relativistic enhancement of the charge density at a nucleus will be quite appreciable, and the magnitude required to give agreement here is not out of line with theoretical estimates [55].

Copper

In the case of copper, we are able to compare modern calculations and experiments, as well as others ten or more years old. Table VI contains calculated form factors based on density functional theory and the LCGO method [27]. The results of Snow [59] who used the Kohn Sham potential and the APW method are given, as are those from a relativistic density functional calculation [60]. These two calculations are in surprisingly good agreement. For comparison, a calculation of form factors by Doyle and Turner [61] based on relativistic Hartree Fock wave functions are shown. The form factors calculated specifically for the solid are slightly smaller for small wave vectors: this difference is significant. We see that the difference between atomic and band structure based form factors is about 2% at most. Also, relativistic effects on the form factors are evidently small.

Two sets of experimental results are given. The first are those of Schneider, Hansen and Kretschmer [62] are obtained using an imperfect (Mosaic) crystal. The other experimental results are those of Temkin, Heinrich, and Racah [63], who used a powder. On the whole, the band calculation form factors agree somewhat better with the powder measurements than with the crystal measurements. In particular; the form factors obtained by Schneider et al seem to be slightly but consistently smaller than the calculated values. However, I would expect that an improved treatment of election correlation in the calculations would probably increase the "attractiveness" of the crystal potential and so increase the form factors. This direction, if correct, is opposite to what would be required to improve agreement with the experiment of Schneider et al. Of course, one can not be certain of this until such a calculation actually is performed.

Covalent Semiconductors.

In Table VII, I compare theory and experiment in regard to some charge density form factors for three covalently bonded semiconductors: silicon, germanium, and gallium arsenide. The calculations involved are those of Wang and Klein [28] who used a Gaussian orbital-density functional approach; those of Racah et al [64] who made self-consistent orthogonalized plane wave calculations using the Kohn-Sham exchange potential; the pseudopotential-density functional calculation of Zunger and Cohen [30]; and the Hartree Fock calculation of Dovesi et al [56]. The experimental results for silicon are from Aldred and Hart [65] and Robert et al [66]. Those for germanium from Matsushita and Kohra [67], and the two sets for gallium arsenide, from Matsushita and Hayashi [68]. Numbers are given as normalized and quoted in Ref. 26.

Again, it is clear that generally, the agreement between theory and experiment is reasonably good. There is some tendency for the calculated values of $\rho(K)$ for the (111) lattice vector from density functional related calculations to be smaller than experiment, but it is not clear whether this is outside possible experimental error. The Hartree Fock calculation in silicon seems to give a too large value. In silicon the pseudopotential form factors for (220) and (331) waves seem to be too small, but it is again not clear that the differences are significant.

Let us consider the case of the (222) lattice vector in a little more detail. As in well known, for this case $\vec{\rho}(K)$ would vanish if the charge density around each of the two atomic sites in the unit cell were spherically symmetric. The non-zero value of this quantity is a measure of angular asymmetry resulting from the formation of covalent bands between atoms. In spite of its small size, the experimental value, which has been confirmed by several measurements is reasonably precise. In fact, the small size of the interesting $\rho(222)$, [of the order of 1% of $\rho(111)$] illustrates the central problem of comparison between theory and experiment in regard to charge form factors: the interesting effects due to the modification of the charge distributions of the outer electrons in solids are very small (order of 1%) compared to the large "uninteresting" contribution from the core electrons plus the unmodified portion of the outer electron distribution. So, unless one can find cases such as this, where the less interesting effects cancel, one must make very precise measurements (better than 1%) in order to discriminate between theories (which can probably get the uninteresting parts reasonably correctly). In the present case, of silicon, I am inclined to regard the disagreement between the Wang-Klein calculation and experiment as meaningful. If so, it is another case in which local density theory has failed to give correct anisotropies. (I think that the Wang-Klein calculation is more likely to be technically correct at the level of accuracy required than the SC-OPW one). However, some caution is required in regard to this conclusion because this discrepancy has almost disappeared in germanium.

IV. Conclusions

I would have liked to use the comparison between theory and experiment to measure the accuracy of local density theory in regard to the most basic quantity of that theory, the density itself. Unfortunately this does not seem to be possible at the present. I do not believe the experimental results are sufficiently reliable. Instead, one may be in a rather different situation in which comparison with theory may indicate which experiment is more reliable.

My personal guess is that the calculated results for transition metals and heavier systems are accurate to about 1%. This is because the cores dominate (at, typically a 90% level), and the band electron densities ought to be good to better than 10%. In lighter systems, where the core is not so overwhelmingly important, the accuracy of computations may be somewhat less. I suggest that emphasis on light systems could be one area in which more interesting contrasts between theory and experiment could develop. Of course many light metals are nearly free-electron like, and these are probably not the ones to emphasize.

When reliable experimental form factor become available at a 1% level of accuracy, my guess it that many interesting discrepancies between theory and experiment will be uncovered. I look forward to this situation.

Acknowledgment

This research was supported in part by the Division of Materials Research of the U.S. National Science Foundation. A portion of this paper was written when I was a visitor in the Department of Physics, University of Western Australia. I wish to thank Professor J. F. Williams for his kind hospitality.

References

1. A. K. Rajagopal, Adv. Chem. Phys. 41, 59 (1981).
2. W. kohn and P. Vashista in "Theory of the Inhomogenous Electron Gas", eds S. Landquist and N. H. Press, (London, Plenum Press, 1982) ch. 2.
3. J. Callaway and N. H. March, Solid State Phys. 38, 135 (1984).
4. P. Hohenberg and W. Kohn, Phys. Rev. 136, B864 (1964).
5. M. Levy, Proc. Natl. Acad. Sci. U.S. 76, 6062 (1979).
6. N. D. Mermin, Phys. Rev. 137, A1441 (1965).
7. W. Kohn and L. J. Sham, Phys. Rev. 140, A1133 (1965).
8. J. F. Janak, Phys. Rev. B18, 7165 (1978).
9. J. Harris and R. O. Jones, J. Phys. F. 4, 1170 (1974); O. Gunnarson and B. I. Lundqvist; Phys. Rev. B13, 4274 (1976); D. C. Langreth and J. P. Perdew, Phys. Rev. B15, 2884 (1977).
10. D. M. Ceperley and B. J. Alder, Phys. Rev. Letts. 45, 566 (1980).
11. G. S. Painter, Phys. Rev. B24, 4264 (1981).
12. U. vonBarth and L. Hedin, J. Phys. C5, 1629 (1972).
13. A. K. Rajagopal and J. Callaway, Phys. Rev. B7, 1912 (1973).
14. L. Lam and P. Platzman, Phys. Rev. B9, 5122 (1974).
15. G. E. W. Bauer, Phys. Rev. B27, 5912 (1983).
16. These methods are reviewed in J. Callaway, "Quantum Theory of the Solid State", (New York, Academic Press, Inc.), ch. 4.
17. C. S. Wang and J. Callaway, Comp. Phys. Commun. 14, 327 (1978).
18. R. E. Watson, A. A. Misetich, and L. Hodges, J. Phys. Chem. Solids 32, 709 (1971).
19. D. Bagayoko, X. Zou, and J. Callaway, Phys. Letts. 89A, 86 (1982); W. Y. Ching and J. Callaway, Phys. Rev. B9, 5115 (1974).
20. W. Y. Ching and J. Callaway, Phys. Rev. B11, 1324 (1975).
21. S. P. Singhal and J. Callaway, Phys. Rev. B16, 1744 (1976).
22. P. Blaha and J. Callaway, Phys. Rev. B32, 7664 (1985).
23. D. G. Laurent, C. S. Wang, and J. Callaway, Phys. Rev. B°7, 455 (1978).

24. D. G. Laurent, J. Callaway, J. L. Fry, and N. E. Brener, Phys. Rev. B$\underline{23}$, 4977 (1981).

25. J. Callaway and C. S. Wang, Phys. Rev. B$\underline{16}$, 2095 (1977).

26. C. S. Wang and J. Callaway, Phys. Rev. B$\underline{15}$, 298 (1977).

27. D. Bagayoko, D. G. Laurent, S. P. Laurent, S. P. Singhal, and J. Callaway, Phys. Letts. $\underline{76A}$, 187 (1980).

28. C. S. Wang and B. M. Klein, Phys. Rev. B$\underline{24}$, 3393 (1981).

29. J. L. Moruzzi, J. F. Janak, and A. R. Williams "Calculated Electronic Properties of Metals" (New York, Pergamor Press, 1978).

30. A. Zunger and M. L. Cohen, Phys. Rev. B$\underline{20}$, 4082 (1979).

31. R. Dovesi, M. Casua, and G. Angona, Phys. Rev. B$\underline{24}$, 4177 (1981).

32. L. Hodges, R. E. Watson, and H. Ehrenreich, Phys. Rev. B$\underline{5}$, 3953 (1972).

33. L. Wilk, W. R. Felner, and S. H. Vosko, J. Phys. $\underline{56}$, 266 (1978).

34. M. Y. Chou, P. K. Lam, and M. L. Cohen, Phys. Rev. B$\underline{28}$, 4179 (1983).

35. R. Dovesi, C. Pisani, F. Ricca, and C. Roetti, Phys. Rev. B$\underline{25}$, 3731 (1982).

36. N. K. Hansen, J. R. Schneider, and F. K. Larsen Phys. Rev. B$\underline{29}$, 917 (1984).

37. P. J. Brown, Phil May $\underline{26}$, 1377 (1972).

38. S. Wakoh and R. Kubo, J. Phys. F$\underline{10}$, 2707 (1980).

39. G. Mazzone and M. Diana, Phil May B$\underline{37}$, 641 (1978).

40. M. V. Linkoaho, Phys. Soc. $\underline{5}$, 271 (1972).

41. O. Terasaki, Y. Uchida, and D. Watanabe, J. Phys. Soc. Japan $\underline{39}$, 1277 (1975).

42. S. Ohba, S. Sato, and Y. Saito, Acta Cryst. A$\underline{37}$, 697 (1981).

43. S. Ohba, Y. Saito, and S. Wakoh, Acta Cryst. A$\underline{38}$, 103 (1982).

44. M. Diana and G. Mazzone, Phys. Rev. B$\underline{5}$, 3832 (1972).

45. U. Von Barth and L. Hedin, J. Phys. C$\underline{5}$, 1629 (1972).

46. W. Kohn and L. J. Sham, Phys. Rev. $\underline{140}$, A$\underline{1133}$ (1965).

47. S. Wakoh and J. Yamashita, J. Phys. Soc. Japan $\underline{30}$, 422 (1971).

48. P. DeCicco and A. Kitz, Phys. Rev. 162, 486 (1967).

49. K. J. Duff and T. P. Das, Phys. Rev. B3, 2294 (1971).

50. T. Paakkari and P. Suortti, Acta Cryst. A22, 765 (1967).

51. M. E. Radchenko and V. P. Tsvetkov, cited by N. Sirota, Acta Cryst. A25, 223 (1969).

52. S. Hoysoya cited by N. Sirota, Ref. 51.

53. B. W. Batterman, D. R. Chipman, and J. J. DeMarco, Phys. Rev. 122, 68 (1961).

54. M. Diana and G. Mazzone, Phys. Rev. B9, 3898 (1974).

55. J. J. DeMarco and R. J. Weiss, Phys. Letts. 18, 92 (1965).

56. T. Shinohara and M. Fujioka, Phys. Rev. B7, 37 (1973).

57. C. E. Violet and D. N. Pipkorn, J. Appl. Phys. 42, 4339 (1971).

58. J. J. Mallow, A. J. Freeman, and J. P. Desclaux, Phys. Rev. B13, 1884 (1976).

59. E. C. Snow, Phys. Rev. 171, 785 (1968).

60. A. H. MacDonald, J. M. Daams, S. H. Vosko, and D. D. Koelling, Phys. Rev. B25, 713 (1982).

61. P. A. Doyle and P. S. Turner, Acta Cryst. A24, 390 (1968).

62. J. R. Schneider, N. K. Hansen, and H. Kretschmer, Acta Cryst. A37, 711 (1981).

63. R. J. Temkin, V. E. Heinrich, and P. M. Raccah, Phys. Rev. B6, 3572 (1972).

64. P. M. Raccah, R. N. Euwema, D. J. Stuckel, and T. C. Collins, Phys. Rev. B1, 756 (1970).

65. P. J. E. Aldred and M. Hart, Proc. Roy. Soc. London A332, 223 (1973).

66. J. B. Roberto, B. W. Batterman, and D. T. Keating, Phys. Rev. B9, 2590 (1974).

67. T. Matsushita and K. Kohra, Phys. Stat. Sol. 24, 531 (1974).

68. T. Matsushita and J. Hayashi, Phys. Stat. Sol. A41, 139 (1977).

Table I

Charge Density Form Factor for beryllium (absolute value)

k(h,k,l)	Theory Ref. [35]	Theory Ref. [36]	Exp Ref. [36]
(002)	3.398	3.330	3.320 (7)
(004)	2.201	2.173	2.196 (10)
(100)	1.892	1.829	1.833 (6)
(101)	2.798	2.816	2.828 (7)
(102)	1.435	1.442	1.455 (11)
(103)	2.132	2.105	2.168 (12)
(200)	1.182	1.168	1.214 (7)
(110)	2.632	2.621	2.659 (18)

The number in parenthesis indicates the uncertainty (one standard deviation) in the last significant figures.

Table II

Charge Density Form Factors for Vanadium

k	Theory			Experiment	
	Ref. 23	Ref. 38	Ref. 39	Ref. 40	Ref. 41
(110)	15.752	15.740	15.75±0.25	15.94±0.08	15.90±0.18
(200)	13.113	13.122		13.16±0.04	13.22±0.17
(211)	11.434	11.459		11.30±0.04	
(220)	10.229	10.193		10.41±0.05	
(310)	9.323	9.280		9.37±0.03	
(222)	8.727	8.718		8.39±0.06	
(321)	8.216	8.204		8.24±0.05	
(400)	7.778	7.746		7.69±0.08	
(330)	7.516	7.520		7.56±0.05	
(411)	7.487	7.474		7.56±0.05	
(420)	7.239	7.240			
(332)	7.046	7.066			
(422)	6.842	6.861			
(431)	6.671	6.693			
(510)	6.636	6.641			

Ratios

	Theory		Experiment
	Ref. 23	Ref. 38	Ref. 42
(330)/(411)	1.0039	1.012	1.010±0.014
(431)/(510)	1.0053	1.016	1.018±0.010

Table III

Charge Density Form Factors for Chromium

K	Theory Ref. [24]	Ref. [43]	Experiment Ref. [44]	Ref. [41]
(110)	16.29	16.26	15.74 (10)	16.30 (6)
(200)	13.39	13.40	13.06 (17)	13.47 (11)
(211)	11.66	11.62	11.37 (15)	
(220)	10.39	10.34	10.10 (14)	
(310)	9.40	9.34		
(222)	8.82	8.81		
(321)	8.27	8.25		
(400)	7.76	7.72		
(330)	7.54	7.54		
(411)	7.48	7.46		

Ratios

Wave vectors	Theory Ref. [24]	Ref. [43]	Experiment Ref. [44]	Ref. [43]
(330/411)	1.008	1.011	1.013 (7)	1.009 (7)
(431/510)		1.015		1.008 (5)
(433/530)		1.008		1.006 (6)
(442/600)	1.014	1.019	1.014 (7)	1.012 (8)
(532/611)		1.012_5		1.006 (6)

The number in parenthesis is the estimated experimental uncertainty.

Table IV

Charge Density Form Factors for Iron

	Wave Vector K			Asymmetry
Calculations	(110)	(200)	(211)	(330/411)
VBH[a]	18.295	15.105	13.074	1.0025
KS[a]	18.267	15.087	13.061	1.0026
Other Authors				
Wakoh amd Yamashita[b]	18.34	15.12	12.99	1.0062
DeCicco and Kitz[c]	18.371	15.124	13.093	1.0032
Duff and Das[d]	17.59	14.28	12.55	1.0076
Experimental				
Terasaki et al[e]	18.34±0.07	15.19±0.07		
Paakkari and Suorti[f]	18.19	15.19±0.08	13.01±0.08	
DeMarco and Weiss[g]				1.011±0.003
Radchenko and Tsvetkov[h]	18.38±0.44	15.23±0.15	13.09±0.17	
Hosoya[i]	18.38	15.13	13.18	
Batterman et al[j]	17.63±0.20	14.70±0.23	12.62±0.21	
Diana and Mazzone[k]	17.54±0.25	14.55±0.20		1.010

a) Ref. 25; b) Ref. 47; c) Ref. 48; d) Ref. 49; e) Ref. 41
f) Ref. 50; g) Ref. 55; h) Ref. 51; i) Ref. 52; j) Ref. 53;
k) Ref. 54.

Table V

Charge and Spin Densities at the Nuclear Site

State	$\rho_\uparrow(0)$	$\rho_\downarrow(0)$	$\rho_\uparrow(0) + \rho_\downarrow(0)$	$\rho_\uparrow(0) - \rho_\downarrow(0)$
1s	5195.787	5195.828	10391.615	-0.041
2s	472.074	472.815	944.899	-0.741
3s	68.661	68.152	136.813	+0.509
3p*	0.003	0.003	0.006	0.000
Band	2.673	2.805	5.478	-0.133
Total	5739.198	5739.604	11478.802	-0.406

Experiment

Band	5.53±0.46[a]	
Total		-0.6466±0.0006[b]

* Resulting from s hybridization into the 3p band.

a) Ref. 56;   b) Ref. 57
All quantities are in atomic units (a_0^{-3}).

Table VI

Charge Density Form Factors for Copper

K	Ref. 27	Ref. 59	Ref. 60	Ref. 61	Ref. 62	Ref. 63
(111)	21.68	21.63	21.73	22.08	21.51 (5)	21.93 (15)
(200)	20.35	20.40	20.39	20.72	20.22 (4)	20.36 (15)
(220)	16.62	16.64		16.78	16.45 (5)	16.70 (16)
(311)	14.70	14.68		14.78	14.54 (4)	14.71 (17)
(222)	14.17	14.16	14.25	14.23	14.07 (5)	14.18 (17)
(400)	12.42	12.42		12.46	12.29 (6)	12.23 (20)
(331)	11.41	11.43		11.46		
(420)	11.13	11.14		11.17	11.02 (6)	
(422)	10.16			10.19	10.08 (6)	
(511/333)	9.58		9.67/9.66	9.61	9.53 (6)	

The numbers in parentheses are the quoted experimental uncertainties in the last two significant figures.

Table VII

Charge Density Form Factor for Some Covalent Semiconductors

Calculation	Wave Vector			
	(111)	(220)	(311)	(222)
		Silicon		
Expt[a]	15.19	17.30	11.35	0.38
Wang, Klein[b]	15.11	17.26	11.37	0.25
SC-OPW[c]	15.12	17.28	11.33	0.34
Pseudopotential[d]	15.13	17.14	11.02	0.38
Hartree-Fock[g]	15.55	17.58	11.46	0.22
		Germanium		
Expt[e]	39.42	47.44	31.37	0.26
Wang, Klein[b]	38.83	47.23	31.29	0.22
SC-OPW[c]	38.95	47.26	31.21	0.48
		Gallium Arsenide		
Expt (1)[f]	39.44	47.18	31.37	
Expt (2)[f]	39.06	46.66	31.04	
Wang-Klein[b]	38.85	47.18	31.25	1.214

a) Ref. (65); (222) value from Ref. (66); b) Ref. (64); c) Ref. (51); d) Ref. (30); e) Ref. (67); f) Ref. (68), two different correlations for anomalous dispersion were used in reducing the data; g) Ref. (31).

PITFALLS IN THE NUMERICAL INTEGRATION

OF DIFFERENTIAL EQUATIONS*

Ronald E. Mickens
Department of Physics
Atlanta University
Atlanta, Georgia 30314 U.S.A.

I. Introduction

This presentation will differ in a significant manner from most papers given at this conference. It will be concerned with the analysis of problems which can arise in the numerical integration of differential equations. The results obtained can be directly applied to the differential equations which occur in the analysis of analytical techniques for material characterization and the construction of related physical theory to explain and predict possible experimental results [1, 2].

In general, there are three methods for determining solutions to differential equations; they are methods which give exact analytical solutions [3], procedures for calculating approximations to exact analytical solutions [4] and numerical solutions [5, 6]. As stated above, we will investigate the pitfalls that can occur in the use of numerical techniques. Our emphasis will center on finite-difference procedures [7, 8]. To proceed, we will briefly discuss

---

*This research is supported in part by grants from ARO and NASA.

several problems which arise in the modeling of differential equations by finite-difference schemes.

The first issue concerned how the derivatives should be modeled. For example, in a given differential equation, the first-derivative can be modeled by any one of the following three expressions [6, 7]:

$$\frac{dx}{dt} \to \begin{cases} \frac{x_{k+1} - x_k}{h}, \\ \frac{x_{k+1} - x_{k-1}}{2h}, \\ \frac{-x_{k+2} + 8x_{k+1} - 8x_{k-1} + x_{k-2}}{12h}, \end{cases} \quad (1)$$

where $h = \Delta t$ is the time-step size. There is, in principle, an unlimited number of possible finite-difference expressions that can be used to represent dx/dt. A similar situation holds for the second and higher order derivatives [6, 7]. The second issue is related to the modeling of nonlinear terms. As an illustration of this problem, consider the modeling of x and x^2 by finite-difference expressions; any of the following will suffice:

$$x \to x_k, \quad x_{k+1}, \quad \frac{x_k^2}{(x_k + x_{k+1})/2}, \quad (2)$$

$$x^2 \to x_k^2, \quad x_{k+1}^2, \quad x_{k+1}x_k, \quad \frac{x_k^2 + x_{k+1}^2}{2} \quad (3)$$

(Note that in the limit $h \to 0$, $k \to \infty$, $hk = t =$ fixed, all the terms on the right-sides of Eqs. (2) and (3) reduce to the indicated functions on the left-sides.) The results given in Eqs. (1)-(3) show the almost total ambiguity in the procedures to construct finite-difference schemes for differential equations. This holds true for both ordinary and partial differential equations.

It is possible for "extra" or "spurious" solutions to occur in the numerical integration of differential equations. In general, they arise whenever the difference equation model is of higher-order than the corresponding differential equation [9]. Another indication of the existence of spurious solutions is when the stability properties of the steady-state solutions for the difference equation model differs from those of the differential equation. Finally, a given difference equation model may have a dependence on the step-size h such that for $0 < h < h_T$, where h_T is a positive constant determined by the given finite-difference scheme, the solution to the difference equation is "close to" and "similar" to the solution of the differential equation. However, for $h \geq h_T$, the solution to the difference equation may become unstable, oscillate wildly and diverge. This fact places strong limitations on the use of particular finite-difference models for making long-time predictions where the device of taking the time-step size large is utilized [10].

Finally, consider a first-order differential equation

$$\frac{dx}{dt} = f(x,t,\alpha), \quad x(t_0) = x_0 = \text{given}, \qquad (4)$$

where $f(x,t,\alpha)$ has the necessary properties such that a solution $x(x_0,t_0,\alpha,t)$ exists and α is a set of N-parameters that define the system. Let the finite-difference model to Eq. (5) be

$$x_{k+1} = g(x_k,k,\alpha,h). \qquad (5)$$

Note that the parameter space of the difference equation is of dimension N+1, while that of the differential equation is of dimension N. This arises from the fact that the step-size h is a parameter for the finite-difference equation. Now suppose for a fixed value of h, say $h = h_1$, the solution to the difference equation, $x_k(h_1)$ is exactly equal to the solution of the differential equation $x(h_1 k)$ at $t = h_1 k$, i.e.,

$$x_k(h_1) = x(h_1 k). \qquad (6)$$

If now h is changed to a new value $h = h_2$, the solution to the difference equation may not equal the solution to the differential equation, i.e.,

$$x_k(h_2) \neq x(h_2 k). \qquad (7)$$

This general result is to be expected from catastrophe theory [11]. Thus, we conclude that the solutions of difference equation models can have a larger number of possible behaviors than the solution to the corresponding differential equation.

The remainder of the paper is organized as follows: In Section II, we give several examples of how ordinary differential equations may be modeled by finite-difference schemes. There stability properties are investigated. Section III presents a fundamental theorem on the relationship between ordinary differential equations and certain difference equation representations. The theorem is illustrated by means of several examples. In Section IV, we briefly discuss the situation with regard to partial differential equations. Section V presents a number of rules that can be used in the construction of finite-difference models of differential equations. Finally, in Section VI, we end the paper with a discussion of other related problems that need to be investigated.

One last comment is in order before we proceed with the calculations. The examples of this paper are, on the whole, rather elementary. However, it should be emphasized that they all show the great difficulties that can arise in the numerical integration of more complicated systems of differential equations.

II. Examples

Consider the first-order equation

$$\frac{dx}{dt} = -x, \quad x(0) = x_0. \tag{8}$$

Its exact solution is

$$x(t) = x_0 \exp(-t). \tag{9}$$

The use of a central difference approximation [5, 6] for the derivative gives

$$\frac{x_{k+1} - x_{k-1}}{2h} = -x_k, \qquad (10)$$

where x_k is an approximation to x(t) at t = hk and h = Δt is the time-step size. Equation (10) can be written in the form

$$x_{k+1} + 2hx_k - x_{k-1} = 0. \qquad (11)$$

This last equation is a second-order, linear difference equation whose general solution is easily found to be [8]

$$x_k = A(r_+)^k + B(r_-)^k \qquad (12)$$

where A and B are arbitrary constants and

$$r_{\pm} = -h \pm \sqrt{1 + h^2}. \qquad (13)$$

An analysis of Eqs. (12) and (13) shows, that for all positive step-sizes h, x_k has a behavior such that for sufficiently large k the solution is oscillating divergent. This means that the difference scheme of Eq. (10) cannot represent the differential equation (8).

The use of the forward Euler procedure [5, 6] gives

$$\frac{x_{k+1} - x_k}{h} = -x_k. \qquad (14)$$

This is a linear, first-order difference equation and has the solution [8]

$$x_k = x_0(1-h)^k. \qquad (15)$$

Examination of Eq. (15) shows it to have the following possible solution behaviors:

$0 < h < 1$: x_k is decreasing monotonic.

$h = 1$: $x_k = 0$.

$1 < h < 2$: x_k decreases to zero with an oscillating amplitude.

$h = 2$: $x_k = x_0(-1)^k$, oscillations with a constant amplitude.

$h > 2$: x_k diverges with an oscillating amplitude.

Note that only for $0 < h < 1$ does the behavior of x_k even agree qualitatively with the solution $x(t)$ of the differential equation. The nature of the solution depends on the value of the step-size!

The logistic differential equation

$$\frac{dx}{dt} = x(1-x), \quad x(0) = x_0, \qquad (16)$$

has the exact solution

$$x(t) = \frac{x_0}{x_0 + (1-x_0)\exp(-t)}. \qquad (17)$$

The application of the forward Euler procedure to Eq. (16) gives

$$\frac{x_{k+1} - x_k}{h} = x_k(1-x_k). \qquad (18)$$

Now, the change of variables

$$y_k = \left(\frac{h}{1+h}\right)x_k, \quad \lambda = 1 + h, \qquad (19)$$

transforms Eq. (18) to the form

$$y_{k+1} = \lambda y_k(1-y_k). \tag{20}$$

This last equation is known as the "logistic difference equation" and can have a variety of solution behaviors depending upon the value of λ [12]. These solutions include steady-state behavior, periodic oscillations and chaotic motions. Only for the situation where h is small will the solutions of Eq. (18) have qualitatively the same behavior as the solutions of Eq. (16).

We can also model Eq. (16) by the central difference scheme; this gives

$$\frac{x_{k+1} - x_{k-1}}{2h} = x_k(1-x_k). \tag{21}$$

However, this procedure has serious difficulties. It has been shown by Ushiki [13] that Eq. (21) has chaotic solutions for all positive time-steps h. This means that none of the solutions of Eq. (21) have the correct behavior expected for the solutions of Eq. (16).

Now consider the following nonlocal, forward Euler scheme for Eq. (16);

$$\frac{x_{k+1} - x_k}{h} = x_k(1-x_{k+1}). \tag{22}$$

The change of variable $x_k = (u_k)^{-1}$ transforms Eq. (22) into the form

$$u_{k+1} - \left(\frac{1}{1+h}\right)u_k = \frac{h}{1+h}, \qquad (23)$$

which can be solved exactly [8] to give

$$x_k = \frac{x_0}{x_0 + (1-x_0)(1+h)^{-k}}. \qquad (24)$$

Comparison of Eqs. (17) and (24) shows that for small values of h, the two solution behaviors are similar. In fact in the limit $h \to 0$, $k \to \infty$, $hk = t =$ fixed, the two solutions are equal.

In summary, we have demonstrated that the straight forward application of finite-difference techniques to the modeling of differential equations, for the purposes of numerical integration, does not always lead to results that can be used to understand either the qualitative or quantitative behavior of the required solutions. Now the two equations considered above have been rather "elementary." If such problems arise for such simple cases, then we should expect no less for the more complex differential equations which occur in the analysis of realistic dynamic systems [2]. In the next section, we show that certain important constraints exist for ordinary differential equations.

III. Theorem for Ordinary Differential Equations

For the results to follow, the variable x can represent either a scalar variable or an N-dimensional vector. Thus, the obtained results apply to the general case of an N-th order differential equation.

Consider the first-order differential equation

$$\frac{dx}{dt} = f(x,t), \quad x(t_0) = x_0, \tag{25}$$

where $f(x,t)$ is such that a solution exists, i.e., $x(t) = \phi(x_0,t_0,t)$. Let the difference equation

$$x_{k+1} = g(x_k,k,h), \tag{26}$$

have the solution $x_k = \psi(x_0,k_0,k,h)$. Equations (25) and (26) are said to have the <u>same general solution</u> if and only if

$$x_k = x(hk), \tag{27}$$

for arbitrary constant values of h [9, 14]. A <u>best difference scheme</u> is one for which the solution to the difference equation has the same general solution as the associated differential equation.

<u>Theorem</u>: Every ordinary differential equation has a best difference scheme.

<u>Proof</u>: We prove this theorem by direct construction of such a difference scheme [9]. Let h be an arbitrary, positive constant. The "group property" of the solutions to Eq. (25) gives [15]

$$x(t+h) = \phi[x(t),t,t+h]. \tag{28}$$

Make the following definitions: $t_k = hk$, $x_k = x(hk)$. Therefore, Eq. (28) can be rewritten as

$$x_{k+1} = \phi[x_k, hk, h(k+1)]. \qquad (29)$$

Thus, by construction, the difference equation of Eq. (29) has the "same general solution" as Eq. (25); consequently, it is the "best difference scheme."

Note that a given ordinary differential equation has only one "best difference scheme." Also, our theorem is only an existence theorem and, in practice, cannot be used directly to help in the construction of a "best difference scheme" for a particular differential equation. However, we can apply this theorem to differential equations with known solutions. This exercise will allow us to study the various possible modeling behaviors of derivatives and nonlinear terms. In this way, we can hopefully construct a set of rules for situations where exact solutions are not known.

The following examples illustrate the use of this theorem. In each case, the exact general solution can be easily calculated and the theorem can then be used to determine a "best difference scheme."

First, consider Eq. (8) and its solution Eq. (9). Application of the theorem gives

$$x(t) = x_0 \exp[-(t-t_0)] \rightarrow x_{k+1} = x_k e^{-h}. \qquad (30)$$

Finally, this last expression can be written as

$$\frac{x_{k+1} - x_k}{(1 - e^{-h})} = -x_k, \qquad (31)$$

which is the "best difference scheme." This is to be compared to the results of Eqs. (10) and (14). Note that Eq. (31) has the solution (see, Eq. (30))

$$x_k = x_0 \exp[-(t_k - t_{k0})], \tag{32}$$

where t_k = hk, etc., and consequently, x_k = x(hk) as expected. An important point is the fact that in the modeling of the derivative term the numerator term is [1 - exp(-h)] rather than the simple expression h used in the forward Euler scheme. Also, for h > 0, Eq. (31) has no stability problems.

In a similar fashion, the "best difference schemes" can be found for the following differential equations:

$$\frac{d^2x}{dt^2} + x = 0 \rightarrow \frac{x_{k+1} - 2x_k + x_{k-1}}{4 \sin^2(\frac{h}{2})} + x_k = 0, \tag{33}$$

$$\frac{d^2x}{dt^2} = a\frac{dx}{dt} \rightarrow \frac{x_{k+1} - 2x_k + x_{k-1}}{\left(\frac{e^{ah}-1}{a}\right)h} = a\left[\frac{x_k - x_{k-1}}{h}\right], \tag{34}$$

$$\frac{dx}{dt} = x(1-x) \rightarrow \frac{x_{k+1} - x_k}{(1 - e^{-h})} = x_{k+1}(1-x_k), \tag{35}$$

$$2\frac{dx}{dt} + x = \frac{1}{x} \rightarrow 2\left[\frac{x_{k+1} - x_k}{(1 - e^{-h})}\right] + \frac{x_k^2}{\left(\frac{x_{k+1}+x_k}{2}\right)} = \frac{1}{\left(\frac{x_{k+1}+x_k}{2}\right)}. \tag{36}$$

We leave until Section V a discussion of what can be learned from the above constructions.

IV. Partial Differential Equations

For the case of partial differential equations the

situation is complicated by the fact that there does not exist an unambiguous definition of the general solution to such equations [16]. However, we do expect to construct "best difference schemes" for special classes of partial differential equations. For such schemes, functional relations are found to hold among the space- and time-step sizes. The following equations illustrate some of the results obtained with these investigations [9, 17]; in each case $u(x,t)$ and $u_m^n = u[(\Delta x)m,(\Delta t)n]$, and the situation is the initial value problem with $u(x,0) = f(x)$ given:

$$u_t + u_x = 0, \tag{37a}$$

$$\frac{u_m^{n+1} - u_m^n}{\Delta t} + \frac{u_m^n - u_{m-1}^n}{\Delta x} = 0, \quad \Delta t = \Delta x, \tag{37b}$$

or simplifying

$$u_m^{n+1} = u_{m-1}^n; \tag{37c}$$

$$u_t + u_x = \lambda u, \tag{38a}$$

$$\frac{u_m^{n+1} - u_m^n}{\left(\frac{e^{\lambda \Delta t} - 1}{\lambda}\right)} + \frac{u_m^n - u_{m-1}^n}{\left(\frac{e^{\lambda \Delta x} - 1}{\lambda}\right)} = \lambda u_{m-1}^n, \quad \Delta t = \Delta x = h \tag{38b}$$

or simplifying

$$u_m^{n+1} = e^{\lambda h} u_{m-1}^n; \tag{38c}$$

$$u_t + u_x = \lambda u^2, \tag{39a}$$

$$\frac{u_m^{n+1} - u_m^n}{\Delta t} + \frac{u_m^n - u_{m-1}^n}{\Delta x} = \lambda u_m^{n+1} u_{m-1}^n, \qquad \Delta t = \Delta x = h \qquad (39b)$$

or simplifying

$$u_m^{n+1} = u_{m-1}^n / (1 - \lambda h u_{m-1}^n). \qquad (39c)$$

Again we postpone to the next section any discussion of these interesting results.

V. Rules for Constructing Difference Schemes

When confronted with an arbitrary ordinary or partial differential equation, how should we proceed to construct a finite-difference scheme? The results of Sections III and IV indicate that we cannot start (at least at the present moment) from a priori arguments. However, we can list the various requirements that would be desirable in any workable finite-difference scheme; they are as follows:

(a) The finite-difference scheme should be explicit [2, 5, 6]. This requirement is for purposes of ease of calculation. In practice, for a particular equation, this may not be possible.

(b) The difference scheme should lead to numerical solutions which have both the qualitative and quantitative properties (within certain error limits) of the solutions to the differential equation of interest. In particular, this requirement means that for ordinary differential equations, the difference scheme should have exactly the same number of steady states (fixed points, limit cycles, etc.)

as the differential equation and that the stability properties of the corresponding steady states be the same. Otherwise, a situation exists that can give rise to "spurious" or chaotic solutions.

(c) If special (not general) solutions are known for the differential equation, then the difference scheme should also have these special solutions in its solution set.

(d) If at all possible, a "best difference scheme" should be used. In general, such will not be the case, since, no a priori rules exist for constructing such schemes. However, if the requirements (b) and (c) are satisfied, then from a practical viewpoint the "best working difference scheme" will be obtained. At the present time, this is the best that can be achieved.

The following rules for the construction of difference equation models for differential equations are based on both analytical and numerical studies of a large number of linear and nonlinear, and ordinary and partial differential equations. When reviewing these rules the reader should refer to the examples of Sections III and IV and the results presented in Section II.

Modeling Rules

(i) The order of the difference scheme should be equal to the order of the differential equation. A corollary to this rule is that the modeling of derivatives will require the use of more complicated expressions for denominator functions, i.e., sinh may need to be used for h or $4\sin^2(\frac{h}{2})$ for h^2, etc. As shown in Section II and III, these seemly minor changes lead to significant changes in the solutions to the corresponding difference schemes.

(ii) Nonlinear terms should be modeled nonlocally on the computational lattice, i.e., x^2 should not be replaced by x_k^2, but, by $x_{k+1}x_k$.

(iii) The difference scheme should be constructed in such a manner that special solutions of the differential equations are also the "same general solutions" of the difference scheme. This rule includes the fact that corresponding stability properties should be the same for both the differential equation and the difference scheme.

We now illustrate the use of these rules by means of three examples.

First, consider the case of the diffusion-free Burgers' equation [18]

$$u_t + uu_x = 0. \tag{40}$$

This equation has the special solution

$$u(x,t) = \frac{A_2 - \alpha x}{A_1 - \alpha t}, \tag{41}$$

where A_1, A_2 and α are arbitrary constants. It can be shown that both of the following difference schemes have special solutions that are exactly equal to Eq. (41):

$$\frac{u_m^{n+1} - u_m^n}{\Delta t} + \frac{u_m^{n+1}(u_{m+1}^n - u_m^n)}{\Delta x} = 0, \qquad (42a)$$

$$\frac{u_m^{n+1} - u_m^n}{\Delta t} + \frac{u_m^n(u_{m+1}^{n+1} - u_m^{n+1})}{\Delta x} = 0. \qquad (42b)$$

In Eqs. (42), the step-sizes Δt and Δx can be arbitrary. These two schemes correspond, respectively, to explicit and implicit difference schemes for Eq. (40). Note that both schemes follow from the modeling rules given above and are not what would have been obtained from application of the standard "textbook" procedures [2, 5, 6].

The nonlinear diffusion equation

$$u_t = u u_{xx}, \qquad (43)$$

can be modeled by the explicit difference scheme

$$\frac{u_m^{n+1} - u_m^n}{\Delta t} = u_m^{n+1}\left[\frac{u_{m+1}^n - 2u_m^n + u_{m-1}^n}{(\Delta x)^2}\right], \qquad (44)$$

where Δt and Δx can take on arbitrary values. It can be shown that rational solutions of Eqs. (43) and (44) are exactly equal to each other, i.e., $u_m^n = u[(\Delta x)m,(\Delta t)n]$. Note that the above difference scheme is explicit.

Then nonlinear, steady-state Burgers' equation is [18]

$$u u_x = \nu u_{xx}, \qquad (45)$$

where ν is a constant. The use of central difference approximations for both the first and second derivatives gives

$$u_k \left[\frac{u_{k+1} - u_{k-1}}{2h} \right] = \nu \left[\frac{u_{k+1} - 2u_k + u_{k-1}}{h^2} \right], \qquad (46)$$

which is an explicit model. This equation can be solved exactly [18]; using this solution, one can show that for $0 < h \ll 1$ it has both the qualitative and quantitative properties of the exact solution to Eq. (45).

VI. Summary

The main purpose of this paper was to indicate some pitfalls that can occur in the modeling of differential equations by difference equations for the purposes of numerical integration. The advantage of constructing best difference schemes is that problems concerning consistency, stability and convergence [5, 6, 7] will not arise. However, best difference schemes for an arbitrary difference equation are difficult to construct. We have shown that certain modeling rules (as presented in Section V) do lead to difference schemes having a number of desirable properties. While these are not exact difference schemes, they may lead to difference equations that minimize a number of problems related to consistency, stability and convergence.

Finally, it should be indicated that all of the above results fit easily into the general framework of mathematical problems encountered by both theorists and experimenters in their analysis of the complex ordinary and partial differential equations that arise in many-body problems, molecular structure calculations, the determination of transport properties, etc. [2]. The results of this paper indicate that non-standard rules of constructing difference schemes can be used with great profit. The actual application of these techniques to the solving of realistic physical problems will provide the real proof of their value.

References

1. M. Gitterman and V. Halpern, *Qualitative Analysis of Physical Problems* (Academic Press, New York, 1981).

2. D. Potter, *Computational Physics* (Wiley, New York, 1973).

3. D. W. Jordan and P. Smith, *Nonlinear Ordinary Differential Equations* (Clarendon Press, Oxford, 1977).

4. A. H. Nayfeh, *Perturbation Methods* (Wiley, New York, 1973).

5. J. Lambert, *Computational Methods in Ordinary Differential Equations* (Wiley, New York, 1973).

6. L. Lapidus and G. F. Pinder, *Numerical Solution of Partial Differential Equations in Science and Engineering* (Wiley, New York, 1982).

7. R. D. Richtmyer and K. W. Morton, *Difference Methods for Initial-Value Problems* (Wiley-Interscience, New York, 1967, 2nd edition).

8. R. E. Mickens, *Difference Equations* (Van Nostrand Reinhold, New York, 1987).

9. R. E. Mickens, "Difference Equation Models of Differential Equations Having Zero Local Truncation Errors," in *Differential Equations*, I. W. Knowles and R. T. Lewis, Eds., (North-Holland, Amsterdam, 1984), pp. 445-449.

10. C. W. Horton, Jr., L. E. Reichl and V. G. Szebehely, *Long-Time Prediction in Dyanmics* (Wiley, New York, 1983).

11. R. Gilmore, *Catastrophe Theory for Scientists and Engineers* (Wiley-Interscience, New York, 1981).

12. R. May, "Simple Mathematical Models with very Complicated Dynamics," Nature **261**, 459-467 (1976).

13. S. Ushiki, "Central Difference Scheme and Chaos," Physica **D4**, 407-424 (1982).

14. R. B. Potts, "Differential and Difference Equations," Am. Math. Monthly **89**, 402-407 (1982).

15. V. V. Nemytski and V. V. Stepanov, *Qualitative Theory of Differential Equations* (Princeton University Press; Princeton, NJ, 1969).

16. E. Zauderer, Partial Differential Equations of Applied Mathematics (Wiley-Interscience, New York, 1983).

17. R. E. Mickens, "Exact Finite Difference Schemes for the Non-Linear Unidirectional Wave Equation," Journal of Sound and Vibration 100, 452-455 (1985).

18. R. E. Mickens, "Exact Solutions to Difference Equation Models of Burgers' Equation," Numerical Methods for Partial Differential Equations 2, 123-129 (1986).

II. SHORT CONTRIBUTIONS

Structural Data from Solid-State Deuterium NMR Spectroscopy: A Karplus-Type Relationship for ^{2}H-C-C-X Torsion Angles.

Leslie G. Butler

Department of Chemistry, Louisiana State University, Baton Rouge, Louisiana 70803 USA

Herein, the structural data available from solid-state deuterium NMR spectroscopy is reviewed. Major applications have been to the study of hydrogen bonding. Recently, metal hydrides have been studied. Structural studies of C-^{2}H bonds, as compared to dynamical studies, have been nonexistent. Based upon the observed unusual asymmetry in the methyl deuterium electric field gradient in thymine, a research project has been initiated into the possible applications of neighboring atom effects for the measurement of torsion angles in amorphous systems.

Introduction

The objective in this research project is the development of a new method for measuring torsion angles in amorphous samples. Torsion angles are difficult to measure by any conventional methods, yet are important to the solid-state properties of polymers, biomaterials, and catalysts intermediates. In solution proton NMR spectroscopy, a method was developed early on which utilizes the ^{1}H-^{1}H J-coupling between hydrogens bound to adjacent carbon centers to establish a torsion angle. Development of the method has lead to the Karplus equation, which has the general form[1]

$$^3J(^1H,^1H) = A + B\cos(\theta) + C\cos(2\theta) \qquad [1]$$

where the torsion angle is defined as shown in the Newman projection, Figure 1. Until the recent work in our laboratory, an analogous solid-state NMR method did not exist.

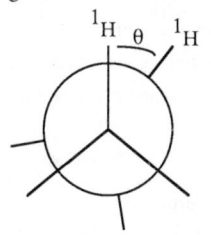

Figure 1. Newman projection defining the ^{1}H-C-C-^{1}H torsion angle used in the Karplus equation (solution NMR).

Theory

In solid-state deuterium NMR spectroscopy, structural information comes about from the fact that the deuterium NMR spectrum is determined by the electric field gradient at the deuterium nuclear site. The electric field gradient is a tensor quantity with a trace of zero. In reports of solid-state deuterium NMR data, the common convention is to refer to an axis system that diagonalizes the the electric field gradient tensor; the larges element of the tensor, eq_{zz}, determines the quadrupole coupling constant, $e^2q_{zz}Q/h$, where e is the charge of the electron, Q is the nuclear electric quadrupole moment, and h is Planck's constant. The electric field gradient is a direct function of the charge distribution in the close vicinity of the deuterium nucleus, as shown in equation [2].

$$eq_{zz} = \sum_n K_n \frac{3z_n^2 - r_n^2}{r_n^5} - e \langle \Psi^* | \sum_i \frac{3z_i^2 - r_i^2}{r_i^5} | \Psi \rangle \qquad [2]$$

where K_n is the nuclear charge of the n-th nucleus. We note the the eq_{zz} is a measure of local charge density; the influence of neighboring charges falls off with a $1/r^3$ distance dependence. We also note that the electronic contribution, the second term in equation [2], is summed only over the occupied molecular orbitals. Since excited states do not affect the value of eq_{zz}, the calculation of eq_{zz} from the results of *ab initio* molecular orbital calculations is straightforward. The utility of deuterium quadrupole coupling constants is shown in applications to hydrogen bonding (O-^{2}H$\cdots$O)[2,3] and the study of metal hydrides.[4,5]

In a C-^{2}H bond, the nuclear charge of the carbon nucleus causes the major axis of the electic field gradient tensor to lie along the C-^{2}H bond axis. Also, the nuclear contribution to eq_{zz} is larger than the shielding afforded by the valence electrons in the C-^{2}H bond. Therefore, most aliphatic C-^{2}H bonds have a positive deuterium quadrupole coupling constant of about 170 kHz. We now consider the effect of neighboring charged atoms. This is pictorially represented in Scheme I.

Multiple contributions to EFG at Deuterium

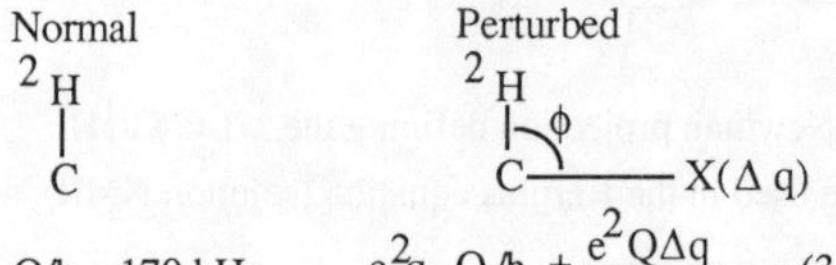

Normal
^{2}H
|
C

$e^2q_{zz}Q/h \sim 170$ kHz

Perturbed
^{2}H
\\ϕ
C————X(Δ q)

$e^2q_{zz}Q/h + \dfrac{e^2 Q \Delta q}{d(^2H\text{----}X)^3}(3\cos^2\phi - 1)$

Scheme I

In the case of thymine-*methyl-d₃*, the solid-state deuterium NMR spectrum shows an unusual asymmetry. Based upon the results of electric field gradients obtained from *ab initio* molecular orbital calculations, the cause of the unusual asymmetry has been traced to the charge resident on an exocyclic oxygen atom ortho to the deuterated methyl group.[6]

Experimental

ADLF Spectrospcopy. Adiabatic demagnetization in the laboratory frame (ADLF) spectroscopy is one of a set of techniques used to obtain high-resolution solid-state deuterium NMR spectra. The ADLF technique, pioneered by Hahn and Edmonds,[7] uses the pulse sequence shown in Figure 2.

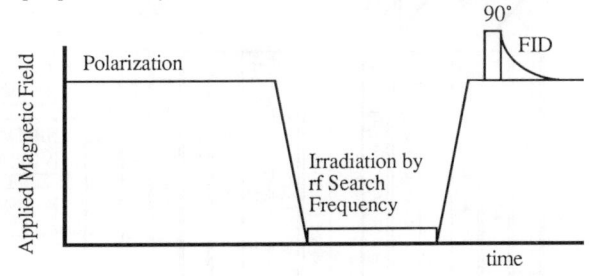

Figure 2. Field Cycling Sequence in ADLF Spectroscopy.

The ADLF spectra are obtained at LSU on an instrument built by the Butler group. A block diagram of the instrument is shown in Figure 3. This instrument is controlled by an IBM 9000 computer (68000 CPU, 1 Mbyte RAM) with about 5000 lines of software written in Pascal. The instrument design is highly flexible due to the use of a CAMAC crate holding twelve CAMAC interface modules. CAMAC is a nuclear physics standard and makes available to the instrument builder many different interface modules at reasonable cost. Special CAMAC modules in this instrument include dual 20 MHz 8-bit digitizers coupled to 24-bit by 8k signal averagers and a home-made pulse programmer with 100 ns resolution and 1024 steps. The CAMAC crate is controlled by the computer over an IEEE-488 bus. The high-field rf equipment consists of a wideband Novex transceiver, preamplifier, and 15 MHz 400 watt power amplifier. In the rather large coil required in ADLF (the sample is moving inside a transport tube immersed in liquid nitrogen), the ^{1}H 90° is 3 μs and the probe ringdown time is 10 μs at 15 MHz. The high-field magnet is a Varian XL100 electromagnet with a 0-120 ampere power supply. Without room temperature shims and flux stabilizer, the pole face gap is 38 mm which is adequate for a glass liquid nitrogen dewar. The zero-field rf irradiation frequency and amplitude are under computer-control. The instrument is usually operated with the sample cooled to 77 K in liquid nitrogen and at a field strength of 0.3 T (a ^{1}H frequency of 15 MHz); these conditions are used to obtain a ^{1}H T_1 long enough ($T_1 > 10$ s) for the field cycling experiment to be successful. The recovered ^{1}H magnetization is measured with a Ostroff-Waugh pulse sequence set to generate 128 spin echoes that are individually digitized.

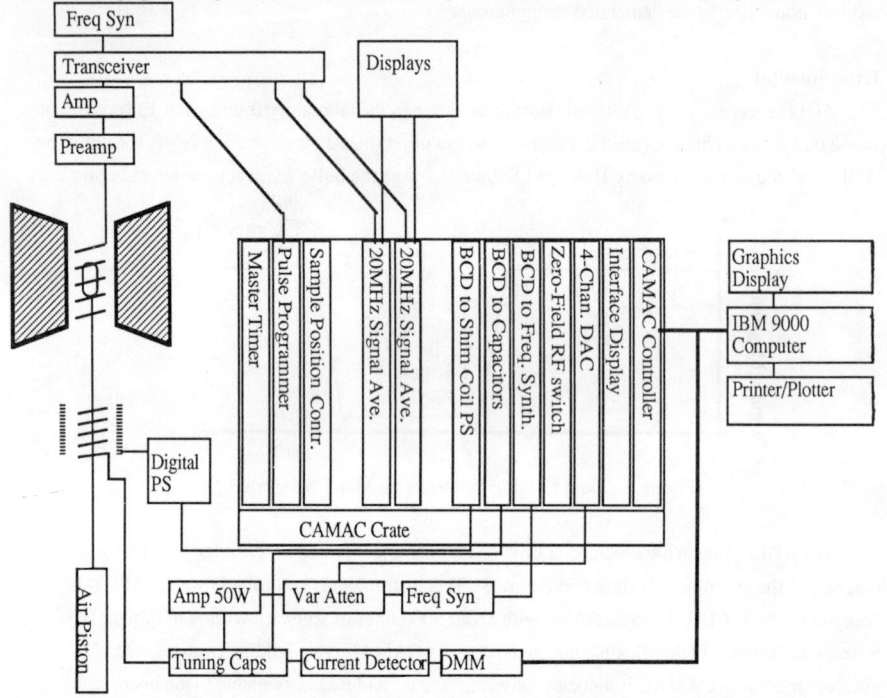

Figure 3. The LSU ADLF spectrometer.

Molecular System. Acetic acid and substituted phenylacetic acids are convenient molecules for establishing the relationship between torsion angles and the deuterium quadrupole coupling constant. The calculated values of the electric field gradient tensor in the acetic acid dimer showed a clear relationship between the ^{2}H-C-C-OH torsion angle and the deuterium quadrupole coupling constant. Experimental confirmation of the calculated results was obtained from the ALDF spectrum of *para*-substituted phenylacetic acids.

Results

Table 1 list the results obtained to date.[8,9] There is a clear correlation between the ^{2}H-C-C-OH torsion angle and the deuterium quadrupole coupling constant. Based on the calculated results for the acetic acid dimer (B, C parameters) and the experimental results of the substituted phenylacetic acids (A parameter), preliminary values for a Karplus-type relationship are

$$e^2q_{zz}Q/h \text{ (kHz)} = 170.14 - 0.55 \cos(\theta) - 1.79 \cos(2\theta). \quad [3]$$

This work represents the first application of solid-state deuterium NMR to determining structural properties.

Table 1. Deuterium Quadrupole Coupling Constants and Structural Data for Substituted Phenylacetic Acids.

Acid	"C-^{2}H Normal"	"C-^{2}H Perturbed"
(4-Chlorophenyl)(2,2-^{2}H)acetic acid		
$e^2q_{zz}Q/h$, kHz	174.0(9)	168.0(9)
θ, deg[a]	89	31
(4-Bromophenyl)(2,2-^{2}H)acetic acid		
$e^2q_{zz}Q/h$, kHz	173.3(11)	168.6(14)
θ, deg	90	30
(4-Nitrophenyl)(2,2-^{2}H)acetic acid		
$e^2q_{zz}Q/h$, kHz	171.5(61)	166.7(12)
θ, deg	69	52
(4-Fluorophenyl)(2,2-^{2}H)acetic acid		
$e^2q_{zz}Q/h$, kHz	171.3[b]	171.3[b]
θ, deg	60	60

[a] θ is the torsion angle between the hydroxyl oxygen ant the deuteron similar to that defined in Figure 1.
[b] Multiplet analysis in progress.

References
1. Becker, E. D. "High Resolution NMR", Academic Press: New York, 1980.
2. Butler, L. G.; Brown, T. L. *J. Am. Chem. Soc.* **1981**, *103*, 7796.
3. Brown, T. L.; Butler, L. G.; Curtin, D. Y.; Hiyama, Y.; Paul, I. C.; Wilson, R. G. *J. Am. Chem. Soc.* **1982**, *104*, 1172.
4. Jarrett, W. L.; Farlee, R. D.; Butler, L. G. *Inorg. Chem.* **1987**, in press.
5. Guo, K.; Jarrett, W. L.; Butler, L. G. *Inorg. Chem.* **1987**, in press.
6. Hiyama, Y.; Roy, S.; Guo, K.; Butler, L. G.; Torchia, D. A. *J. Am. Chem. Soc.* **1987**, *109*, 2525.
7. Edmonds, D. T. *Phys. Rep.* **1977**, *29C*, 223.
8. Jackisch, M.; Jarrett, W. L.; Guo, K.; Butler, L. G. *Polym. Prepr., Am. Chem. Soc., Div. Polym. Chem.* **1987**, *28* (1), 204.
8. Jackisch, M.; Jarrett, W. L.; Guo, K.; Fronczek, F. R.; Butler, L. G. *Polym. Prepr., Am. Chem. Soc., Div. Polym. Chem.* **1987**, *28* (2), in press.

CHARACTERIZATION OF COATING MATERIALS AND COATED SURFACES

CHAND PATEL* AND C. COOK**
* Materials Evaluation Laboratory, Inc., 17695 Perkins Road, Baton Rouge, LA 70810
** Louisiana State University, Mechanical Engineering Department, Baton Rouge, LA 70803

ABSTRACT

This paper describes the properties of coatings obtained by several processes, in particular, the ion-implantation and Conforma Clad processes. A detailed study of these coatings as well as surface modification due to, is examined as a function of time and temperature.
One of the main objectives of using coatings is to improve the wear resistance of a material, and the hardness of a material is an indication of its wear resistance. Results of microhardness measurements taken both on the coated surface and on a section perpendicular to this surface are presented. The latter of these measurements indicates the diffusion mechanism that occurs at the interface of the coating and the substrate during application of the coating. In addition, surface analytical techniques such as SEM/EDAX, AES/ESCA are used to characterize the coating and the interface. The results are in good agreement with predictions.
A further study of hardness of the coating at and just below the surface shows a decrease in hardness over time. A mechanism to explain this phenomenon is suggested.

INTRODUCTION

There has been a recent upsurge in interest in hard, tough materials now that prospects of using ceramics in engines, machines, tools and wear-resistant components are being realized as a result of increased toughness obtained with new compositions and improved processing. Higher specifications for surface properties can be met by improving existing surface hardening processes or by applying refractory coatings. {1-2}
The decision on whether a hard material should be used in solid or coated layer form is rarely determined by a hardness requirement alone, but is more usually decided by other properties which stem from the specific intrinsic properties of the atomic bonding of these hard materials.
In many applications, it is necessary for a component to have bulk properties different from those of its surface. That is, for example, the material is required to have adequate strength or toughness, while the surface must resist corrosion and wear. Such a combination of properties is possible by the use of proper coatings.

TECHNIQUES

In recent years, several new surface probes techniques for determining the elemental and chemical composition of solid surfaces and thin films have been developed. All of these techniques exhibit fundamental strengths and weaknesses, but some offer greater practical utility than others. X-ray photoelectron spectroscopy, also known as (XPS or ESCA), auger electron spectroscopy, (AES), and scanning electron microscopy, (SEM), and numerous other techniques, have proven to be particularly valuable and are now being utilized to investigate a wide range of material problems. Since for total characterization, the needed information cannot be provided by any one technique alone, data from two or more techniques is required to solve many real-world problems. Full details of these techniques and their application are given in {3-4}.

RESULTS

The sample was cut to convenient sizes by using a low speed diamond wheel cutter. This was done to prevent any changes in the sample that a high speed, high temperature cutting operation might produce. The samples were cut into four approximately equal pieces with the dimensions of 0.5 in. x 0.3 in. x 0.3 in. See Photo 1 and Figure 1A, which show coating-substrate interface.
Figure 1A-G shows the morphology and structure of the interface. Furthermore, Figure 1C-E, F shows the changes in the microstructure below the interface as function of depth. The dark contrast observed in 1G is due to a substantial amount of interdiffusion effects.
Figure 2 shows the AES spectrum of interdiffused interface, major elements present are marked.
Figure 3 shows the variation in hardness number measured from bottom of substrate region to top of coating, value of Vickers hardness number at the interface is indicated by the arrow. The hardness of the substrate is approximately a constant 200 VHN. However, at a position about 0.02 inches from the substrate-coating interface, there is a sharp increase in the hardness values, Figure 3ABE. From the interface, the hardness continues to increase with a gradient of approximately 16,000 VHN/in. The fact that there is an increase in hardness before and after the interface is reached shows that there is a transition region about the interface. This transition is the result of a diffusion process that occurs when the coating is applied.
Because of the high temperatures required to apply the coatings, the formation of a strong bond between the substrate and coating is inevitably accompanied by a diffusion process in the zone of contact between layers. This diffusion process is shown schematically in Figure 4.
The results of the heat treatment tests are shown in Figure C-D. These show the hardness profile of the sample 1 hour, 2 hours, respectively, after heating to 600°C for 1 hour. As before, the hardness increases across the interface. However, the values shown in these graphs are higher than values obtained before.
At 600°C it is not likely that any significant diffusion or phase transformations occurred. It is possible, however, that the increased hardness was the result of atomic rearrangement caused by the movement of chemical complexes. The matrix of the coating consists of a Ni-Cr-B-Fe-C alloy. Thus, Ni-B and carbon-boron complexes are possible. Complexes containing boron can be moved relatively easy with the aid of thermal energy. At 600°C it is most likely that the Ni-B complex would move easiest, causing the increased hardness.
The results from the heat treatment tests help also to confirm the aging process described before. The sample had been exposed to the atmosphere for several months prior to the heat treatment, and was not polished before or after the test. As in Figure 3A, the hardness decreases very close to the surface, demonstrating the aging effect.
In addition, the average values of the coated surface after heat treating are very similar to those of the unpolished, aged sample. This suggests that though the aging process increases the peak hardness above the interface, it does not affect the hardness of the coating itself.
Figure 5 shows the results of the chemical analysis performed using SEM/EDX for base material, interface and the coating. The diffusion mechanism is very much in evidence from the charts. The composition analysis is shown in the table.

CONCLUSIONS

The conclusions that can be reached from the results discussed above are as follows:

1. By examining the hardness of the sample across the interface, the diffusion process occurring during manufacturing can be identified. This helps in the characterization of the coating-substrate bonding.
2. The coatings are subject to a decrease in hardness due to the aging process described.
3. Heat treating the coating causes the movement of complexes in the coating matrix resulting in an increase in the maximum hardness of the coating near the interface. This treatment, however, does not affect the hardness of the coating surface.

Additional experiments involving SEM/EDAX, ESCA, and X-ray diffraction need to be done in order to fully characterize this coating. The results from such experiments and the results presented here could completely describe the chemical, mechanical, and crystallographic characteristics of the diffusion, aging, and heat treatment processes discussed in this paper.

REFERENCES

1. Duret, C. and Pichoir, R., "Protective Coatings for High Temperature Materials: Chemical Vapour Deposition and Pack Cementation Processes," Coatings for High Temperature Application, Applied Science Publishers, Essex, England, 1983.

2. Shewell, D. E., "New Method of Applying Wear Resistant Coatings," Metal Progress, v124 Nov. 1983, pp. 41-42, 45-46, 48-49.

3. Ho, P. S., G. W. Rubloff, J. E. Lewis, V. L. Moruzzi and A. R. Williams, Phys. Rev., B22, 1781 (1980).

4. Patel, C., Journal de Physique, Coll. C2 Suppl., au no. 3, Tome 47 (1986).

ACKNOWLEDGEMENT

My (CP) deepest thanks to the staff of M.E.L. for their continued help technically, and long hours. I also wish to thank Metal Product Division of Imperial Clevite, Inc. for supplying the samples and also their help.

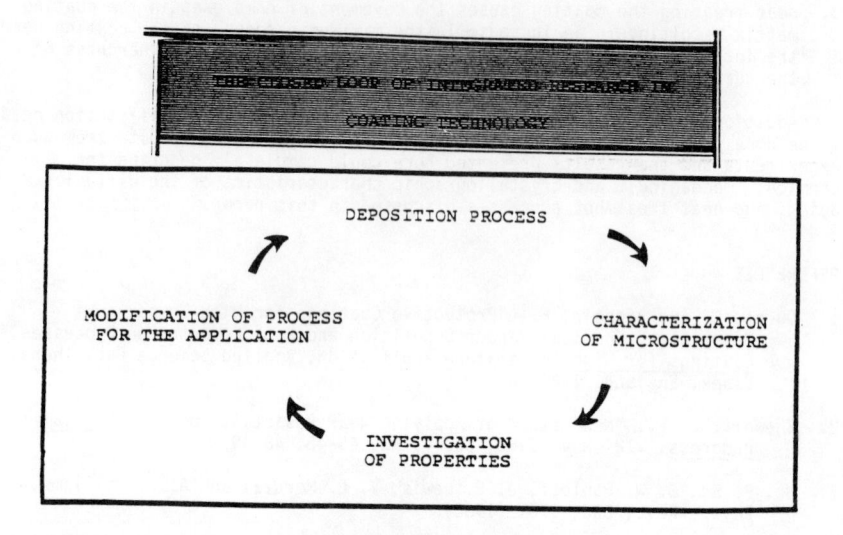

THE CLOSED LOOP OF INTEGRATED RESEARCH IN
COATING TECHNOLOGY

```
                    DEPOSITION PROCESS
                   ↗                ↘
  MODIFICATION OF PROCESS          CHARACTERIZATION
     FOR THE APPLICATION           OF MICROSTRUCTURE
                   ↖                ↙
                    INVESTIGATION
                    OF PROPERTIES
```

TABLE: 1

THE PROCESS BY WHICH A COATING FOR AN APPLICATION IS DEVELOPED IS ILLUSTRATED BY TABLE I. THE MICROSTRUCTURE AND PROPERTIES OBTAINED BY A NEW OR EXISTING PROCESS ARE STUDIED, AND THE RELATIONSHIP BETWEEN THE TWO IS ESTABLISHED. THE COATING PROCESS CAN THEN BE REFINED ACCORDING TO THE REQUIREMENTS OF THE APPLICATION.

Table 2
Common Coating Techniques

Method	Advantages	Disadvantages	Applications
PVD	-very versitile -purity, structure, and adhesion can be controlled		
Vacuum Evaporation	-relatively fast -coating can be optimized with ultra high vacuum	-harder to control	-semi-conductors
Sputtering	-easy to control alloy composition -easy to control rate of deposition	-relatively slow	-ceramic coatings for drills and gear cutters
Ion Plating	-good adhesion on temperature sensitive substrate -covers complicated shapes -enhanced reactivity	-high voltage required -high gas press. required -difficult to control -not for internal surfaces	-corrosion resistant fasteners
CVD	-various kinds of coatings -high rates of deposition -complex components can be coated	-high temperatures required -low pressures required -substrate must be heated	-semi-conductors -magnetic film -optical film
Ion Implantation	-no distortion(low temperature) -adaptable to automation -no adhesion problems	-shallow penetration -equipment only recently on market	-industrial tooling -punches -dies

from Ref. 1.

TABLE: 2

158

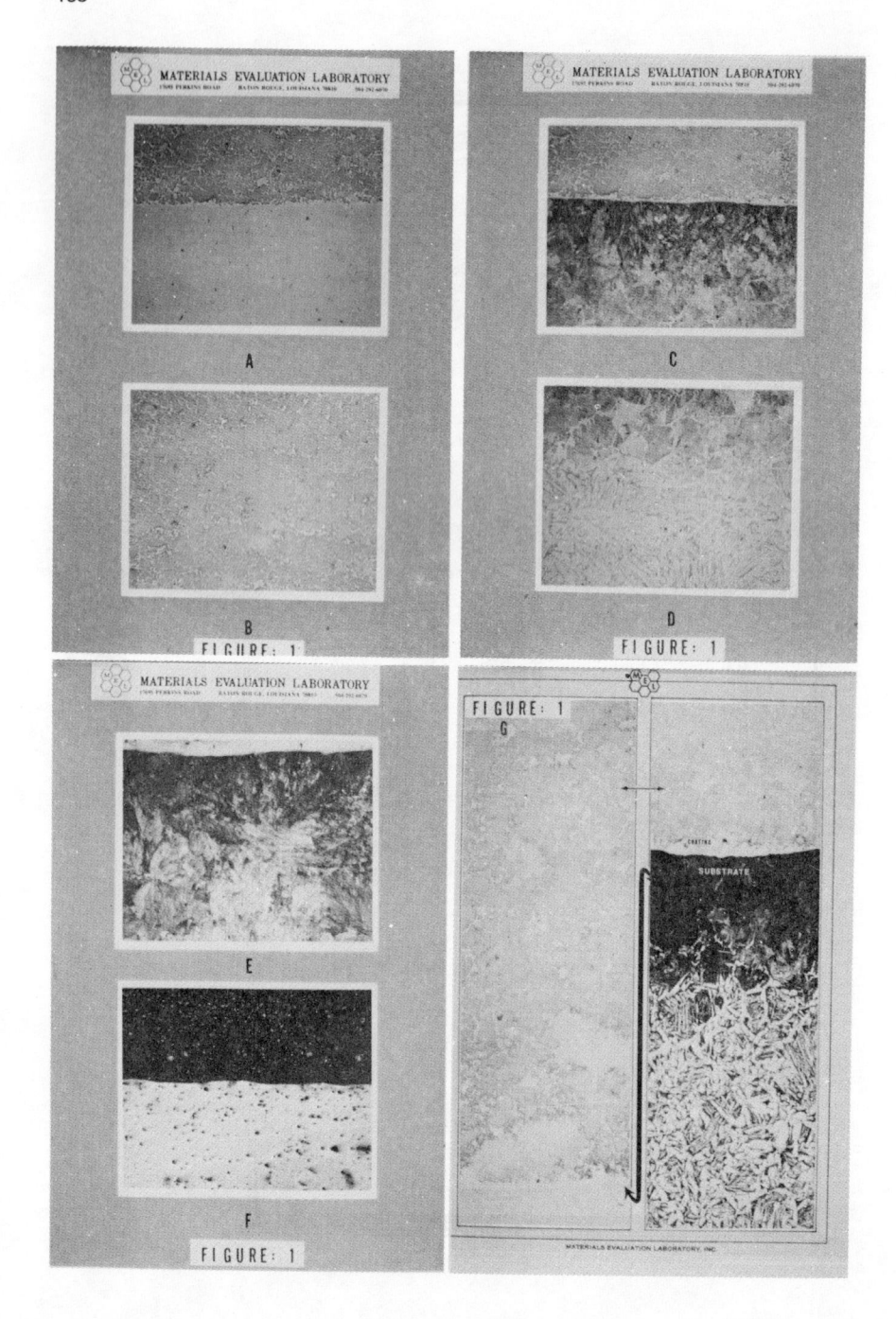

FIGURE: 1

159

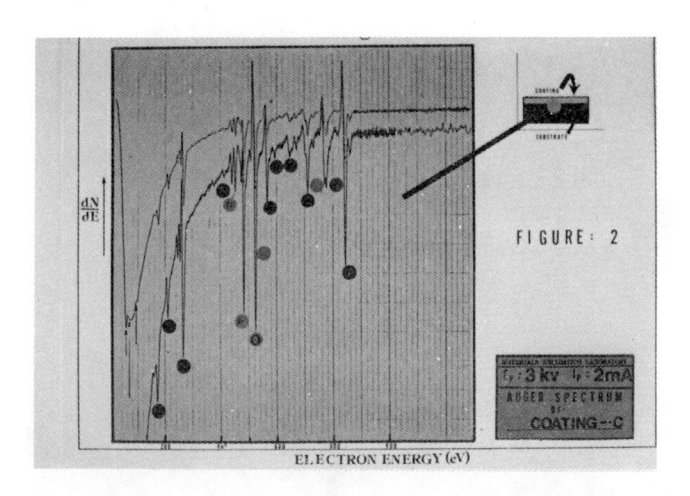

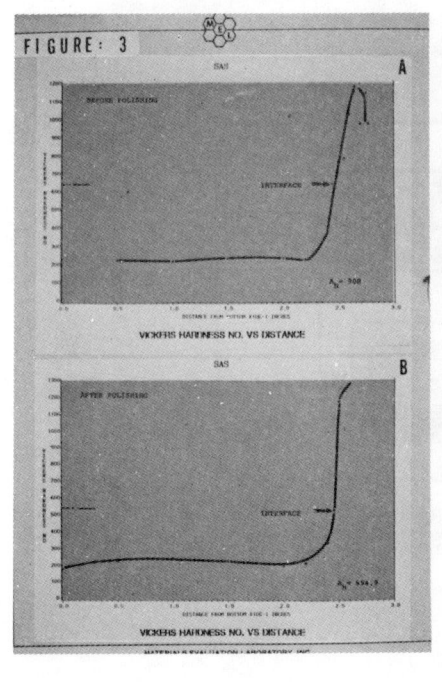

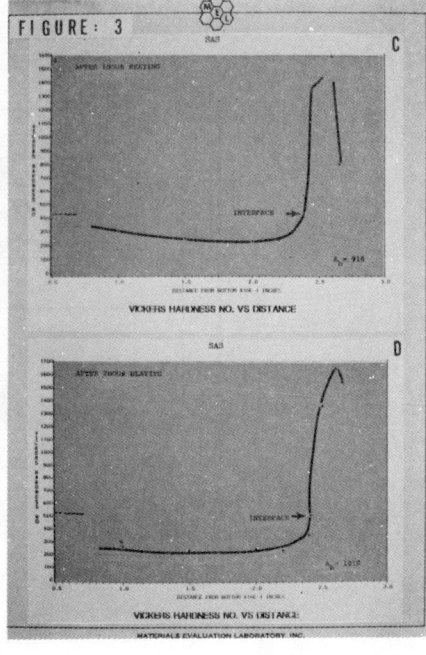

FIGURE: 2

FIGURE: 3

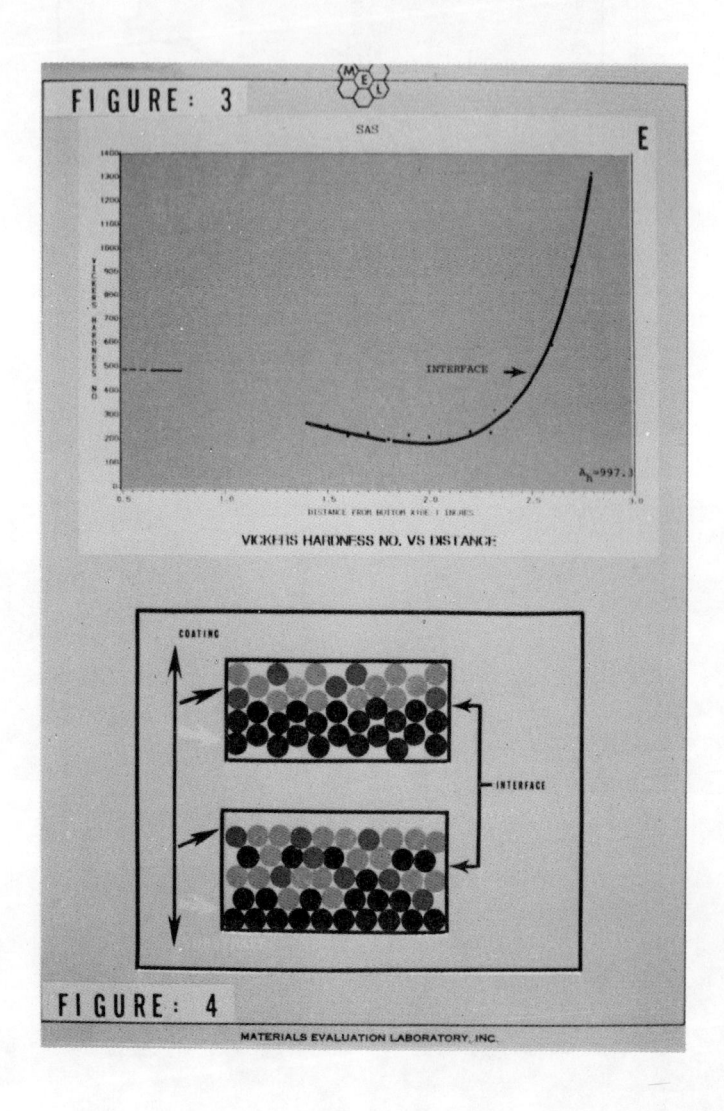

FIGURE: 3

VICKERS HARDNESS NO. VS DISTANCE

FIGURE: 4

MATERIALS EVALUATION LABORATORY, INC.

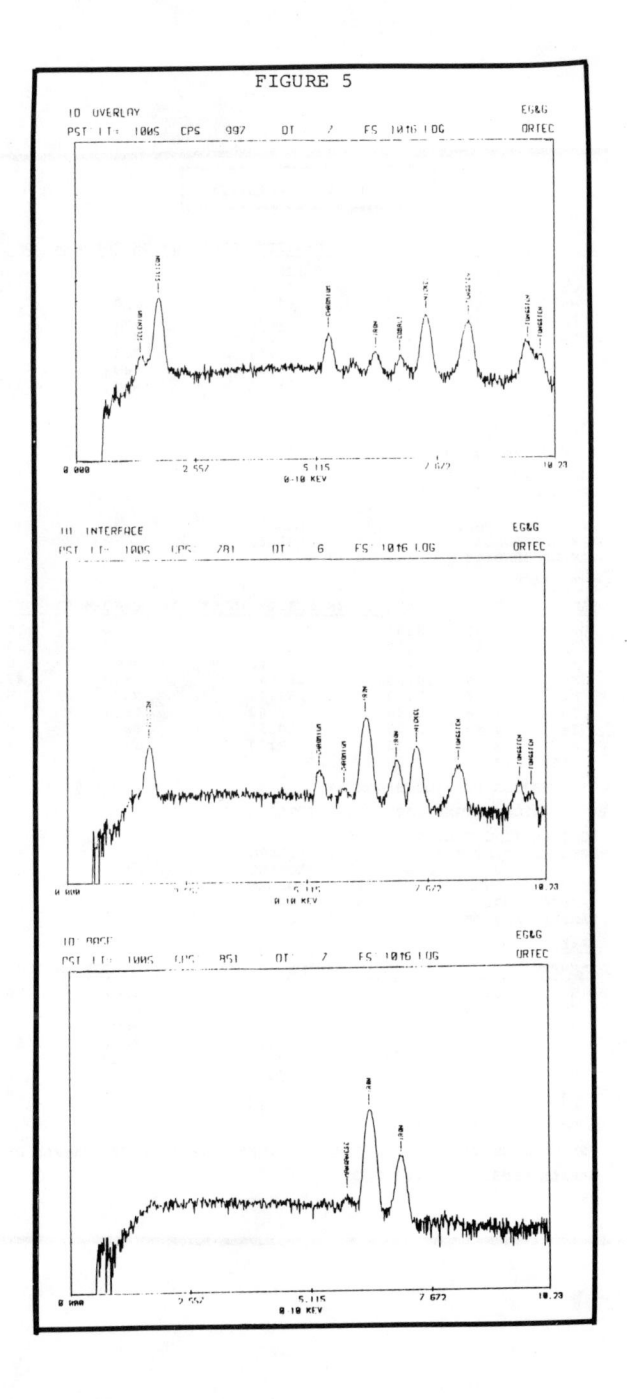

FIGURE 5

COMPOSITION ANALYSIS

OVERLAY

ELEMENT	WT. %	ATOMIC %	INTENSITY (CPS)	OXIDE FORMULA	OXIDE %
Si	17.10	19.73	121.71	SiO_2	36.57
Cr	3.28	2.04	28.94	Cr_2O_3	4.79
Mn	0.00	0.00	-0.39	MnO	0.00
Fe	1.45	0.34	10.96	Fe_2O_3	2.07
Ni	13.59	7.50	33.11	NiO	17.29
Co	1.08	0.60	7.14	CoO	1.38
Se	2.60	1.07	4.60	SeO_3	4.18
W	29.83	5.26	56.71	W_2O_3	33.72
O**	31.08	62.97			

** COMPUTED BY STOICHIOMETRY
# OF OXYGEN ATOMS BASED ON 0 TOTAL

ACCELERATING VOLTAGE: 25.0 KV
SPECIMEN TILT X-AXIS: 0.0 DEGREES
              Y-AXIS: -25.0 DEGREES
INCIDENCE ANGLE   : 94.63 DEGREES
TAKEOFF ANGLE   : 35.37 DEGREES
SAMPLE DENSITY   :   2.57 G/CC

INTERFACE

ELEMENT	WT. %	ATOMIC %	INTENSITY (CPS)	OXIDE FORMULA	OXIDE %
Si	7.45	8.64	33.58	SiO_2	15.94
Cr	1.87	1.17	13.76	Cr_2O_3	2.74
Mn	0.23	0.13	1.42	MnO	0.29
Fe	33.19	19.35	190.46	Fe_2O_3	47.46
Ni	12.76	7.08	53.48	NiO	16.24
Co	1.56	0.86	7.69	CoO	1.98
Se	0.59	0.25	0.57	SeO_3	0.96
W	12.73	2.26	17.50	W_2O_3	14.40
O**	29.61	60.26			

** COMPUTED BY STOICHIOMETRY
# OF OXYGEN ATOMS BASED ON 0 TOTAL

ACCELERATING VOLTAGE: 25.0 KV
SPECIMEN TILT X-AXIS: 0.0 DEGREES
              Y-AXIS: -25.0 DEGREES
INCIDENCE ANGLE   : 94.63 DEGREES
TAKEOFF ANGLE   : 35.37 DEGREES
SAMPLE DENSITY   :   2.77 G/CC

BASE*

ELEMENT	WT. %	ATOMIC %	INTENSITY (CPS)	OXIDE FORMULA	OXIDE %
Si	0.13	0.15	1.06	SiO_2	0.28
Cr	0.18	0.11	3.02	Cr_2O_3	0.26
Mn	0.61	0.36	7.53	MnO	0.79
Fe	67.25	38.59	749.48	Fe_2O_3	96.14
Ni	0.00	0.00	-0.30	NiO	0.00
Co	1.85	1.01	17.91	CoO	2.35
Se	0.00	0.00	-0.31	SeO_3	0.00
W	0.15	0.03	0.40	W_2O_3	0.17
O**	29.83	59.76			

*SETTING CONDITIONS ARE SAME AS THOSE ABOVE (EXCEPT SAMPLE DENSITY).
SAMPLE DENSITY: 2.85 G/CC

A STUDY OF
LASER ACTIVATED GOLD-SILICON INTERFACE CONTACT

BILL DIXON* AND CHAND PATEL

Materials Evaluation Laboratory, Inc.
17695 Perkins Road
Baton Rouge, LA 70810
(504) 292-6070
U.S.A

ABSTRACT:

The use of a flashlamp-excited, tunable dye laser to produce Ohmic contacts between gold and silicon is discussed. Approximately 1000 angsroms of Au is evaporated onto Si <111> p-type wafers and is laser irradiated at operating wavelengths of 585, 600, and 615 nm. Pulse rates of 6, 10, and 15 pulses per second are used and samples are treated for varying periods of time. The damaged areas on the evaporated gold caused by the laser treatment is examined by spectroscopic techniques such as SEM, EDAX, ESCA, and AES. Results of the analysis show the change in the chemical properties in the Au or Si and the occurrence of interdiffusion between the elements. Electrical properties of the Si-Au bond are then determined using the interdiffusion and chemical analysis results.

INTRODUCTION:

Ohmic contacts are formed between thin metal films and semiconductor materials by an annealing process. The contacts provide a bonding pad for leads between internal micro-circuitry and external connections from the device.

Metal-semiconductor Ohmic contacts are defined as "interfaces that possess current-voltage characteristics with a linear region for both directions of current flow over a large temperature range" [1, p. 467]. For the contact to be useful, its resistance must be negligible so that device performance will not be impaired and the current-voltage relationship must be linear in the range of application.

The inability to achieve low values of contact resistance is a major problem in Ohmic contact formation. The current method of forming Ohmic contacts is by furnace or thermal annealing. This method requires deposition of a chosen material on a semiconductor surface, then annealing the system in a furnace. Depending on the temperature and the time of annealing, a shallow, heavily doped region will be produced. This region provides a low resistance path from the metal contact to the semiconductor bulk.

The nature of the furnace annealing process requires bulk heating, high temperatures, long heating times, and melt formation. Consequences may be: a) out diffusion of the semiconductor constituents and/or in diffusion of surface impurities, b) formation of high-resistivity compounds of the metal and semiconductor materials, and c) formation of crystallites in various phases which could cause surface roughness [1, p. 468].

To add to these problems, advances in technology are requiring further miniaturization of micro-electronic devices. Presently, micro-electronic devices range in size of 2 by 2 mm to 10 by 250 mm and may contain over 10^5 elements of circuitry [2, p. 126]. Higher circuitry densities and greater performance in the areas of speed and reliability are now being sought [3, p. 30]. This requires smaller physical dimensions of the contact areas and improvements in contact resistance values.

Methods to provide these improvements are currently being investigated. The method must provide very localized heating so that the physical, chemical and electrical properties of the semiconductor material and the circuitry surrounding the point of the contact formation will not be affected. For the same reason, the method must also be easily controlled with respect to processing time and energy transferred to the area.

A method which meets the above criteria is the use of lasers for the formation of Ohmic contacts. Investigations into their application in this area has shown good results [1, 3, 4].

In this project, we examine the surface characteristics as well as the structural arrangement induced during the irradiation of the metal-semiconductor junction. We have employed the surface analytical techniques such as SEM-EDAX, AES, ESCA and Metallographic methods [5-7]. [See also the keynote paper by C. Patel.]

Si-Au CONTACT THEORY:

Metal layers and multiple layers of metals provide an immense number of possibilities for metal-semiconductor and metal-oxide-semiconductor contacts. Requirements such as low contact resistivity, ease of application, good adhesion to semiconductors and semiconductor oxides, ease of bonding, and reliability greatly reduce the number of possibilities. In fact, the number is reduced to two systems which are generally used: aluminum-based and gold-based systems [8, p. 185].

Aluminum adheres well to silicon without additional contact layers or adhesion layers. It is also inexpensive, easy to apply and bond, and therefore has widespread applications. Aluminum, however, is susceptible to corrosion and elecromigration. Gold is less sensitive to corrosion and electromigration, but its metallization is more complex. Contact and adhesion layers are required, resulting in an expensive, but high quality metallization. The high quality metallization, combined with gold's electrical properties, make it the most desirable choice for use in contact formation [8, p. 186].

Interfaces of Si-Au may exist in one of three basic forms. Figure 1A shows an ideal interface in which no chemical reactions of diffusion have taken place. Figure 1b shows an impurity layer (typically an oxide) between the Si and Au layers. The impurity layer may react with the substrate (in the case of Si with an oxide overlayer, SiO_2 will be formed). Figure 1c shows an interface of Si-Au in which diffusion or chemical reaction has formed a layer of unknown composition which may or may not be thermodynamically stable [9].

Studies into the growth of gold films on clean Si <111> surfaces have shown that ideal interfaces truly are ideal. At room temperature it is observed that gold and silicon interdiffuse when the film thickness is 1.0 to 3.0 nm [10, p. 10]. Furthermore, the interaction at the Au-Si interface produces a silicide on both sides of the film when film thicknesses are 10 to 100 nm. The silicide layers are separated by thicker gold layers with less silicon content. The silicide layer is suggested to have the composition of $Au Si_5$ [10, p. 10]. Further studies suggest that the interface may be a mixed region consisting of a silicide layer and an alloy [10, p. 10].

Tu [11, p. 554] proposed a reaction mechanism which accounts for interdiffusion of Au with Si at temperatures well below the eutectic temperature of 370 degrees C. Brillson et al. [11, p. 554] has obtained results consistent wih Tu's mechanism. The mechanism as described by Brillson et al., suggests that Au atoms diffuse into the Si as interstitals which weaken the Si-Si bond. The partially dissociated Si atoms are then available for reaction with the Au at lower temperatures than otherwise required.

In an annealing process at 200 degrees C for 45 minutes, the percentage of Si in the Au layer is increased even more [11, p. 554].

The advantages of using a laser with the Si-Au system now becomes evident. At room temperature diffusion between Au and Si is already occurring along with possible alloy formation. Higher temperatures, as in the annealing process used in forming Ohmic contacts, increases the diffusion and may possibly cause the formation of an eutectic compound [10, p. 12]. Laser beams, having concentrated energies, can prevent these high temperature formations from occurring throughout the bulk material. Since laser light is a source of heat, localized formation of compounds will be seen. However, these will be minimal due to the rapid (in the order of nanoseconds) heating and cooling of the irradiated area.

EXPERIMENTAL PROCEDURE:

Sample Preparation

One inch Si(111) p-type wafers with a bulk resistivity of 38-39 Ohm-cm are cleaned using concentrated HF acid. These wafers have an oxide layer on them which contains other atmospherical contaminants. The HF rids the wafers of the contaminated oxide layer to expose a clean oxide layer. The wafers are first cleaved in half and dipped in the

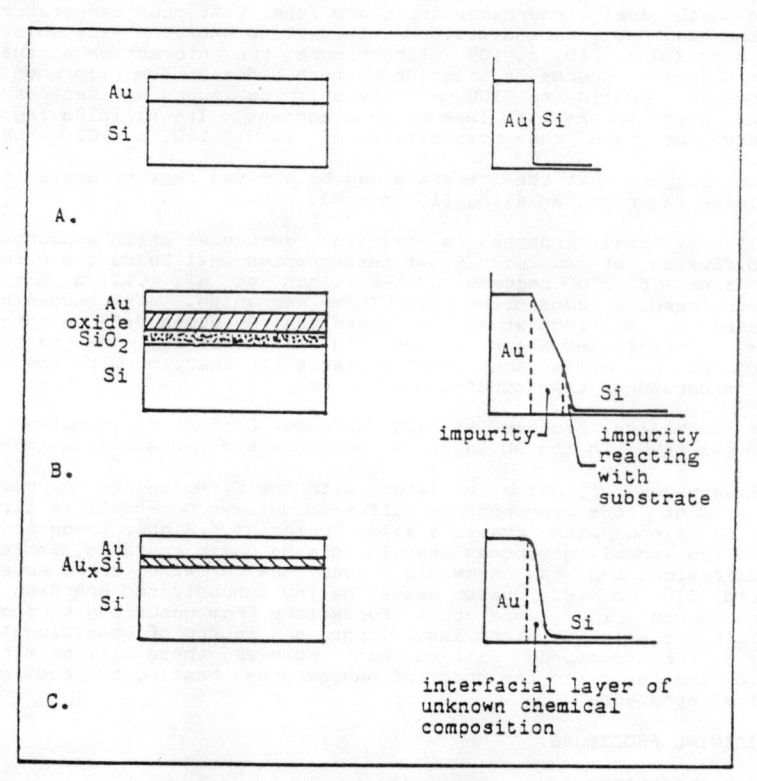

FIGURE 1
Three basic types of interfaces.

HF for 1 minute. The samples are then placed in methanol for 1 minute to stop the HF reaction. Rinsing with deionized water and drying with He gas concludes the process.Figure 2A-B.

The sample is examined using an optical microscope to determine surface cleanliness. If the surface still contains some impurities, the cleaning process is repeated except the time of HF exposure is decreased to 15 seconds.

After the sample has been properly cleaned, a gold film is evaporated onto the Si surface. Figure 2A shows the two mask designs which are placed over the sample during film evaporation to achieve a desired pattern. Evaporation is done for 5 minutes to produce a film of approximately 1000 angstroms thick.

Laser Irradiation of Samples

After the AU is evaporated onto the Si, laser irradiation of the sample is done. The laser used in this experiment is the Chromatix Model CMX-4 flashlamp pumped dye laser. Rhodamine 6G dye (tunable from 577-625 nm with peak gain at 598 nm) [12, p. 3-6] is the amplifying medium for this particular experiment. Figure 3 shows the laser and various electronic support equipment connected to it. A discussion of these components will not be given in this paper.

A lens is used to focus the beam and the sample is placed approximately at the lens' focal point to obtain maximum energy intensity from the laser beam. Each Au spot on the sample is irradiated with varying parameters (wavelength, pulse rate, and irradiation time).

Heating Jig

One of the samples is heated with a heating jig designed for this purpose. Figure 4 is a picture of the jig. A 100 watt heater is inserted into the top section which is a clamp. Two thermocouples are used to monitor the temperature between the jaws of the clamp. Calibration shows that the temperature reaches approximately 320 degrees C after few minutes from the time the heater is switched on.

The sample is heated in an Ar environment to see how the high temperature will affect the Au film. As mentioned in a previous section, Au alone does not adhere well to Si. It is therefore expected that some effects due to heating will be observed.

DISCUSSION OF EXPERIMENTAL PROCEDURE AND SETUP

A few comments on the experiment are necessary to help explain some of the results which follow.

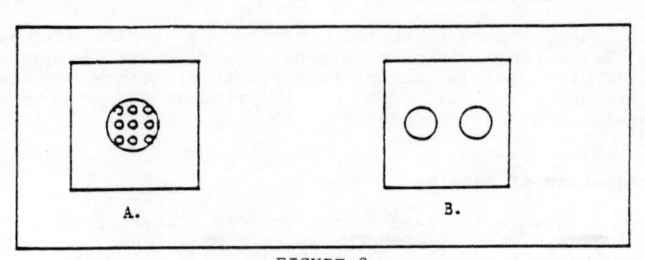

FIGURE 2
(A): Mask designs used while sputtering Au onto Si surface. Mask A has 9-2mm diameter holes contained in a 1cm diameter circle. Mask B has 2-½ in. holes with a center to center distance of ½ in.

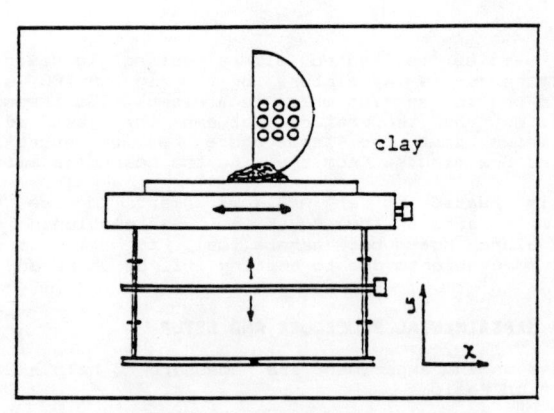

(B): **Apparatus and setup of sample for position manipulation**

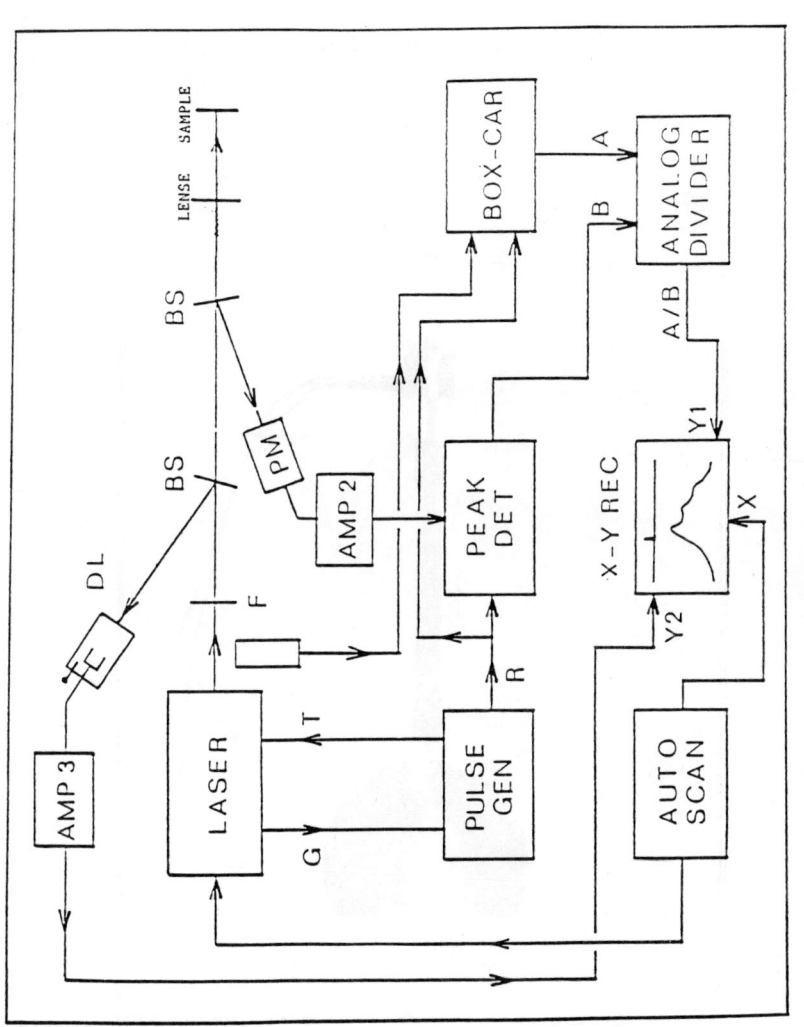

FIGURE 3: EXPERIMENTAL SETUP FOR LASER STUDIES

FIGURE 4: HEATING JIG

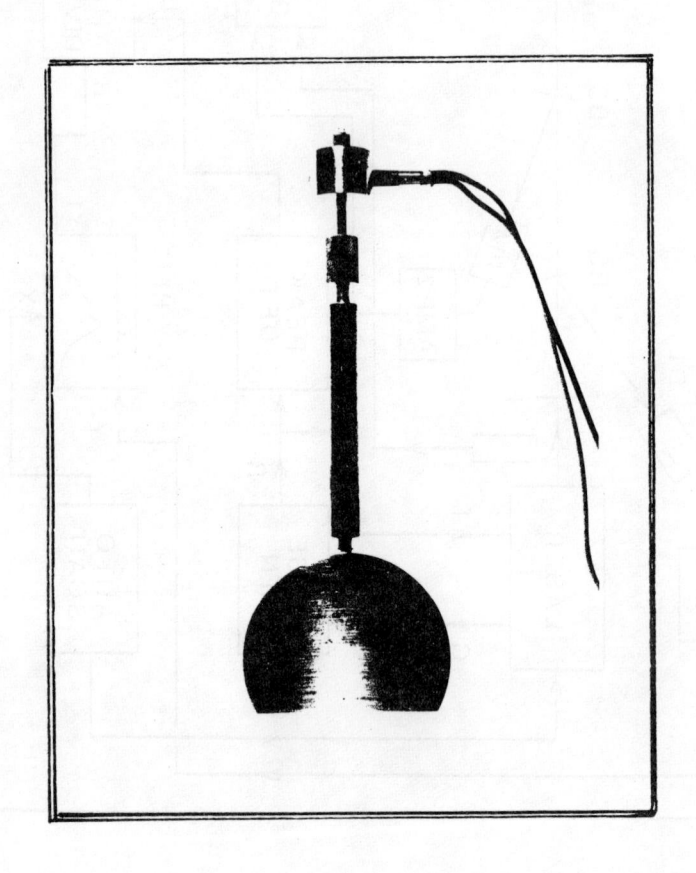

TABLE A

SAMPLE 2

39.4 Ohm-cm cleaned: 2-20-86
(111)   evaporated: 2-20-86
p-type   irradiated: 2-20-86

spot	time sec.	pps	lambda nm.	energy /pulse	total joules
1,1	50	15	615	5.0 mj	3.75
1,2	40	15	615	5.0 mj	3.0
1,3	30	15	615	5.0 mj	2.25
2,1	20	15	615	5.0 mj	1.50
2,2	10	15	615	5.0 mj	0.75
2,3	5	15	615	5.0 mj	0.375
3,1	30	15	585	4.5 mj	2.025
3,2	20	15	585	4.5 mj	1.35
3,3	10	15	585	4.5 mj	0.675

dye: Rhodamine 590 "RoG"

SAMPLE 1

39.4 Ohm-cm cleaned: 2-20-86
(111)   evaporated: 2-20-86
p-type   irradiated: 2-20-86

spot	time sec.	pps	lambda nm.	energy /pulse	total joules
1,1	40	10	600	6.6 mj	2.64
1,2	60	10	600	6.6 mj	3.96
1,3	60	15	600	6.6 mj	5.94
2,1	55	15	600	6.6 mj	5.445
2,2	50	15	600	6.6 mj	4.95
2,3	40	15	600	6.6 mj	3.76
3,1	50	15	600	5.3 mj	3.975
3,2	60	15	600	5.3 mj	4.77
3,3	70	15	600	5.3 mj	5.565

dye: Rhodamine 590 "RoG"

First, the evaporation of the Au onto the Si surface is not done immediately after cleaning of the oxide layer. Approximately 30 minutes elapses between cleaning and evaporation. During this time the sample is in a covered Petri dish being transported to the evaporator. It is quite possible that contamination of the sample occurs during this time.

Second, the repeatability of the irradiation of the samples is quite low. Figure 2B, shows the device used for mounting and positioning of the samples. No method of alignment is used other than eyesight. This makes for uncertainty in the perpendicularity of the sample with respect to the laser pulses. Also, the focal point of the lens is found by noticing where the spot diameter of the laser pulses is smallest. The sample is then located approximately at that point.

Last, it is noted that some of the first EDAX scans taken wear done using an old version of software. Tungsten is shown as the element corresponding to the main Si peak. This was discovered more than midway through the project. Due to time constraints, the scans were not able to be redone using an updated software version, but it was seen that the peak is Si and not W.

RESULTS:

Laser Irradiation of Samples

Three samples were prepared and irradiated by the laser. Table A, gives a tabulation of the parameters used for the irradiation of each spot on the samples. Sample 1 was irradiated without a focusing lens. This results in a relatively larger laser pulse diameter with a low energy per unit area. Consequently, no damage was observed. Repeating the experiment with the lens in place, damage was observed on the irradiated spots (Sample 1).

Sample 2 was also irradiated using the focusing lens. Again, extensive damage was observed. Laser was left on in operation while the sample was being repositioned to a new spot. This procedure makes sure that the laser is striking the correct place. Repositioning is done at a slower pulse rate and lower energy than when the main irradiation takes place. Micrographs in Figure 5 shows the respective damage, together with the EDX charts.Figure 6 and Figure 7

AES/ESCA techniques were applied to the Sample 2 with all the spots exposed to the electron beam. Chart 1 shows the auger spectrum of the results, and Chart 2 gives the spectrum obtained after 2 minutes of sputtering with Ar-gas at 15 mA and 1.5 kv. The contaminant levels due O and C have decreased substantially, and give rise to the low energy peaks. Chart 3 shows the resulting spectrum due to ESCA.

FIGURE 5

SAMPLE 1

Damage due to laser irradiation for 60s. (600x)

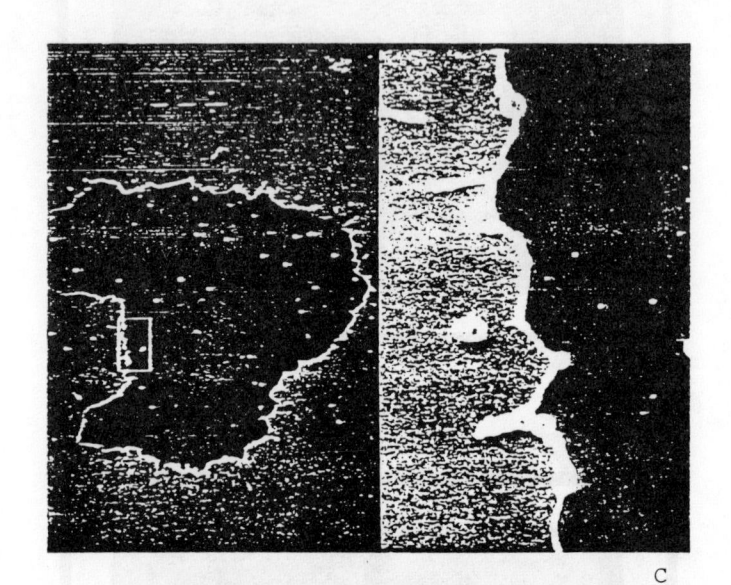

FIGURE 5: Damage due to laser irradiation for 60s. (230x, 2300x)

SAMPLE 1

FIGURE 6: A-B

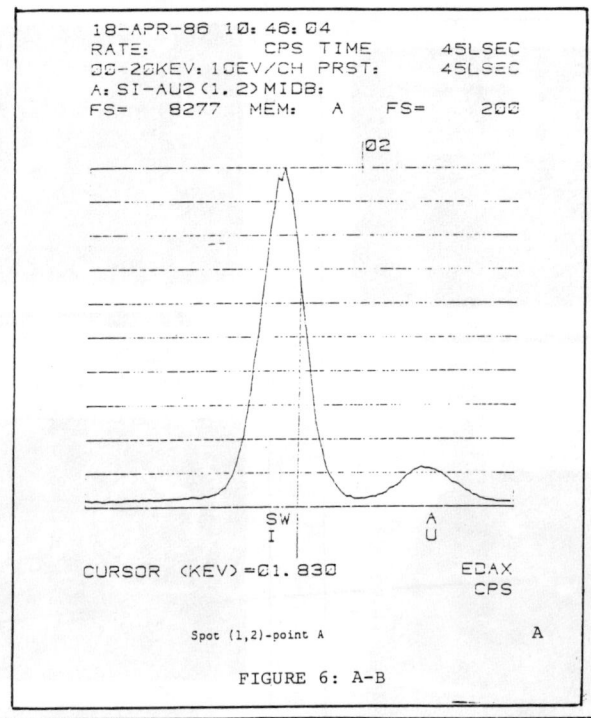

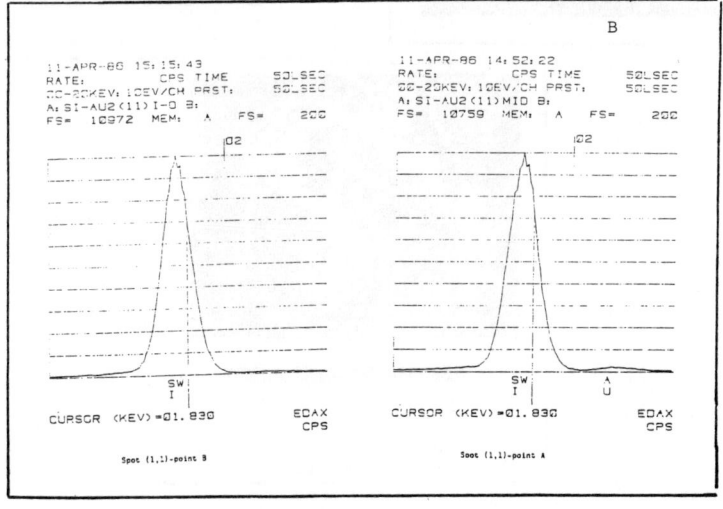

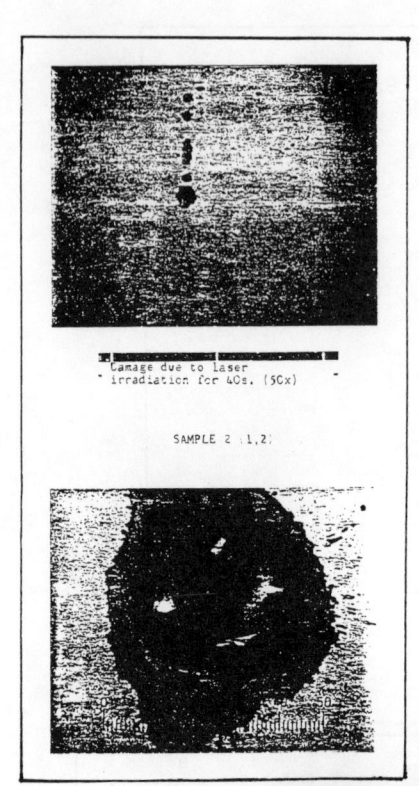

Damage due to laser irradiation for 40s. (50x)
SAMPLE 2 (1,2)

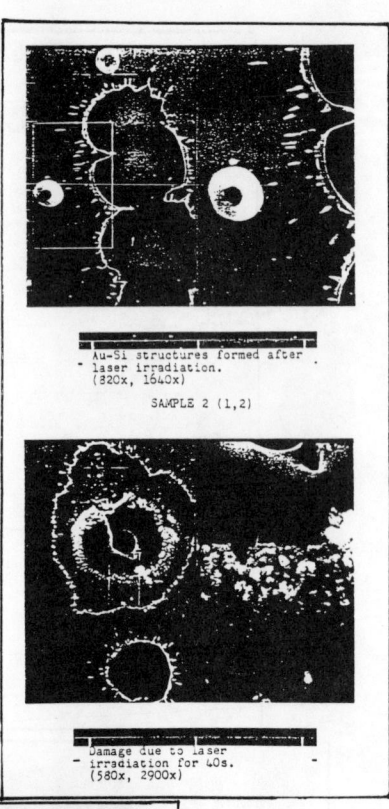

Au-Si structures formed after laser irradiation. (820x, 1640x)
SAMPLE 2 (1,2)

Damage due to laser irradiation for 40s. (580x, 2900x)

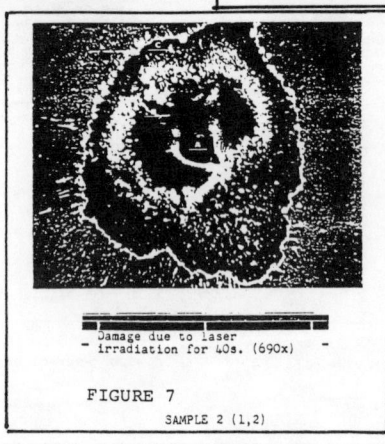

Damage due to laser irradiation for 40s. (690x)

FIGURE 7
SAMPLE 2 (1,2)

FIGURE 7

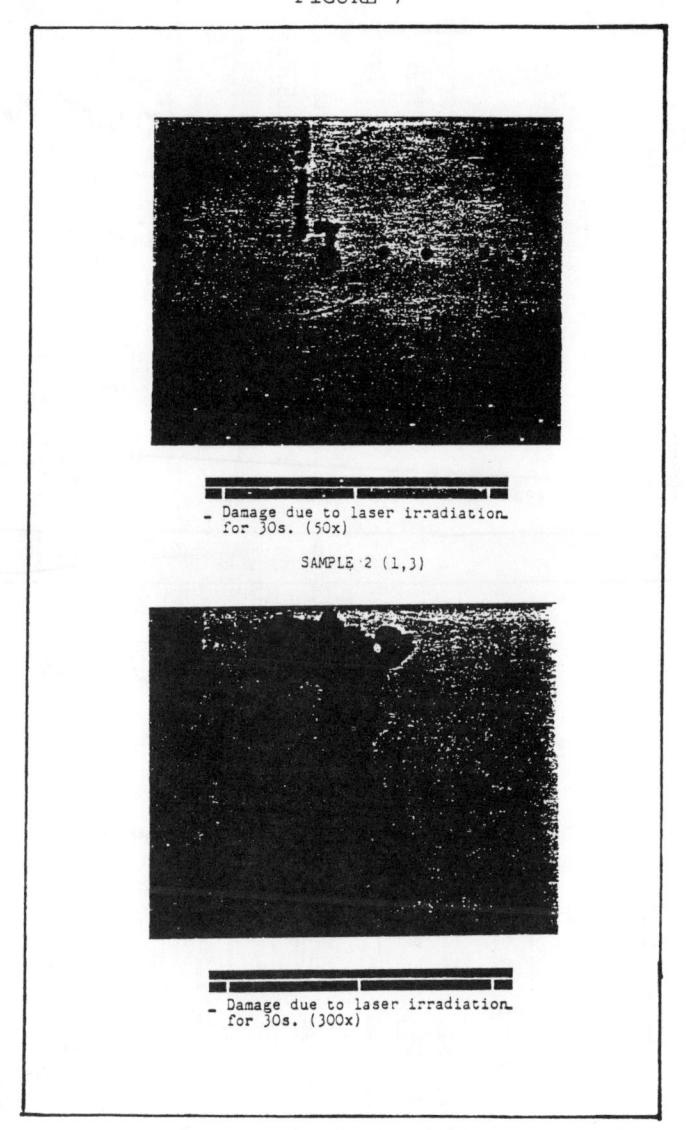

Damage due to laser irradiation for 30s. (50x)

SAMPLE 2 (1,3)

Damage due to laser irradiation for 30s. (300x)

178

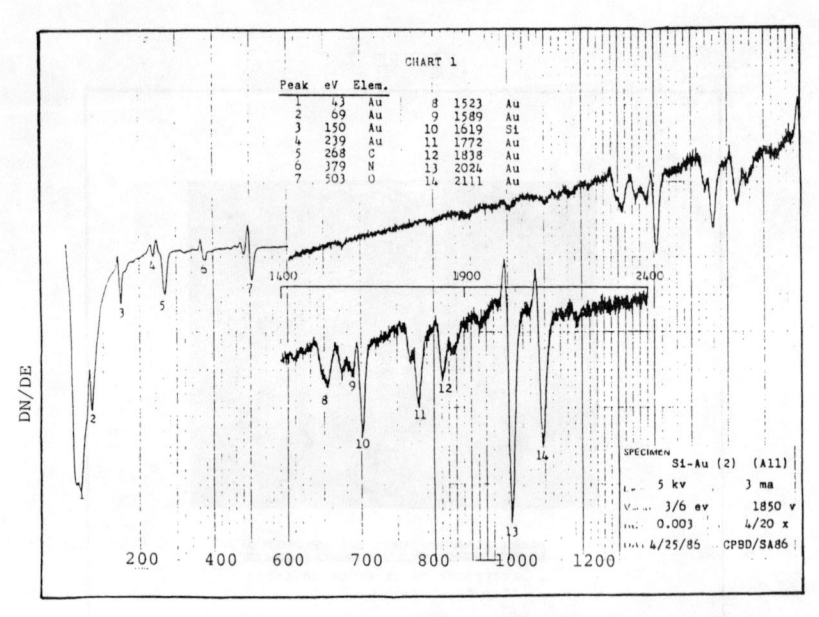

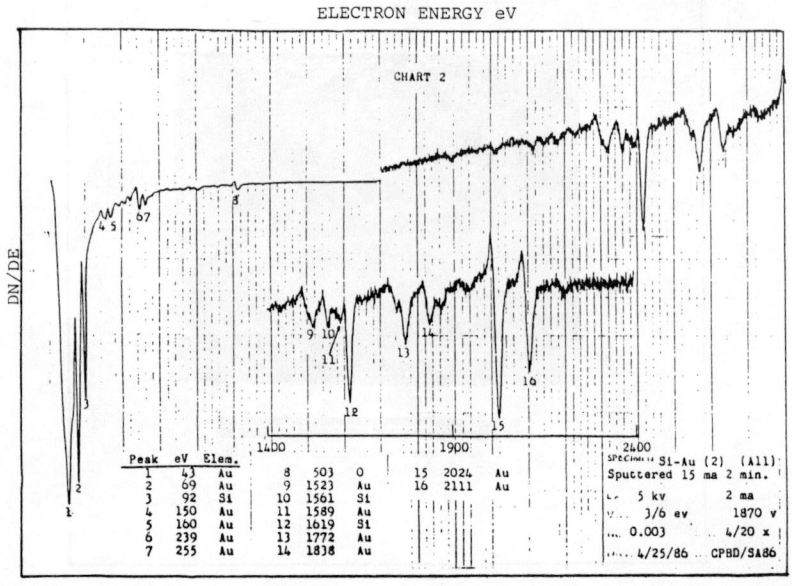

179

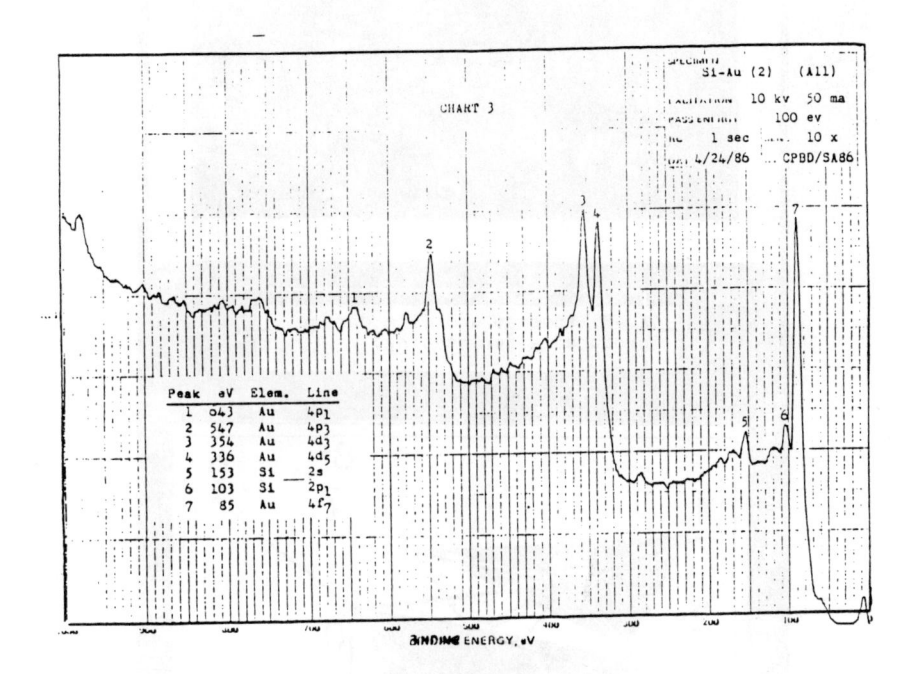

CHART 3

Specimen
Si-Au (2) (All)
Excitation 10 kv 50 ma
Pass Energy 100 ev
 1 sec   10 x
4/24/86   CPBD/SA86

Peak	eV	Elem.	Line
1	643	Au	$4p_1$
2	547	Au	$4p_3$
3	354	Au	$4d_3$
4	336	Au	$4d_5$
5	153	Si	$2s$
6	103	Si	$2p_1$
7	85	Au	$4f_7$

BINDING ENERGY, eV

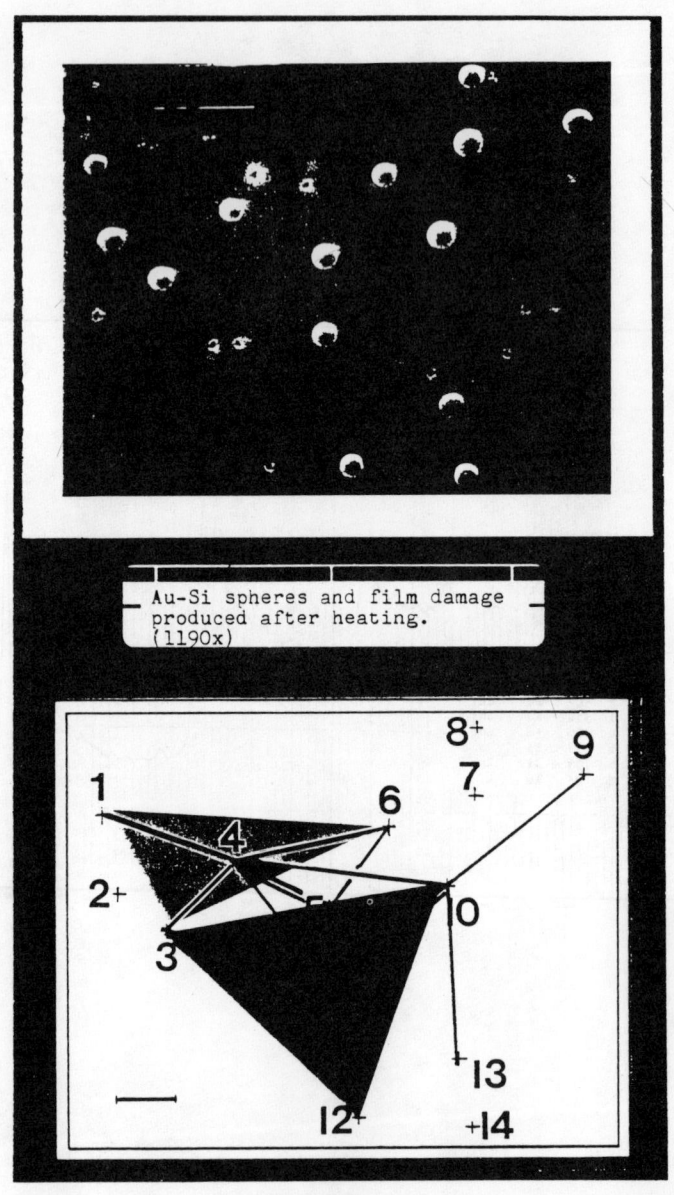

Au-Si spheres and film damage produced after heating. (1190x)

FIGURE 8:A

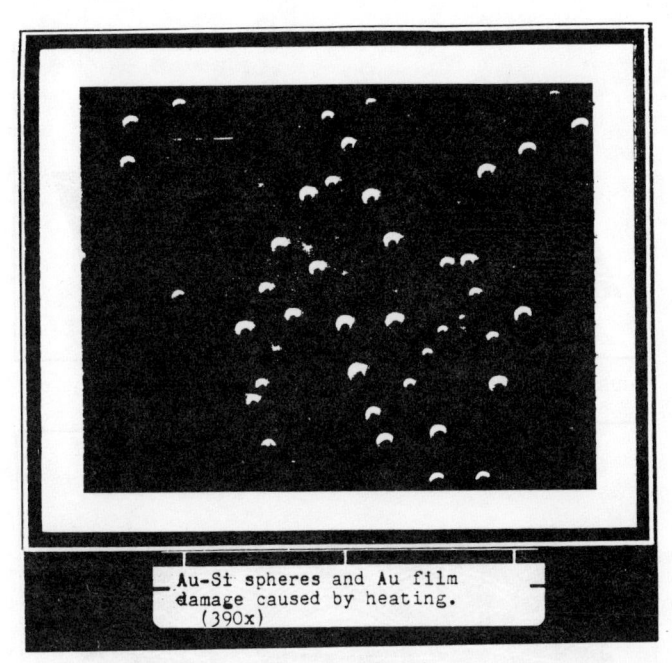

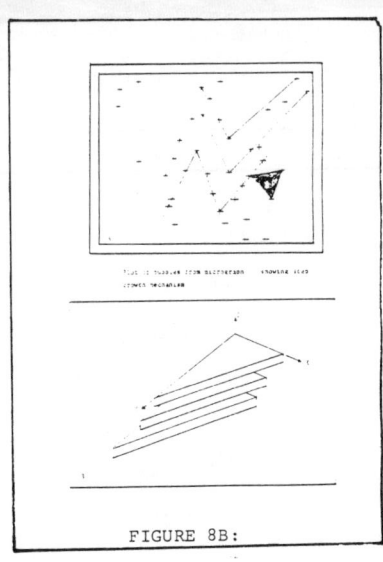

Au-Si spheres and Au film damage caused by heating. (390x)

FIGURE 8B:

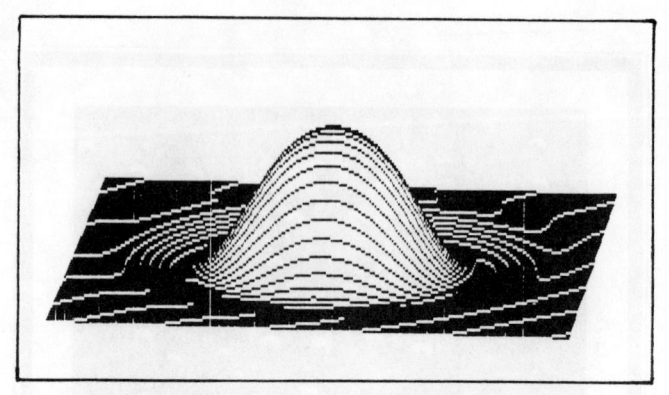

FIGURE 9:COMPUTER SIMULATION OF LASER DAMAGE (A)

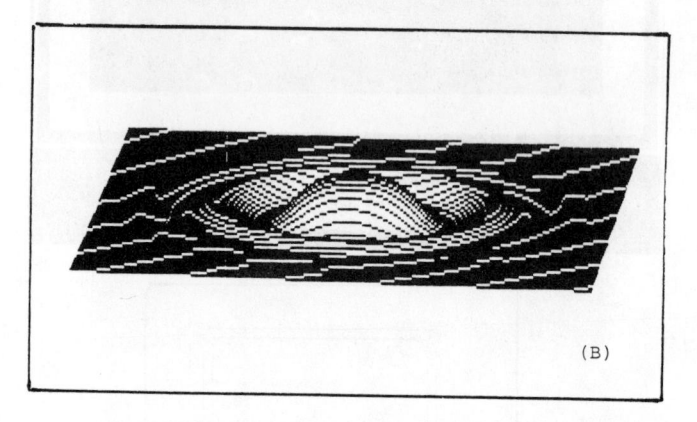

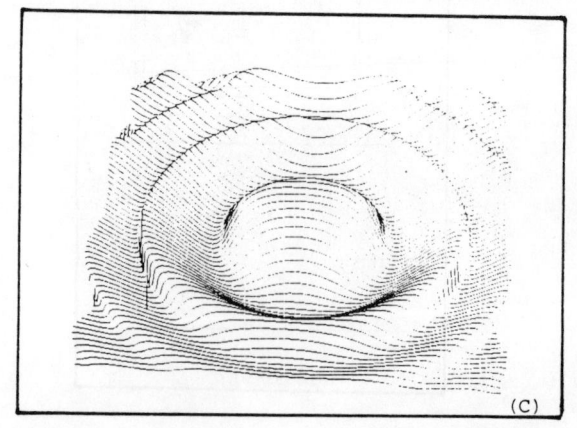

(B)

(C)

Heating Results

After observing the Sample 1 in SEM, it was heated in an argon gas environment using the specially-designed heating apparatus, Figure 4. Heating was done for 25 minutes to achieve a peak steady temperature of 320 deg. C. The resulting microstructures are shown in micrographs, Figure 8A-B with the possible structure schematics indicating <111>.

CONCLUSION:

From the structure formation which results from the Si-Au irradiation as shown in the micrographs, general chemical and interfacial structure appears as predicted. Figure 8A, shows that the film takes on the same structure as the underlying substrate. The AES-ESCA charts does show some interdiffusion effects. Since there are no phase of Si-Au as such documented it is highly plausable to say that SiO-component plays an important role in determining the chemical status of the interface. Electrical properties, and more or less depend on the crystal structure of the material. If such is the case, it seems that very little contribution to the change in electrical properties is made by the phase transformation.

We have included some results of theoretical investigation of structural simulation of the laser damage area, using computer methods, results are depicted in figure 9A-C.(this is based on the Gaussian distribution of the laser power, an extened paper is submitted for publication in surface science).

ACKNOWLEDGEMENTS:

We express our deepest gratitued to Mr.phil Kitchen for invaluable library research,help in designing of the specimen stage. I (C.P),Futher thank P.K for his loyalty to the project.

(*) BILL DIXON IS A GRADUATE OF L.S.U and NOW WORKS AT PRATT WHITNEY ,JUPITOR ,FL.,U.S.A

REFERENCES:

1. G. Eckhart, in *Laser and Electron Beam Processing of Materials*, C.W. White and P.S. Peercy, Eds. New York, Academic Press, 1980, pp. 467-480.

2. H.M. Muncheryan, *Laser Technology*, Indianapolic, Indiana, Howard W. Sams Co., 1979.

3. L.D. Hess, S.A. Kolorowski, G.L. Olson, and G. Yaron, in *Laser and Electron-Beam Solid Interactions and Materials Processing*, J.F. Gibbons, L.D. Hess, and T. W. Sigmon, Eds. New York, North-Holland, 1981, pp. 307-319.

4. O. Eknoyan, W. Van Der Hoeven, T. Richardson, and W.A. Porter, in *Laser and Electron-Beam Solid Interactions and Materials Processing*, J.F. Gibbons, L.D. Hess, and T.W#. Sigmon, Eds. New York, North-Holland, 1981, pp. 289-294.

5. C. Patel, Journal de Physique, Coll. C2, Suppl. au No. 3 Tome 47, (1986), p. c2-53.

6. M.T. Postek, K.S. Howard, A.H. Johnson, and K.L. McMichel, Scanning Electron Microscopy, (1980).

7. J.P. Jones and C. Patel, Surface Science, 80, (1979), P. 265.

8. P.A. Hart, in *Handbook on Semiconductor Materials, Vol. 4: Device Physics*, T.S. Moss, series editor, C. Hilsum, volume editor, New York, North-Holland Publishing Co. pp. 87-250.

9. C. Patel, private communication.

10. T.G. Anderson, in *Gold Bull*, 1982, 15,(1), pp. 7-18. (and references therein.)

11. L.J. Brillson, A.D. Katnani, M. Kelly, and G. Margaritondo, *J. Vac. Sci. Technol. A 2 (2)*, Apr.-June 1984, pp. 551-555 (and references therein.)

12. *Handbook of Auger Electron Spectroscopy*. Physical Electronics Industries, U.S.A.

A STUDY OF GOLD-STRONTIUM FLUORIDE INTERFACE
USING SURFACE PROBE TECHNIQUES

CHAND PATEL[*], R.DESBRANDES and M.E.McCONNAUGHHAY[*]

ABSTRACT:
Very accurate resistivity measurements in metal films have been undertaken for a better understanding of the electron-phonon interaction. A knowledge of the interaction of the film with the substrate is essential and has prompted the present work.

We report our preliminary findings with gold deposited in ultra-high vacuum on strontium-fluoride crystals. The surface characterization was perfomed using AES-ESCA and SEM-EDAX. The results indicate an interfacial reaction with change in the properties of the gold overlayers.

(*)Louisiana State University, Department of petroleum engineering, Baton Rouge, Louisiana, 70803, U.S.A

INTRODUCTION:

A systematic investigation of resistivity variations in metal films deposited on various substrates with temperature and different excitations such such as electron, X-ray and Laser beam has been started. The type and quality of the bonding of the film to the substrate is essential for interpreting the recorded variations of resistivity. The work reported here concerns gold films deposited on strontium fluoride crystals. Other substrates such as calcium fluoride, quartz, calcite, anhydrite and gypsum as well as other metals such as copper, aluminum, silver and zinc are being investigated.

In recent years, several new surface probe techniques for determining the elemental and chemical composition of solid surfaces and thin films have been developed. All of these techniques exhibit fundamental strengths and weaknesses, but some offer greater practical utility than others. Recent research has exploited a number of new techniques to gather information on interface electronic, atomic structure, composition, reactivity and electrical properties. Surface spectroscopies such as AES, (Auger Electron Spectroscopy) [1], Ultraviolet photoemission spectroscopy (UPS), ESCA (XPS) and field emission microscopy (FEM), [2], measure the interface electronic structure and thereby give specific information about chemical bonding. When used in conjunction with depth profiling, composition and diffusion data are generated. Scanning electron microscopy (SEM) [3] is used to study surface and interface morphology.

Due to their optical properties and the natural availability of most fluoride crystals, they are widely used in applied physics. Strontium Fluoride (SrF_2) is a crystal whose physical, as well as the mechanical properties, fall between CaF_2 and BaF_2 in the infrared. Furthermore, its behaviour is so similar to CaF$_2$ in the visible region that its often used in place of CaF$_2$. Being a strongly ionic crystal [4], it has attractive universal index characteristics.

EXPERIMENT:

In the present study samples were 1 cm. x 1 cm. x 0.3 cm. (supplied by Optovac. Inc.). No polishing was performed, and SrF$_2$ crystals were used in 'as received' condition. Samples were cleaned in vapor of acetone and ultrasonically cleaned for about 5-10 sec. Crystals were then dried using spectroscopic pure N$_2$ or Ar. Using the technique of vacumm thermal evaporation, gold was deposited in a vacumm of about 8 x 10^{-10} Torr, thickness of the gold was approximately 1000A. Figure 1 shows the schematic of the sample and the gold evaporation. SrF$_2$-Au is then mounted in the AES chamber for analysis. Literature shows no data on SrF$_2$-Au using AES-ESCA. Furthermore, our results show some interesting features on the Auger Spectrum for the sample with Gold without any cleaning and after sputtering using Ar$^+$ ions.

RESULTS:

Although a spectrum (AES) on SrF$_2$ and on Au can be found individually literature articles, no data exists on SrF$_2$-Au to this date. Figure 2 shows a AES scan of the sample. The low as well as the high Auger energy peaks are distinct, though some interfacial reaction is obvious in the high energy region.

In our experimental setup we have dual techniques, consequently, an ESCA Scan is also made. This is shown in Fig. 3. Some information on the binding at the interface of the system can be concluded from the ESCA scans. The occurrence of doublets at 350ev does support the hypothesis of the interdiffusion or surface reaction. The scanning electron micrograph in Fig. 4 also shows a complex microstructure, and we observe no surface damage or charging effects. The interfacial phenomena is schematically shown in Fig. 6.

INTERPRETATION:

The chemical inertness of SrF_2 suggests that it is likely to form an ideal interface when a metal film is adsorbed on it. We believe that in practice no ideal interface can be produced as shown in figure 6 (a), hence the non-ideal and total diffused state conditions do appear experimentally and can be systematically investigated using surface probe techniques.

Physically there is a likelihood of charge or ion transfer, which dramatically effects the resistivity value of the system. Since there may be no direct chemical interaction possible between Au and SrF_2 it is most likely to be effected by a charge exchange mechanism across the interface. This argument is represented by the non-ideal case as shown in Fig. 6 (b) and is substantiated by the AES spectrum in the lower energy range of Fig. 2. The higher energy side indicates some interaction between Au-SrF_2 (1400-1600 ev). The ESCA chart shows chemical activities in the energy range of 300 ev - 400 ev with a doublet peak to peak height. The SEM micrographs shows abrupt as well as mixed regions suggesting that the interface is more likely to be distorted and ill-defined. We are at present systematically checking our preliminary findings using more stringent and well-defined experiments such as in detail depth profiling and sputtering of the interface.

CONCLUSION:

The surface probe techniques show some real effects occurring at the SrF_2-Au interface. If the interface formed does have any of the geometry shown in Fig 6 (b) or (c), then it is highly likely that the charge exchange across the interface or interaction mechanism can be explained. From these models, some meaningful estimate can be made about the change in resistivity due to these mechanisms. Our ongoing work shows that using the depth profile technique we can further map the interfacial resistivity effects accurately [4].

ACKNOWLEDGEMENT:

We wish to thank Dr. Z. Bassiouni for his support of this project and the Department of Petroleum Engineering for its funding of this research. We also wish to thank the Department of Mechanical Engineering for the use of their facilities.

REFERENCES:

1. P. S. Ho, G. W. Rubloff, J. E. Lewis, V. L. Moruzzi and A. R. Williams, Phys. Rev., B22 1781, (1980).

2. C. Patel, "Field Emission Microscopy of Gallium Arsenide", Journal de Physique, Vol. 47, (1986).

3. A. G. Gullis, S. M. Davidson and G. R. Brooker, eds., Proc. of Inst. Phys. Conf. on "Microscopy of Semiconducting Materials", Inst. of Phys., London (1983).

4. Optovac, "Optovac Crystal Handbook", Optovac, Inc. (1982).

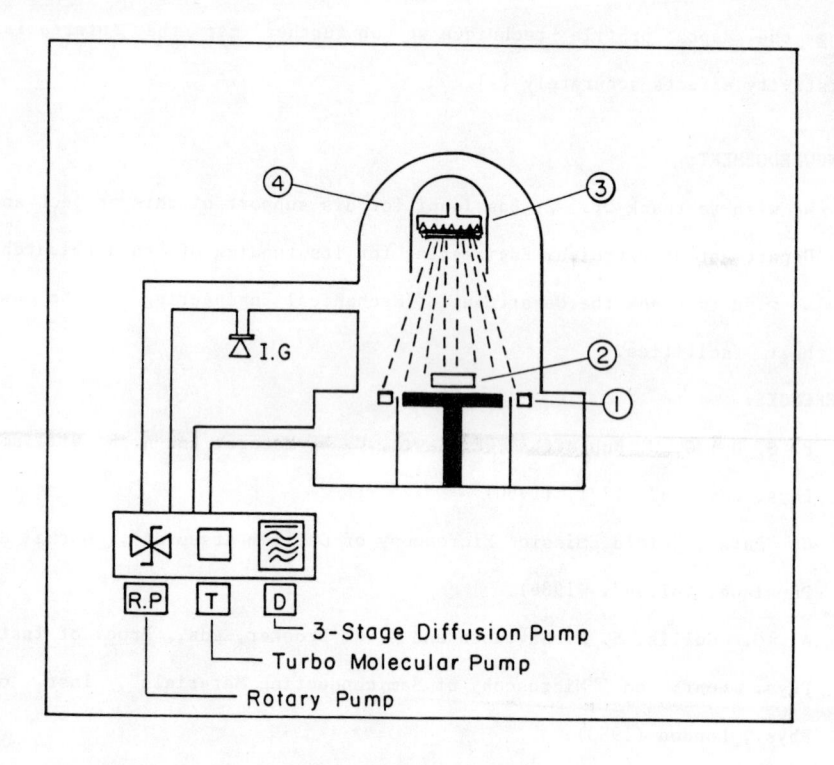

Figure 1:   Schematic Gold Thin Film Deposition Setup.

(1) Thickness monitor system

(2) SrF_2 - sample

(3) Gold source

(4) Vacuum better than 10^{-10}

IG: Ionization Gage

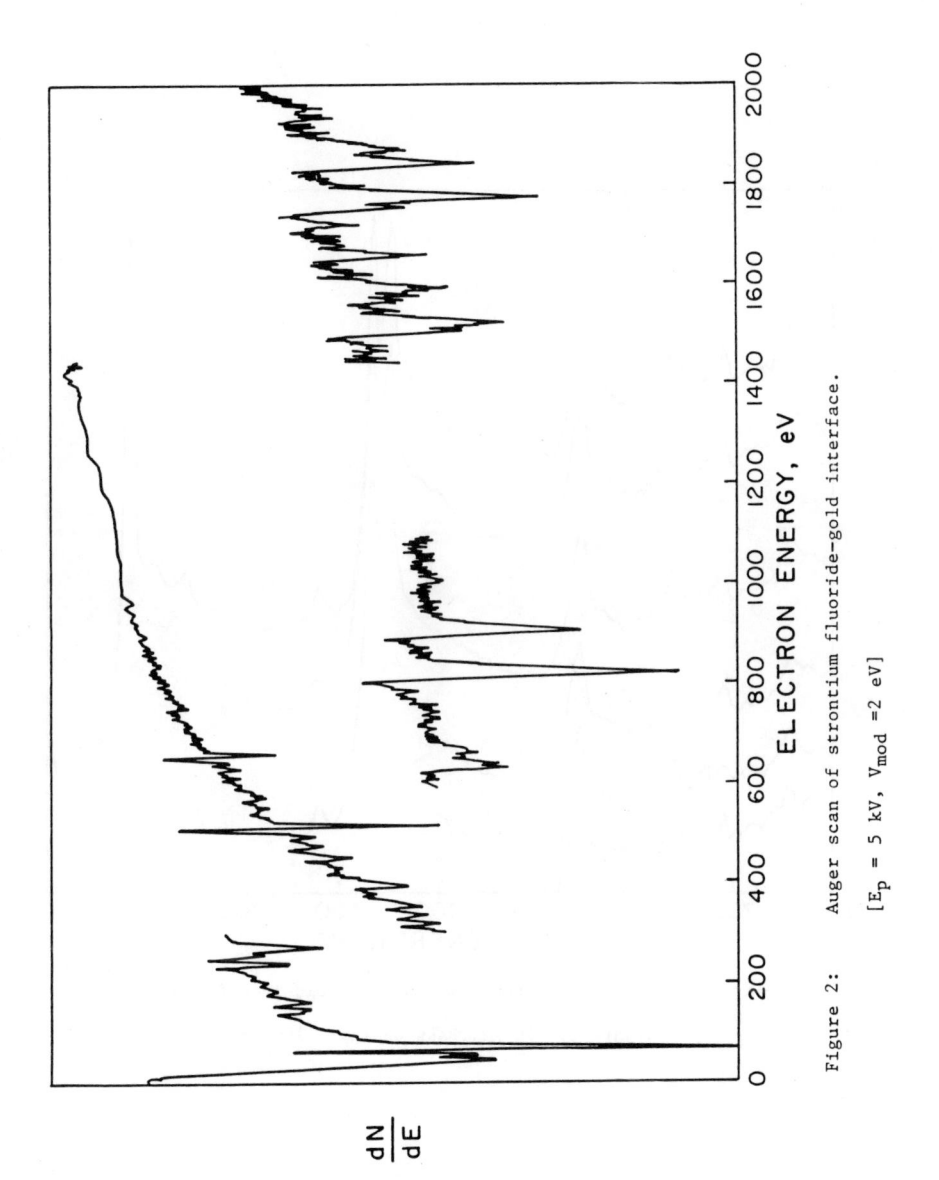

Figure 2: Auger scan of strontium fluoride-gold interface.
[E_p = 5 kV, V_{mod} = 2 eV]

(E)

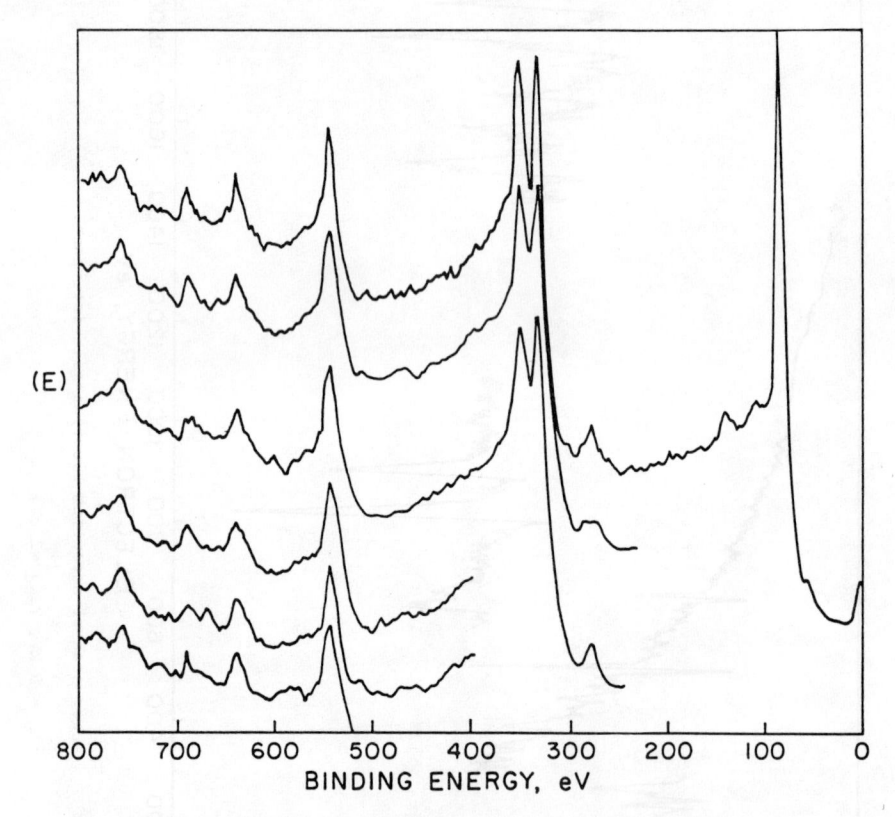

Figure 3: ESCA scan of strontium fluoride-gold interface.
[Excitation: 10 kV, Pass energy: 100 eV]

193

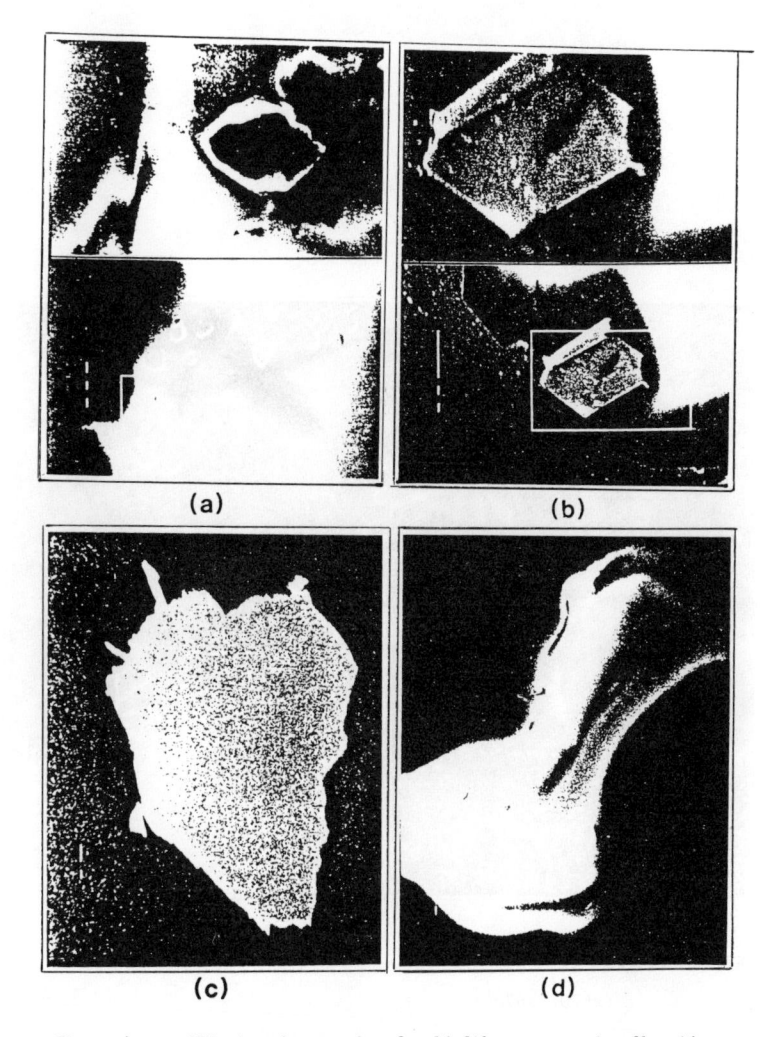

Figure 4:   SEM microphotographs of gold film on strontium fluoride.

(a) Dual magnified image of gold interface

[x210, x1050]

(b) Dual magnified image of gold crystal

[x1170, x2340]

(c) Section of gold film [x3600]

(d) Section of gold film [x21]

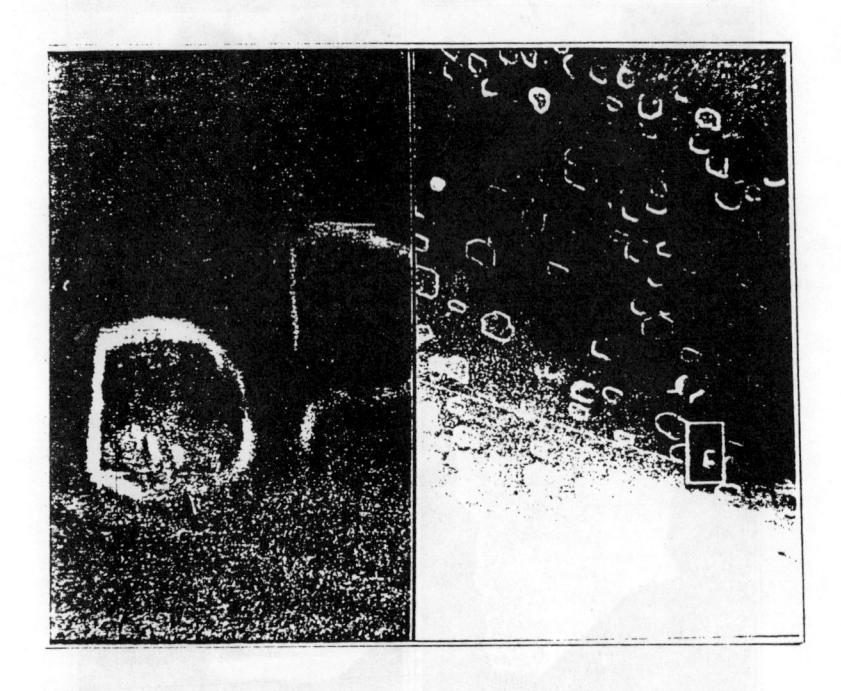

Figure 5: SEM microphotograph of stroutium fluoride crystal. [x670, x6700]

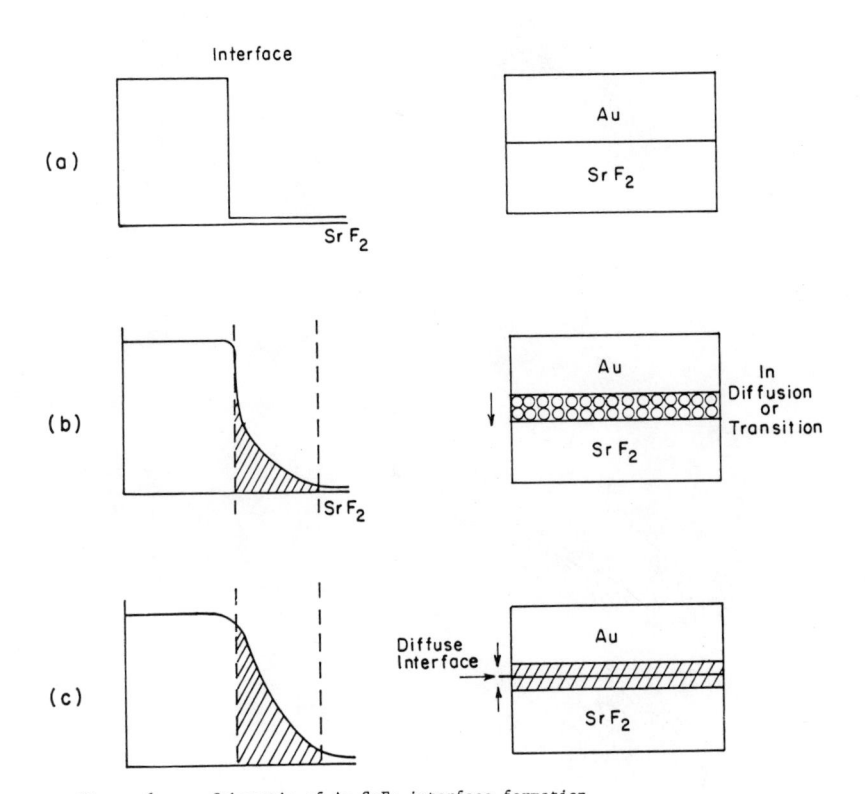

Figure 6: Schematic of Au SrF$_2$ interface formation.

    (a) ideal case

    (b) non-ideal case

    (c) totally diffused state

AN XPS STUDY ON THE STRUCTURE OF Mg(PO$_3$)$_2$-BaF$_2$-AlF$_3$ GLASS SYSTEM

Chen Binjiang, Mi Qingzhou, Wang Shizhuo

Changchun Institute of Optics and Fine Mechanics, Academia Sinica

An X-ray photoelectron spectroscopy(XPS) study has been carried out on the Mg(PO$_3$)$_2$-BaF$_2$-AlF$_3$ glass system with the aim of investigating the structure characterization. In this fluorophosphate glass system, the O$_{1s}$ line can be separated into bridge oxygen(O$_b$), nonbridge oxygen (O$_{nb}$) and double bonded oxygen(O$_d$) by useing curve fitting procedure. As fluoride content increases, the areas of O$_b$, O$_d$ peaks decrease and the area of O$_{nb}$ peak increases progressively, indicating the decrease of "branching units" and "middle units" and the increase of "end units" and "isolated units". The O$_d$ peak's disappearance implies the delocalization effect of π electron of P=O bond. The F$_{1s}$ peak has been resolved into two, which supports the hypothesis of the existence of network forming and network modifying fluorine atoms in fluorophosphate glasses. Two identifiable peaks of Al$_{2p}$ line prove the existence of Al(O,F)$_4$ and Al(O,F)$_6$ units. The variations of the chemical shifts of all atom's characteristic energy levels coincide with those of the properties of related glasses.

INTRODUCTION

It has been shown that X-ray photoelectron spectroscopy(XPS) is a powerful tool not only for analytical purpose but also for determining the electronic structure of solids. Although this technique is widely used recently, there are relatively few papers as far as glasses are concerned. This is because there are serious experimental difficulties when XPS is applied to glasses. Most of the works concerning glasses are limited in discriminating the structural states of oxygen atoms and investigating the relations between the "chemical shifts" and properties of glasses. It has been known till now that this technique is the only one available for determining unambiguously the concentrations of oxygen atoms of different structural states.

As far as we know, all XPS studies which have been reported are confined in silicate[1-4], borosilicate[5], germanosilicate[6], phosphate[2,7], and fluoride[8] glasses. No report on fluorophosphate glass has been seen.

The purpose of this paper is to extend the XPS measurements to fluorophosphate glass of $Mg(PO_3)_2-BaF_2-AlF_3$ system with systematically varied compositions to see whether or not O_{1s} photoelectron consists of three states (bridge oxygen O_b, non-bridge oxygen O_{nb}, and double bonded oxygen O_d), F_{1s} consists of two states (network forming fluorine and network modifying fluorine) and Al_{2p} consists of two states ($Al(O,F)_4$ and $Al(O,F)_6$) in this fluorophosphate glass system. For this purpose, the best-fit procedure was employed by useing VGS5000 software. The second aspect is to investigate the structure of the glasses from the different states of atoms and their relative concentration. Of further interest is to study the relations between the variations of the characteristic energy levels of all atoms and properties of related glasses.

EXPERIMENTAL

The glasses were prepared from reagent grade $Mg(H_2PO_4)_2$, $AlF_3 \cdot 3\frac{1}{2}H_2O$ and BaF_2. The target compositions were $(1-x-y)Mg(PO_3)_2 - xBaF_2 - yAlF_3$ (see Table 1). After the compounds had been mixed, the 100g batches were melted in Pt crucibles in an electric furnace at temperatures ranging from 1100-1400°C and then poured into preshaped moulds. The anneal process was carried out in a muffle furnace.

Table 1: Glass compositions in this work

No	1	2	3	4	5	6	7	8	9	10
1-x-y	1	0.8	0.6	0.4	0.2	0.2	0.2	0.2	0.2	0.2
x	0	0.2	0.4	0.6	0.8	0.7	0.6	0.5	0.45	0.4
y	0	0	0	0	0	0.1	0.2	0.3	0.35	0.4

The XPS measurements were carried out with an ESCA LAB MARK II spectrometer. A Mg Kα source was used for excitation($h\nu=1253.6$ eV). The sample preparation was useing "in situ" splity method. During the experiments, the pressure inside the analyser chamber was about 5×10^{-10} mbar. The analyser energy was 20 eV. The spectra were calibrated by utilizing the C_{1s} peak of the contamination of the pumping oil. The O_{1s}, F_{1s}, Al_{2p} spectra were analyzed by means of VGS 5000 best-fit computer program. The peak positions(E_B), full widths at half maximum (FWHM) and intensity(peak area) ratios can be determined with this program.

The refractive index was determined with V-prism refractometer and the density was measured by the Achimeden method.

RESULTS AND DISCUSSION

O_{1s} spectra are shown in fig.1. In fig.1 (a), (b) and (c), O_{1s} spectra can be well separated into three peaks. In fig.1 (d), O_{1s} spectra can only be separated into two peaks. This proves that there are P=O double bond in fluorophosphate glass. Their peak positions and area percentages are shown in table 2.

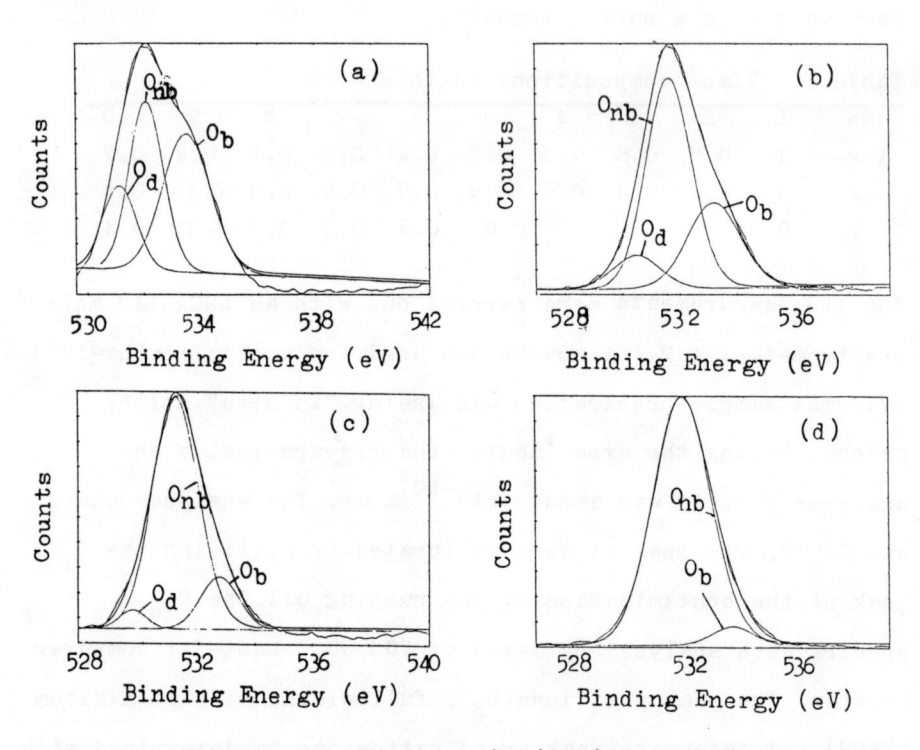

Fig.1 O_{1s} XPS spectra of $(1-x-y)Mg(PO_3)_2-xBaF_2-yAlF_3$ glass.
Here: (a): x=0 y=0; (b): x=0.4 y=0; (c): x=0.8 y=0;
(d): x=0.5 y=0.3.

According to the ratio of the three oxygen content(the ratio of the peak areas), the following conclusions can be given

Table.2 Peak position and area percentage of O_b, O_{nb} and O_d.

	peak position(E_B) in eV				area percentage(%)			
	a	b	c	d	a	b	c	d
O_b	533.50	533.20	532.75	533.60	42.6	29.3	17.2	5.7
O_{nb}	532.10	531.65	531.20	531.85	37.2	59.4	76.7	94.3
O_d	531.15	530.50	529.65		20.2	11.2	6.1	

out. The structure of sample 5(fig.1 a) is mainly composed of "middle units"(fig.2 a) with small amount of "branching units" (fig.2 b). That is to say the structure is mainly composed of polyphosphate or cyclic metaphosphate with little three dimensional structure groups. As the value of x increases, the content of O_b decreases and the content of O_{nb} increases, indicating the amount of middle unit and branching unit decreases and the amount of "end unit"(fig.2 c) and "isolated unit"(fig.2 d) increases. The structure of sample 5(fig.1 c) is all composed of end and isolated units mixture. This

$$\begin{array}{cccc} O & O & O & O \\ \| & \| & \| & | \\ -O-P-O- & -O-P-O- & F-P-O- & F-P-O^- \\ | & | & | & | \\ O_- & O_- & O_- & O_- \end{array}$$

Fig.2   a.middle unit   b.branching unit   c.end unit   d.isolated unit

means the glass structure has transformed into pyrophosphate and monofluorophosphate mixture.

In fig.1, the content of P=O bond decreases as the value of x increases. This is due to the delocalization effect of π electron in the P=O double bond.[9] As AlF_3 have been added into glasses the P=O bond vanishes. This may be explained

that a tetrahedral coordination of aluminum in the direct
neighbourhood of phosphorus tetrahedra causes a strong
delocalization effect on the P=O double bond. The π electron
of P=O bond transfers to aluminum.[9] The three dimensional
network of AlP(O,F)$_4$ is formed.

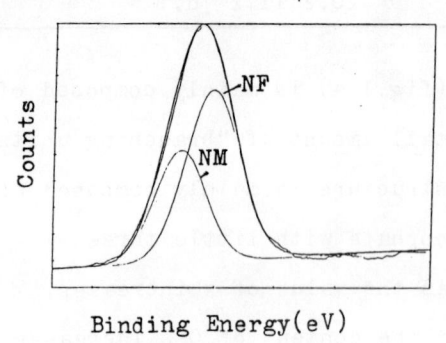

Fig.3 F$_{1s}$ XPS spectrum of 0.2 Mg(PO$_3$)$_2$-0.5BaF$_2$-0.3AlF$_3$ glass

Fig.4 Al$_{2p}$ XPS spectrum of 0.2Mg(PO$_3$)$_2$-0.5BaF$_2$-0.3AlF$_3$ glass

The F$_{1s}$ spectrum has been separated into two peaks in this
fluorophosphate glass system(see fig.3). The distance between
the two separated peaks is 1.05eV. This confirms unambiguously
the existence of the structural double role of fluorine in
fluorophosphate glasses, i.e. the network forming fluorine and
the network modifying fluorine.[9,10]

The Al$_{2p}$ spectrum can also be fit with two peaks. This is due
to the existence of the Al(O,F)$_4$ and Al(O,F)$_6$ two states.[11]
The distance of the two peaks is 1.15eV and the area ratio of
Al(O,F)$_4$ and Al(O,F)$_6$ peaks is 1:2.625.

The chemical shifts of characteristic energy levels of all

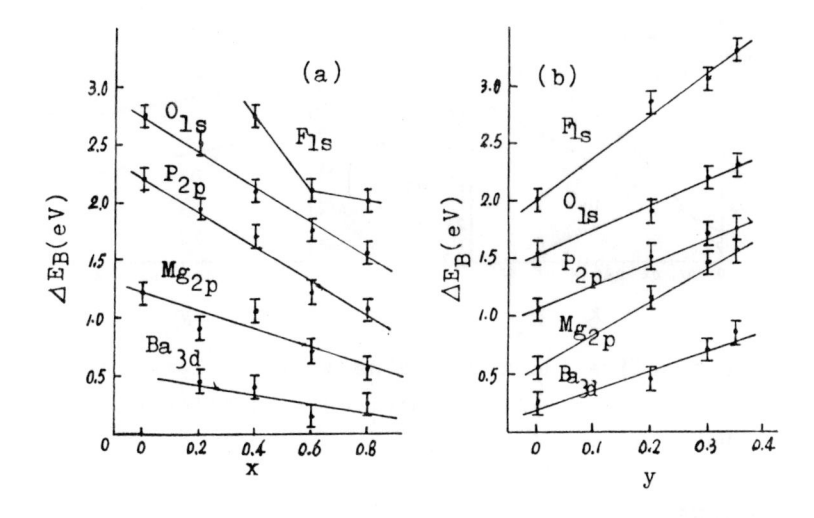

Fig.5 Chemical shifts of characteristic energy levels of all atoms as a function of composition.
Here: a. y=0   b. x+y=0.8

atoms are shown in fig.5. In fig.5 (a), binding energies of all atoms decrease as the value of x increases. It indicates that a continuous breakdown of the network structure takes place. Atoms in glass become closer. The electronic densities of all atoms increase[1] and the average polarizability of the glass increases. In fig.5(b), binding energies of all atoms increase as the value of y increases. It shows that some of Al atoms act as network forming atoms and the $AlP(O,F)_4$ network is formed. Atoms in the glass become looser. The electronic densities of all atoms decrease and the average polarizability of the glass decreases. All the facts above coincide well with the refractive index and the density of related glasses(see fig.6).

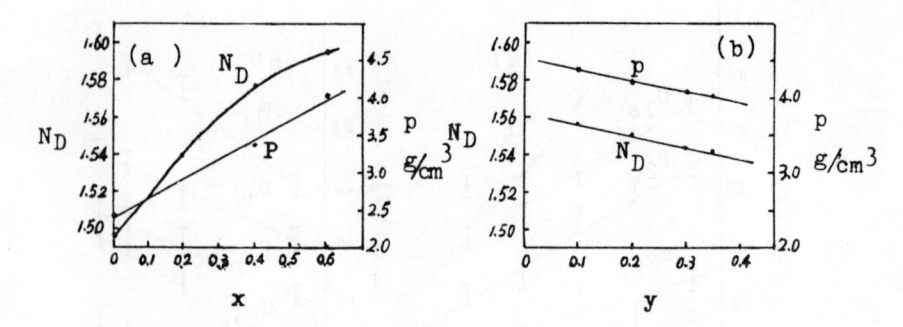

Fig.6 Refractive index N_D and density of $(1-x-y)Mg(PO_3)_2$-$xBaF_2$-$yAlF_3$ glass system as a function of composition
Here: a. y=0   b. x+y=0.8

CONCLUSIONS

It has been shown that valuable information concerning the structure of $Mg(PO_3)_2$-BaF_2-AlF_3 fluorophosphate glass system can be derived from X-ray photoelectron spectroscopy data.

(i) In high phosphate region, the glass structure is mainly composed of middle units with small amount of branching units. The amount of branching and middle units decreases and the amount of end and isolated units increases as the content of BaF_2 increases. The glass structure transforms from polyphosphate into pyrophosphate and monofluorophosphate mixture.

(ii) The P=O double bond exists in fluorophosphate glass. The delocalization effect of π electron occurs in P=O bond and P=O bond disappears as the content of fluoride increases.

(iii) The fluorine plays two role(one for network forming and the other for network modifying) in this fluorophosphate glass system.

(iv) Aluminum is present in both four and six-fold coordination $(Al(O,F)_4$ and $Al(O,F)_6)$. The three dimensional network of $AlP(O,F)_4$ is formed as aluminum is added into the glass.

(v) The chemical shifts of characteristic energy levels of all atoms coincide with refrective index and density of the glasses.

REFERENCES

1. R.Brückner et al, Glastechn.Ber. 51 (1978) 1-7.
2. D.J.Lam et al, J.Non-cryst Solids 42 (1980) 41-48.
3. R.Brückner et al, J.Non-cryst Solids 42 (1980) 49-60.
4. B.W.Veal et al, J.Non-cryst Solids 49 (1982) 309-320.
5. B.M.J.Smets,T.P.A.Lommen, Phys Chem Glasses 22(6) (1981) 158-162.
6. B.M.J.Smets,T.P.A.Lommen, J.Non-cryst Solids 46(1981)21-32.
7. R.Gresch et al, J.Non-cryst Solids 34 (1979) 127-136.
8. R.M.Almeida et al, J.Non-cryst Solids 69(1984) 61-165.
9. M.Sammet,R.Bruckner,XIV Intl Congr on Glass (1986) 102.
10. A.Bertoluzza et al, Can.J.Spectrosc 27(6)(1982)171-177.
11. J.J.Videau et al, J.Non-cryst Solids 48 (1982) 385-392.

AN INVESTIGATION OF GLASS STRUCTURE ON THE $Mg(PO_3)_2$-BaF_2-AlF_3 SYSTEM BY VIBRATIONAL SPECTROSCOPY

Chen Binjiang, Mi Qingzhou, Wang Shizhuo

Changchun Institute of Optics and Fine Mechanics,
Academia Sinica

The structure of the $Mg(PO_3)_2$-BaF_2-AlF_3 fluorophosphate glass system has been investigated by infrared and Raman spectroscopy. The experimental data provide evidences for the existence of double bonded oxygen (P=O) and mixed network of $Al(O,F)_4$ and $P(O,F)_4$ tetrahedra. The v_{Ba-F}, v_{Mg-F}, and v_{Al-F} vibrations occur at 480-510 cm^{-1}, 530-550 cm^{-1} and 580 cm^{-1} respectively in Raman spectra. Changes in both IR and Raman spectra reveal that with increasing of fluoride content the glass structure transforms gradually from polyphosphate to pyrophosphate and monofluorophosphate mixture. The P=O bond vanishes simultaneously.

INTRODUCTION

The interest in fluorophosphate glasses was increased recently by its fine optical properties, such as: relatively broad spectral transparency; high Abbe value; wide range of refractive indexes; low or negative temperature coefficient of refractive index[1] and non-linear refractive index; etc. These special optical constants make it possible to use fluorophosphate glasses successfully for eliminating residual chromatism. It has also been used as a laser host materials for high energy lasers[2]. As far as we know, informations about the structure of fluorophosphate glasses are very limited. No report about the structure of $Mg(PO_3)_2-BaF_2-AlF_3$ glass system has been seen at present.

The purpose of this paper is to investigate the structures of this glass system by IR and Raman spectroscopy with systematically varied compositions.

EXPERIMENTAL

The glasses were prepared from reagent grade $Mg(H_2PO_4)_2$, BaF_2 and $AlF_3 \cdot 3\frac{1}{2}H_2O$. The target compositions were $(1-x-y)Mg(PO_3)_2-xBaF_2-yAlF_3$ (see table.1). After the compounds had been mixed, the 100g batches were melted in Pt crucibles in an electric furnace at temperatures ranging from 1100-1400°C and then poured into preshaped steel moulds. The anneal process was carried out in a muffle furnace.

Table.1: Glass compositions in this work

No	1	2	3	4	5	6	7	8	9	10
1-x-y	1	0.8	0.6	0.4	0.2	0.2	0.2	0.2	0.2	0.2
x	0	0.2	0.4	0.6	0.8	0.7	0.6	0.5	0.45	0.4
y	0	0	0	0	0	0.1	0.2	0.3	0.35	0.4

Infrared spectra were measured on a Perkin-Elmer 783 spectrometer, useing the KBr pellet technique.

Raman spectra were obtained on a JY-T800 triple-grating spectrometer, useing the 514.5um line of a Spectra-Physics 265 exciter Ar^+ laser. A 90° scattering geometry was employed. Slit width was 1000 um. The exciting power was kept at about 50 mW. The spectra were recorded at $100cm^{-1}$/min.

RESULTS AND DISCUSSION

The IR spectra of sample 1 and it's heat-treated multi-crystal product are shown in fig.1. The crystal spectrum is a typical spectrum of Cyclic Metaphosphate(CMP) and Linear Polyphosphate (LPP) crystal mixture.[3,4] The correspondence between absorption peaks and vibrational modes are as follow:

Wavemumbers(cm^{-1})		Vibrational mode	Basic units	
1338		$v_{P=O}$		
1325	1293	$v_{as}\ PO_2$	CMP	LPP
1180		$v_{as}\ PO_3$	End group	
1125	1110	$v_s\ PO_2$	CMP	LPP
1050	940	$v_{as}\ P-O-P$	CMP	LPP
746	720	$v_s\ P-O-P$	CMP	LPP

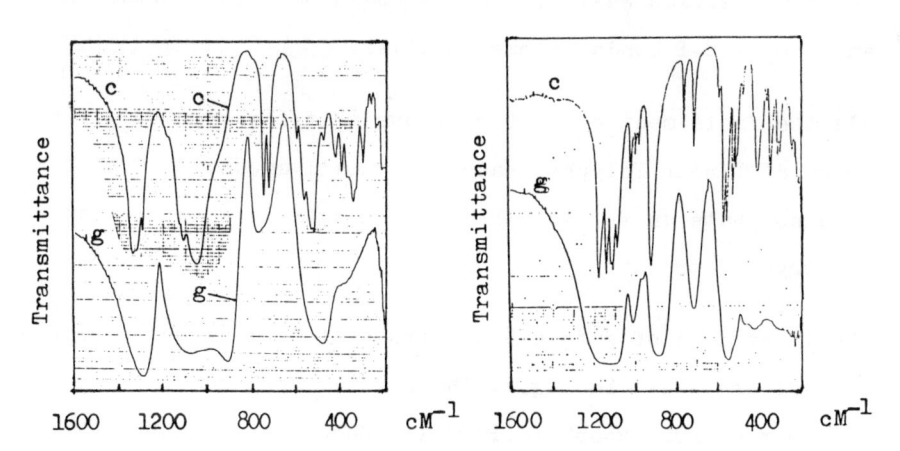

Fig.1 IR spectra of sample 1.   Fig.2 IR spectra of sample 5.
   g—glass   c—crystal   g—glass   c—crystal

The spectrum of glass coincide with that of its crystal product, but there are some shifts and broading of the peaks so that some peaks join together becoming undistinguishable.

The 1290 cm^{-1} band of glass corresponds to the three peaks (1338, 1325, 1293 cm^{-1}) of crystal and 1080 cm^{-1} broad band consists of $v_{as} PO_3$, $v_s PO_3$, and v_{as} P-O-P of CMP absorption peaks. The $v_{as} PO_3$ peak of glass is stronger than that of crystal(1180). The 910 cm^{-1} (v_{as} P-O-P) peak of glass is much stronger than that of crystal(940 cm^{-1}). The 1050 cm^{-1} (v_{as} P-O-P) peak becomes smaller so that it is very difficult to identify. From crystal to glass, the peaks higher than 900 cm^{-1} shift toward low frequency and 746, 720 cm^{-1} peaks shift toward high frequency to 770, 730 cm^{-1}. According to formula $v=1303(K/u)^{\frac{1}{2}}$ and Corbridge relation[4], it indicates that the strength of P-O bond becomes weaker, the P-O chain becomes shorter and the angle of P-O-P bonds becomes smaller.

All the facts mentioned above prove that the sample 1 is composed of Linear Polyphosphate with small amount of Cyclic Metaphosphate units, but the chain length is shorter than the crystal product.

The IR spectra of sample 5 and its heat-treated multi-crystal product are shown in fig.2. The crystal spectrum is a typical spectrum of Pyrophosphate(PYP) and Monofluorophosphate(MFP) crystal mixture.[2-4] The assignments are: 1183, 1146, 1123, 1096 cm^{-1} for $v_{as} PO_3$; 1040, 1018, 1000 cm^{-1} for $v_s PO_3$; 930 cm^{-1} for v_{as} P-O-P; 783, 733 cm^{-1} for v_s P-O-P.

The spectrum of glass coincides well with that of its crystal product. The absorption bands are assigned as follows: 1140 cm^{-1} for $v_{as}PO_3$; 1015 cm^{-1} for $v_s PO_3$; 890 cm^{-1} for v_{as}P-O-P; 730 cm^{-1} for v_sP-O-P. These prove that the structure of sample 5 glass is composed of pyrophosphate and monofluorophosphate mixture.

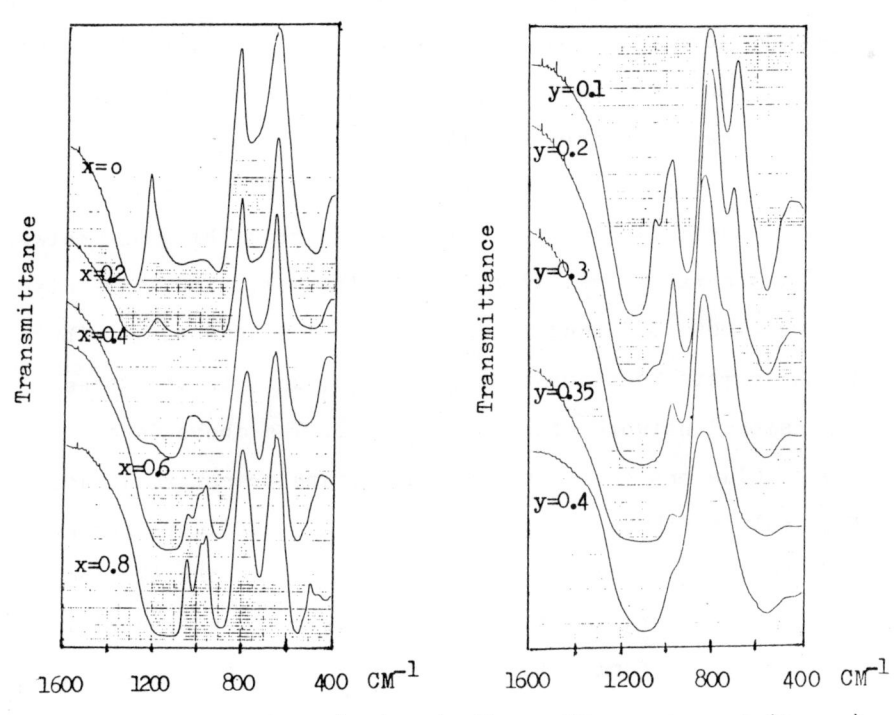

Fig.3 IR spectra of $(1-x-y)Mg(PO_3)_2$ $-xBaF_2$ $-yAlF_3$ glass system. Here: y=0

Fig.4 IR spectra of $(1-x-y)Mg(PO_3)_2-xBaF_2-yAlF_3$ glass system. x+y=0.8

The spectra of IR from sample 1 to 10 are shown in fig.3 and 4. In fig.3, the relative intensity of 1290-1260 cm^{-1} ($v_{P=O}$ and $v_{as}PO_2$) band becomes weaker as the value of x increases, at last it vanishes. The frequency of this band decreases. At the same time, the relative intensity of $v_{as}PO_3$ (1140-1170 cm^{-1}) band and $v_s PO_3$ (around 1020 cm^{-1}) band becomes stronger. That is to say a continuous breakdown of P-O chain takes place. The content of PO_2 unit decreases and the content of PO_3 unit increases. As the value of x increases, the v_sP-O-P(770 cm^{-1}) band vanishes, due to the disappearence of ring units. On the other hand, the relative intensity of v_{as}P-O-P(900 cm^{-1}) and v_sP-O-P(730 cm^{-1}) bands becomes weaker, indicating the content of P-O-P unit decreases and P-O chains become shorter. In fig.4, a new shoulder band at 610-630 cm^{-1} appears as the value of y increases. This is associated with the vibration of $Al(O,F)_4$ unit, indicating the Al atoms join the glass network as tetrahedral coordination[5]. At the same time, the intensity of 1030-1060 cm^{-1} band rises rapidly. This can be attributed to the stretching vibration of $P(O,F)_4$ unit in the three dimensional network of $AlP(O,F)_4$[5,6]. The above facts lead to the conclusions that the glass structure transforms gradually from polyphosphate to pyrophosphate and monofluorophosphate mixture as the value of x increases. Some of the Al atoms join the glass network and the three dimensional network structure of $AlP(O,F)_4$ is formed as the AlF_3 content increases.

The agreement between the results of IR spectra and Raman spectra (see fig.5 and 6) is excellent. The most important absorption bands, vibrational modes and related units in Raman spectra are listed in table 2. In fig.5, the relative intensity of band A and band B becomes weaker as the value of x increases, at last they disappear. At the same time, the relative intensity of band C and band D becomes stronger and

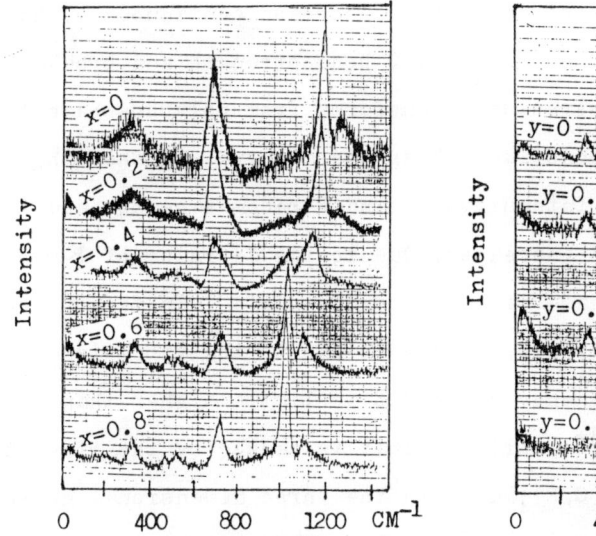

Fig.5 Raman spectra of (1-x-y) Mg(PO$_3$)$_2$-xBaF$_2$-yAlF$_3$ glass system. Here: y=0

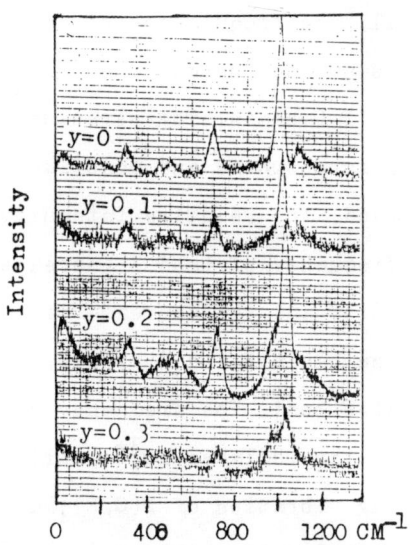

Fig.6 Raman spectra of (1-x-y) Mg(PO$_3$)$_2$-xBaF$_2$-yAlF$_3$ glass system. Here: x+y=0.8

Table.2: Positions of Raman vibrational bands and their structural interpretation

No	Frequencies (cm-1)	Vibrational modes	Related units	References
A	1280	v_{as} PO$_2$	CMP LPP	7, 8
B	1150-1210	v_s PO$_2$		
C	1110-1120	v_{as} PO$_3$		
D	1030-1060	v_s PO$_3$	PYP MFP	4,9,10
E	990-1000	v P(O,F)	PO$_2$F$_2$ POF$_3$	9,11
F	700- 750	v_s P-O-P		8,9
G	580	v_{Al-F}	Al(O,F)$_4$, Al(O,F)$_6$	12
H	530- 550	v_{Mg-F}	MgF$_6$	
I	480- 510	v_{Ba-F}	BaF$_6$	
J	330- 350	δ_{P-O}		10

all the frequencies of four bands decrease continuously. This is due to the decrease in the P-O bond strength and breakdown of P-O chain. The structure transforms from polyphosphate to pyrophosphate and monofluorophosphate mixture. On the other hand, the relative intensity of band F becomes weaker and the frequency of band F increases, due to the decrease of P-O-P units and the decrease of P-O-P band angle[4]. In fig.6, two new bands E and G appear. Band G is due to the vibration of Al-F bond. The frequency of band E is little higher than that of the isolated PO_4^{3+} unit[13]. That is caused by the stretching vibration of $P(O,F)_4$ unit in the three dimensional network of $AlP(O,F)_4$. All the bands in fig.6 shift toward high frequency as the value of y increases, indicating that the strength of all bands increase.

CONCLUSIONS

The effects of structure on composition of $Mg(PO_3)_2-BaF_2-AlF_3$ glass system have been investigated by IR and Raman spectroscopy. The agreement between IR and Raman spectra is excellent. In high phosphate region, the glass is mainly composed of polyphosphate and exists P=O bond. As the BaF_2 content increases, the P=O bond disappears and the glass structure transforms into pyrophosphate and monofluorophosphate mixture. As the AlF_3 added, some of Al atoms join the glass network and make up the three dimensional network of $AlP(O,F)_4$ so that the strength of all bonds in the glass increases.

REFERENCES

1. B. Kumar, R. Harris, Phys Chem Glasses 25 (1984) 155.
2. Gan Fuxi et al, J. Non-cryst Solids 52 (1982) 263.
3. D. E. C. Corbridge, E. J. Lowe, J. Chem Soc, (1954) 493.
4. D. E. C. Corbridge, Topics in Phosphorus Chemistry. Vol 6, 237.
5. M. Sammet, R. Brückner, XIV Intl Congr on Glass, (1986) 102.
6. John R. Van Wazer, Phosphorus and Its Compounds, Vol.1, Interscience Publishers, INC., New York (1958) 76.
7. C. Garrigou-Lagrange, Proceedings of the Eighth International Conference on Raman Spectroscopy (1982) 551.
8. Gan Fuxi et al, J. Non-cryst Solids 52 (1982) 203.
9. J. J. Videau, J. Portier, J. Non-cryst Solids 48 (1982) 385.
10. F. L. Galeener, J. C. Mikkelsen, Solid State Comun, Vol. 30, 505.
11. A. Bertoluzza et al, Can. J. Spectrosc. 27(6), (1982) 171.
12. Chen Haiyan, Gan Fuxi, J. Chinese Silicate Soc, 14(1)(1986) 123.
13. C. Nelson, D. R. Tallant, Phys Chem Glasses 26(4)(1985) 119.

Additional Electron Beam Induced Voltage Contrast due to Skin Resistance

Chen Boliang, Fang Xiaoming and Yu Jinbi
Shanghai Institute of Technical Physics,
Academia Sinica
420 Zhong Shan Bei Yi Road, Shanghai, China

Mercury cadmium telluride material has been studied using electron beam induced voltage technique. An additional contrast has been observed which differs from that caused by diffusion and recombination of minority carriers. This contrast can be explained reasonably in terms of skin resistance. The dependence of the contrast upon the skin resistance has been calculated by means of an equivalent circuit. The result is consistent with experimental observation.

Introduction

Electron beam induced voltage (current) technique has been widely used in characterization of semiconductor materials and devices. However, its imaging contrast theory has not been quite perfect yet. In addition to the contrasts that can be reasonably explained by inhomogeneities and electrical active defects in the material, electron beam induced voltage micrographs occasionally contain artifacts whose origin is obscure.[1] It was mentioned that the electron beam "writing" effects[2] and the specimen geometry effects[3] would cause additional contrast. Usually, a metal-semiconductor contact (Schottky barrier) or a shallow p-n (or n-p) junction is made on the surface of the semiconductor material sample to be tested using electron beam induced voltage (current) technique. The electron beam impinges normal to the Schottky barrier or shallow p-n junction upon the sample (plan-view). Or in the section-view geometry it impinges parallel to the Schottky barrier or shallow p-n junction upon the sample. The Schottky barrier or shallow p-n junction collects the minority carriers to form charge collection signal. These structures themselves do not cause any additional contrast.

The authors have observed an additional contrast which differs from those conventional ones caused by the diffusion and recombination of minority carriers in their experiments of electron beam induced voltage characterization with $Hg_{1-x}Cd_xTe$ crystal material at low temperature. Attempt was made to reasonably explain the mechanism of this additional contrast in terms of the conception of skin resistance by means of calculation with an equivalent circuit. Conversely, this additional contrast mechanism has proved helpful for judgement of the surface electrical property of the sample.

Experiment

The experiments were performed on $Hg_{1-x}Cd_xTe$ crystal material with x value from 0.24 to 0.49. After annealing under the mercury atmosphere, the CMT wafers usually changed into N-type, or sometimes they would have a P-type core with a N-type skin. After removing the bottom surface layer, the sample was bonded with indium at bottom side and with indium or silver paste at top side as electrical connections. The incident electron beam was normal to the sample surface. For some samples, another top connection was bonded on a circular (0.5 mm in diameter) Pt-HgCdTe Schottky contact, as shown in dashed frame in Fig.1.

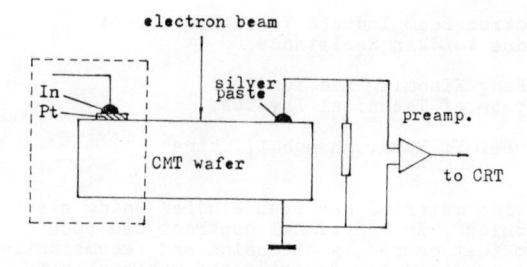

Fig.1. Schematic of the structure for EBIV test

The electron beam induced voltage characterization was performed using a DX-3A SEM (manufactured in China) with a cooling sample stage. The sample was cooled at 90K with liquid nitrogen as cryogen. EBIV signal was amplified and then used to modulate the brightness of the display CRT to form a charge cellection micrograph. Or it might be used to drive the Y axis of a XY recorder to obtain a EBIV profile when the beam was scanned across the sample. The SEM was operated at the accelerating voltage ranging from 15KV to 30KV and the beam current was about 10 nA.

From some samples we got conventional EBIV images which reveal the defects and inhomogeneities in the material. These results had been discussed elsewhere.[4] From other samples we obtained EBIV micrographs with additional contrast. Fig.2 is a typical example. Fig.2(a) is the image formed by the EBIV signal taken from the lower connection which is bonded on a circular Pt-HgCdTe Schottky barrier. It shows clearly that the EBIV signal has a maximum just around the circular contact, and decreases gradually away from the contact. But the contrast extends widely. When we took the EBIV signal from the upper connection which was made by silver paste, we got Fig.2 (b). Similar to Fig.2(a), the EBIV signal has a maximum around the silver paste connection, and decreases gradually as the distance increases. However, it has another maximum as the electron beam reaches just around the Schottky contact. This shows obviously the existence of a Schottky barrier. Fig.3 shows two EBIV profiles obtained when the electron beam scans horizontally towards the two connections. The zero level is also shown in the figure. The one that results from the Schottky contact has a higher peak, but the decay manner of the two profiles is similar. From the section-view EBIV image of this sample, we saw the contrast extend to a quite large area along the surface of the sample. From other samples we got similar contrast profiles.

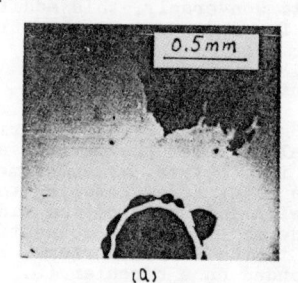

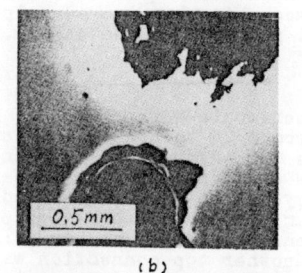

Fig.2. EBIV images with additional contrast, signals taken from the lower connection (a), upper connection (b).

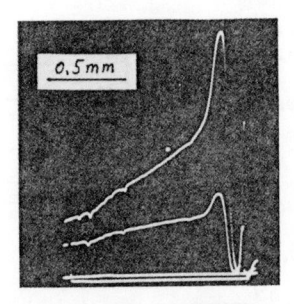

Fig.3. EBIV profiles of the same sample as Fig.2, upper: from lower connection; lower: from upper connection.

In our case, the extension area of EBIV response is quite large. (usually to several hundred microns). This is far more than the range of lateral diffusion of the minority carriers in mercury cadmium telluride which has the typical value of less than a hundred microns. Therefore, it is unreasonable to explain the contrast mechanism here by means of the diffusion and recombination theory of the monority carriers.

Discussion

To interpret the observed additional contrast, we assume that a very thin layer of barrier or space charge region exists at the top surface of the sample, as shown in Fig.4. The dynamic resistance R_0 and the skin resistivity r are the two characteristics indicating the barrier and the skin properties, respectively. The electron beam induced voltage is V_0 at the beam impinging point and V at the connection. V does not equal V_0 because of the existence of the skin resistance. The structure shown in Fig.4 can be described by an equivalent circuit shown in Fig. 5, where R_i is an imagined junction resistance proportional to R_0, and r_i is an imagined skin resistance proportional to r. For convenience of the calculation and discussion, let $R_1 = R_2 = R_i = C_1 R_0 A$, $r_1 = r_2 = r_i = C_2 r$, where A is area of the barrier of the sample and C_1 and C_2 are constants. Futhermore, suppose the resistance between the two points B and A is proportional to the distance d. These simplified assumptions are acceptable for the qualitative discussion here. From the equivalent circuit the depedence of the ratio V/V_0 on the distance of the incident electron beam has been calculated for a number of $R_0 A$ and r values. (See Fig.6 and 7)

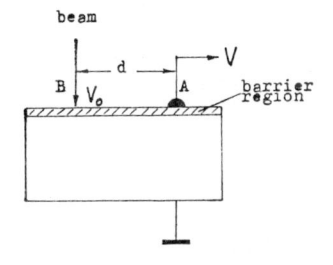

Fig.4. Surface barrier of the sample

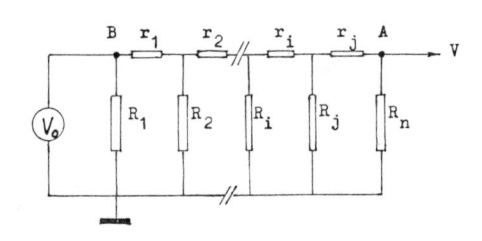

Fig.5. Equivalent circuit of the structure in Fig.4.

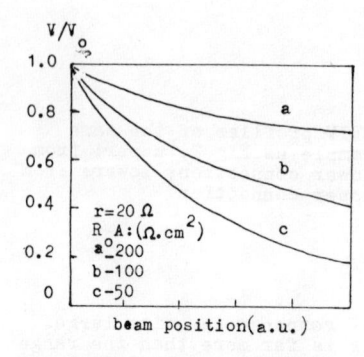

Fig.6. EBIV decay curve for fixed r.

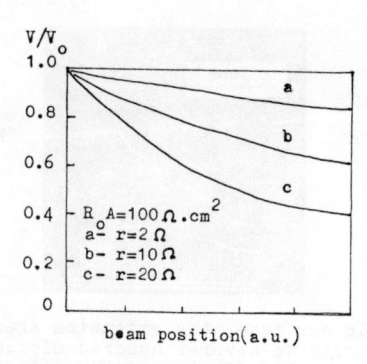

Fig.7. EBIV decay curve for fixed $R_o A$.

From Fig.6 and Fig.7 we have:
For fixed skin resistivity r, the decay of the ratio V/Vo with the beam distance d does slow down as $R_o A$ increases. Hence the extension area of EBIV contrast increases.

For same $R_o A$, if the skin resistivity r increases, the extension area of EBIV contrast will reduce. When $r \rightarrow o$, the EBIV response will extend uniformly towards whole surface of the sample, as in usual cases in which a shallow p-n junction is used to perform a plan-view EBIV experiment.

These are consistent with the experimental results shown in Fig.2 and Fig.3.

Conversely, from the additional EBIV contrast shown in Fig.2 and Fig.3, we deduce that a very thin skin barrier with relatively higher resistivity exists at the surface of the sample. This deduction has been verified by other experiment.

Conclusion

Skin resistance of mercury cadmium telluride crystal wafer can cause additional electron beam induced voltage contrast. The extension of this contrast depends on both skin resistivity and barrier resistance.

Acknowledgement

The authors would like to thank Dr. Shen Jie for providing mercury cadmium telluride crystal material and helpful discussion.

References

(1) H.J. Leamy, J. Appl. Phys., 53, R51 (1982)
(2) Kato et al., J. Appl. Phys., 46, 2288 (1975)
(3) J.H. Tregilgas, J. Vac. Sci. Technol., 21, 208 (1982)
(4) Tong Feiming, Chen Boliang and Chang Ming, Chinese Physics, 4, 934 (1984)

A NOTE ON THE RUTHERFORD BACK SCATTERING TECHNIQUE

W. Eugene Collins
S.P.A.M.S. Laboratory
Physics Department
Southern University
Baton Rouge, La. 70813

Rutherford Back Scattering has shown itself to be an excellent tool for depth profile studies. This paper will be concerned with information that can be obtained on island structure (average height and percentage coverage) in some cases for substrates coated with sufficiently higher mass thin-film material. An excellent general review article on R.B.S has been given by Van Der Veen (1). Both the limitations and advantages of the technique were discussed. The first systematic R.B.S. study of interface formation which made explicit use of high deph resolution, was performed by Van Loenen et al (2). He and his co-workers illustrated the technique for a metal silicon interface as shown in figure 1. Five cases are illustrated in the figure:

a) clean Si susrface with beam alligned with crystal axis

b) amorphous silicide deposited on Si

c) Pure metal deposited on Si

d) metal islands of equal thicknesses

e) Metal islands of varying thicknesses

The results are quite simple to evaluate in such cases (2). In many cases where the roughness of the bottom layer is needed (assuming a smooth top surface), the R.B.S. technique can yield useful information. Standard computer codes that allow for diffusion and/or islands can readily be used to model surfaces as shown in figure 2. For typical roughness domains occuring in practical situations, the R.B.S. spectrum may appear as shown in figure 3. The relative heights of lines a and b yield information on the percentage coverage of the islands or valley areas (A). The distance, t, determine the average height of the islands or depth of the valleys.

In some cases where diffusion is believed to have occured due to ion bombardment or other means, failure to look at the region indicated could lead to eroneous conclusions concerning diffusion.

REFERENCES

1. E.J. Van Loenen, M. Iwami. R.m. Tromp and J. F. van der Veen, Surface Sci. 137 (1984) 1.

2. Van Der Veen, "Surface Science Reports"", vol.5/6, Amsterdam, Elsevier Science Publishers B.V., 1985.

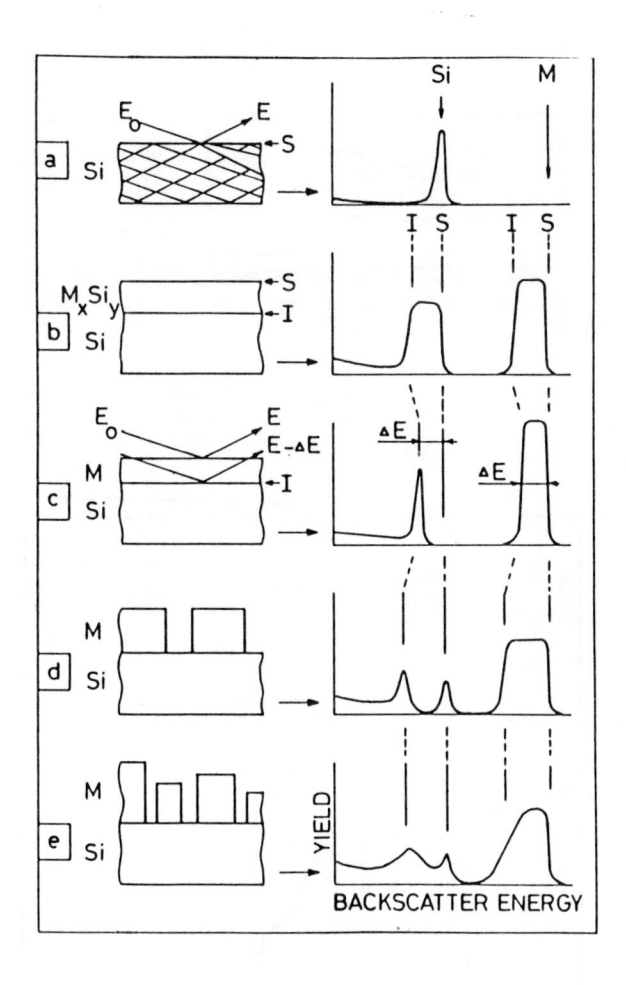

Figure 1. Films of varying composition and morphology on Si, and their corresponding backscattering enegy spectra. (From reference 2)

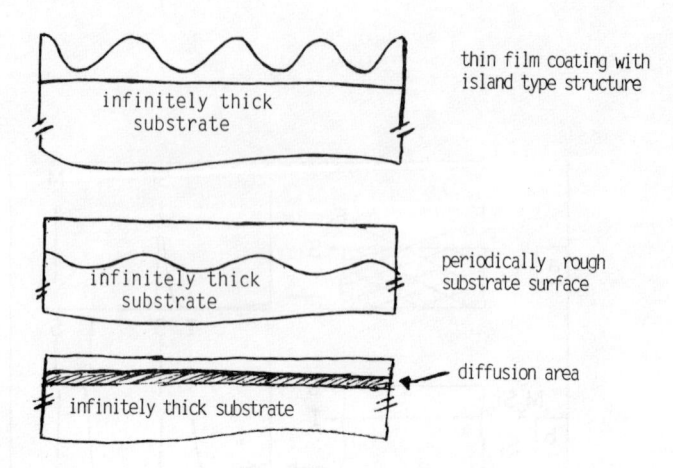

Figure 2. Surface and interface strucures that yield similar R.B.S. spectra.

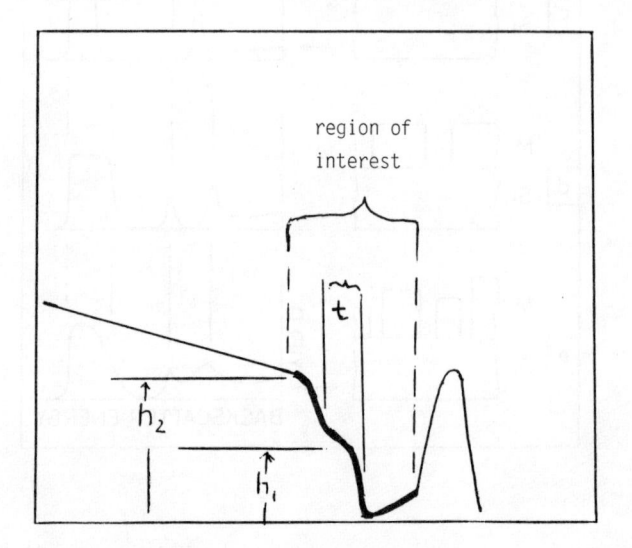

Figure 3. Rutherford Back Scattering Spectra corresponding to model in figure 2. As the roughness increases and or the island heights increase, the spectra begins to approach those of figure 1d and 1e.

CHARACTERISATION OF GAS-NITRIDED STAINLESS STEEL

M F Chung and Y K Lim
Physics Department, National University of Singapore
Singapore 0511

1 INTRODUCTION

Austenitic stainless steels are widely used in industry because of their weldability and strong resistance to corrosion and fatigue. However, they are relatively soft and cannot be hardened by normal heat treatment. In recent years, a technique has been developed to case-harden steels by gas nitriding to improve wear resistance and prolong machine life. The engineering aspects of gas nitriding have been extensively investigated while much of its underlying physical processes is still not well understood. This paper reviews the results of our experiments to characterise gas-nitrided austenitic 304 stainless steel samples.

2 EXPERIMENTAL PROCEDURE

Austenitic 304 stainless steel samples were machined into cylinders 10 mm in diameter and in length. Some with an appendix in the form of a needle of 0.2 mm diameter along the sample axis for powder X-ray diffraction studies. After nitriding the samples were cross-sectionally cut to expose the nitrided region and the bulk. The cut surface was mechanically polished to a mirror finish and then chemically cleaned. Surfaces intended for Auger analysis were further cleaned by light Argon ion bombardment without annealing to remove the surface oxide layer.

Two methods of nitriding were employed: Thermochemical diffusion in which anhydrous ammonia was allowed to enter an alumina tabular vacuum furnace containing the sample which had earlier been evacuated to 1 Pa; and plasma nitriding in which a mixture of nitrogen and hydrogen gases in the pressure ratio of 1:10 was admitted to a stainless steel vacuum chamber which had been pre-evacuated to 1 Pa. In both cases the nitriding temperature was maintained at 520 C. The detailed experimental setup has been published elsewhere[1,2,3]. Figure 1 shows the thickness of the nitrided layer plotted against the square root of nitriding time. A linear relationship shows that both processes are diffusion in nature[3]. To achieve the same thickness of the nitrided layer ammonia nitriding requires a significantly longer time. The samples used for present work had nitrided layers of thickness 110 um and the nitriding times required for this thickness by the NH_3 and plasma nitriding were respectively 25 and 16 hours.

Analysis techniques employed were scanning electron microscopy (SEM) using a JEOL 35CF, wavelength dispersive X-ray analysis (WDX) using a JEOL 35FCS, scanning Auger microscopy (SAM) using a PHI 600, powder X-ray diffraction, X-ray photoelectron spectroscopy (XPS) and microhardness measurements using a Knoop hardness gauge.

3 RESULTS

(a) NH$_3$-nitrided Sample

Figure 2 shows the SEM micrograph of the cut surface of an ammonia-nitrided sample. The nitrided layer can be clearly distinguished from the bulk by a sharp boundary 115 um from the surface. Also shown are the diamond-shape indentation marks from microhardness measurement, the WDX scanning line and the intensity distribution of the nitrogen k_α line obtained by WDX. Figure 3 gives the resulting hardness and nitrogen concentration profiles. It is remarkable that no linear relationship exists between the hardness and nitrogen concentration. An increase of hardness by a factor of 3.5 in terms of the Knoop hardness number may be noted. Figure 4 gives the concentration profiles, also obtained by WDX, of Ni, Cr and Fe for the same region.

More accurate concentration profiles of the elements could be obtained by elemental Auger line scans. These, shown in Fig 5, are in agreement with those obtained by WDX. Several points may be noted: The nitrogen concentration appears to decrease exponentially from the surface and then flatens until it drops sharply at the boundary. The flat part of the profile is present only after prolonged nitriding, > 20 hours. The Fe concentration in the nitrided region is distinctly lower than that in the bulk whereas the concentrations of Ni and Cr show no change on nitriding. Comparing the concentration profiles of the different elements there appears to be no correlation among them as regards the positions of the peaks and valleys.

(b) Plasma-nitrided Sample

Figure 6 shows a SEM micrograph of a cross section of the sample. The nitrided and bulk regions can be clearly distinguished with a sharp boundary between them. Also shown are the diamond-shape indentation marks from hardness measurement, the position of the WDX line scan and the WDX intensity profile of the nitrogen k_α line. Figure 7 shows the hardness profile and the nitrogen concentration profile obtained by WDX. The hardness is seen to have increased by a factor of 3 in terms of the Knoop hardness number, about the same as for the NH$_3$-nitrided sample. The concentration profiles of the various elements for the same region as obtained by Auger electron microprobe are shown in Figure 8. It can be noted that the nitrogen concentration remains constant for the first 70 um from the surface, and then decreases exponentially towards the boundary at 110 um where it drops sharply. The Fe concentration is lower for the nitrided layer while the concentrations of Cr and Ni are not changed by nitriding, again similar to the case of NH$_3$

nitriding. Regarding the peaks and valleys of the different profiles, they are essentially in phase between N and Cr, in antiphase between N and Fe, and not correlated between N and Ni.

Figure 9 show the Auger imaging of Cr and of N respectively of a portion of the area shown in Fig 6 within the first 70 um of the surface where the nitrogen concentration is constant. The bright spots correspond to islands of high concentration, each of about 1 to 2 um in diameter. A careful comparison shows that the two distributions are the same, indicating that the two elements N and Cr precipitate together, probably in the form of CrN. This is confirmed by taking the Auger spectra at a bright and a dark spot as shown in Fig 10. Comparing the heights of the peaks with those of the Fe peaks, we see that the Cr and N peaks are higher at a bright spot than at a dark spot, while the Ni peaks are the same at both spots.

Figure 11 shows the nitrogen spectrum obtained by X-ray photoelectron spectroscopy on a plasma nitrided sample. The twin peaks indicate two bonding states for the nitrogen present.

Powder X-ray diffraction was carried out on the needle samples before and after plasma nitriding. Before nitriding the characteristic FCC γ structure with an average lattice constant a_γ = 0.35862 nm was observed with a set of weak lines corresponding to an FCC γ' structure with $a_{\gamma'}$ = 0.39442 nm. The weak lines could be interpreted as arising from the presence of some impurity $Fe_4N(\gamma')$. After prolonged plasma nitriding four sets of structures - the FCC γ(a_γ = 0.36064 nm), γ'($a_{\gamma'}$ = 0.39564 nm) and B_1 (a_{B_1} = 0.41435 nm), and the BCC α (a_α = 0.28636 nm) were observed. The values of the lattice parameters suggest that the γ and γ' structures correspond respectively to the austenitic iron and the impurity $Fe_4N(\gamma')$ <3>. The slight increase in the lattice parameters may be related to a small volume increase which is also evidenced from the fall of iron concentration on nitriding. The B_1 and α structures may correspond to $CrN(B_1)$ and the ferritic BCC α iron. The presence of two nitrogen compounds (Fe_4N and CrN) has been suggested by the observation of two N peaks in the XPS spectrum. The samples become ferromagnetic after plasma nitriding confirming the presence of BCC α iron.

4 DISCUSSION

From the experiments it can be seen that austenitic 304 stainless steel can be readily nitrided by both thermochemical diffusion in ammonia and plasma nitriding in a mixture of nitrogen and hydrogen. Though the latter process is faster, both produce about the same increase in hardness. In the former case there is no evidence of element segregation or micrograin formation, whereas plasma nitriding produces a region where Cr and N segregate and precipitate together to form grains of diameters 1-2 um<4>.

A remarkable difference between the two processes lies in the resultant nitrogen concentration profiles for the nitrided region. In the NH_3-nitrided layer, the N concentration falls off exponentially until the boundary is reached where it drops suddenly. In the case of prolonged NH_3-nitriding, a flat layer is observed after the exponential fall and before the boundary. We interpret this as the result of diffusion between two uniform media, the nitrided region with a larger nitrogen diffusion coefficient and the bulk with a small diffusion coefficient<5>. The flat region may be due to reflection at the boundary arising because of a large disparity in diffusion coefficients.

The nitrogen concentration profile for the plasma-nitrided region shows a different behaviour. The concentration remains constant for about 70 um, then falls exponentially until the boundary is reached where it drops sharply. The region of constant nitrogen concentration is characterised by the formation of large grains of CrN. The exponentially decreasing part of the profile is similar to the N concentration profile of the NH_3-nitrided layer. It appears therefore that the diffusion takes place through three layers, first a 'porous' layer with a very large diffusion coefficient for nitrogen, then a nitrided layer similar to the NH_3-nitrided layer and finally the bulk. Plasma nitriding may be pictured as follows: Nitrogen is diffused into the steel to produce a layer of exponentially decreasing nitrogen concentration similar to the case of NH_3-nitriding. The nitriding front or the boundary of this layer moves steadily into the bulk at a rate proportional to the square root of time. In this layer nitrogen may combine with Cr and Fe to form finely dispersed crystalline particles with diameters not greater than 0.1 um. With a further continuous supply of active nitrogen, large crystalline grains of CrN, and perhaps fine particles of Fe_4N also, are formed in the high N concentration region near the surface making it porous to further diffusion of nitrogen. As nitriding proceeds, this porous region expands into the already nitrided region, maintaining a thin layer of exponentially decreasing nitrogen concentration in its front.

Reference:

1. A M Staines and T Bell, Thin Solid films, 86, 201 (1981)
2. M F Chung, A K Yap and Y K Lim, Scr. Metall. 19, 415 (1985)
3. M F Chung and Y K Lim, Scr. Metall. 20, 807 (1986)
4. M F Chung and Y K Lim, Scr. Metall. 21, 579 (1987)
5. M F Chung and Y K Lim. "Nitrogen depth profile of nitrided steels", to be published.

229

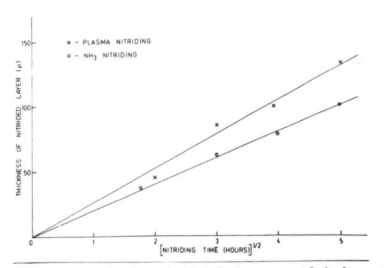

Fig 1 Nitrided layer thickness versus nitriding time. Plasma process is faster than the NH₃ process.

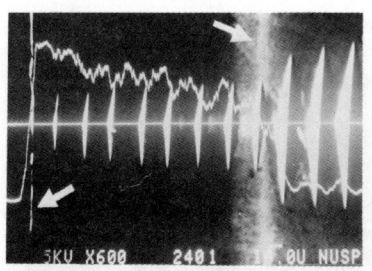

Fig 2 SEM micrograph showing the cross section of a sample after nitriding in NH₃ for 25 hours. Nitriding surface (left arrow) and boundary (right arrow) indicate a thickness of ∼115 μm. Microhardness indentation marks and N$_{k\alpha}$ line scan by WDX are also shown.

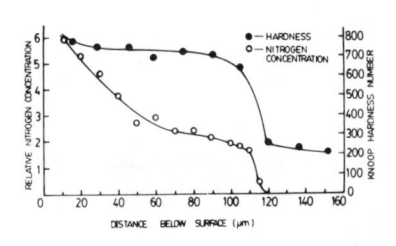

Fig 3 Nitrogen concentration (obtained by WDX) and microhardness depth profiles for the NH₃ nitrided sample indicating a nitrided depth of ∼120 μm.

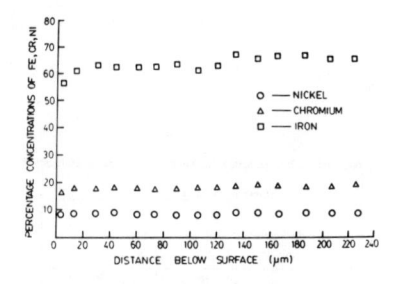

Fig 4 Concentration depth profiles of Fe, Cr and Ni for the NH₃ nitrided sample obtained by WDX.

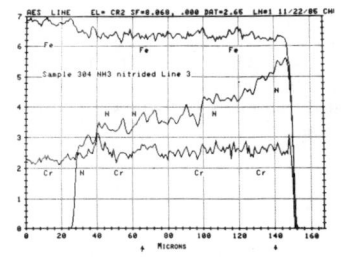

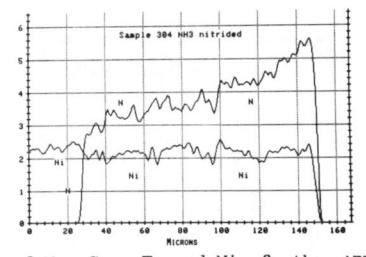

Fig 5 Concentration depth profiles of N, Cr, Fe and Ni of the NH₃ nitrided sample obtained by Auger line scan.

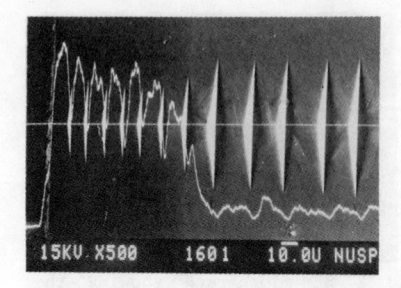

Fig 6 SEM micrograph of the cross section of a sample after plasma nitriding for 16 hours. The nitrided boundary ∼100 μm from the surface can be clearly seen. Microhardness indentation marks and $N_{k\alpha}$ line scan by WDX across the indentation are also shown.

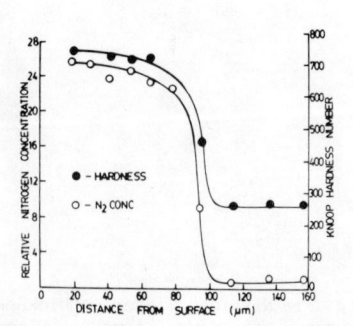

Fig 7 Nitrogen concentration (obtained by WDX) and microhardness depth profiles of the plasma nitrided sample indicates the nitrided depth to be ∼100 μm.

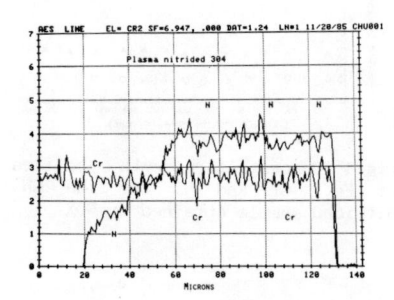

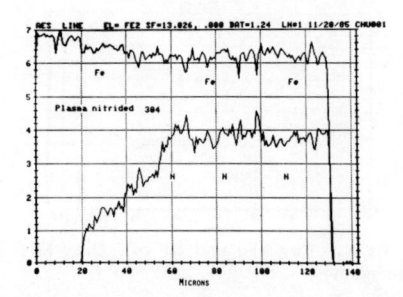

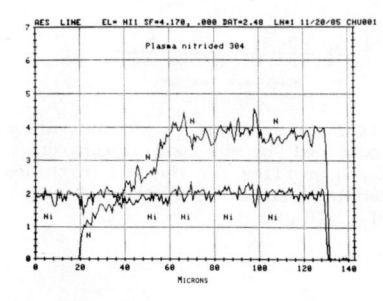

Fig 8 Concentration depth profiles of N, Cr, Fe and Ni of plasma nitrided samples obtained by Auger line scan.

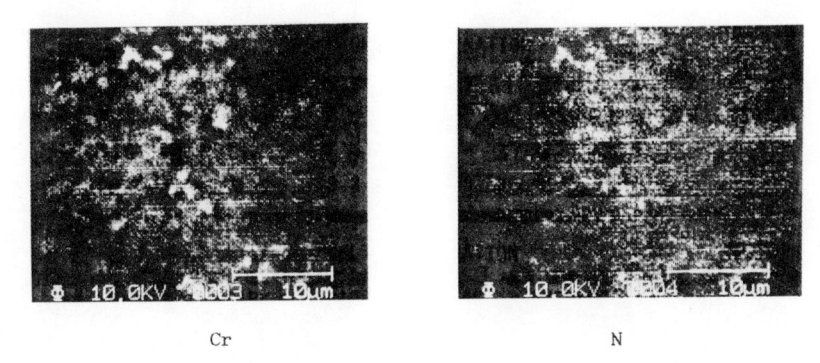

Fig 9 Cr and N distributions obtained by Auger mapping in the constant N concentration region of the plasma nitrided sample.

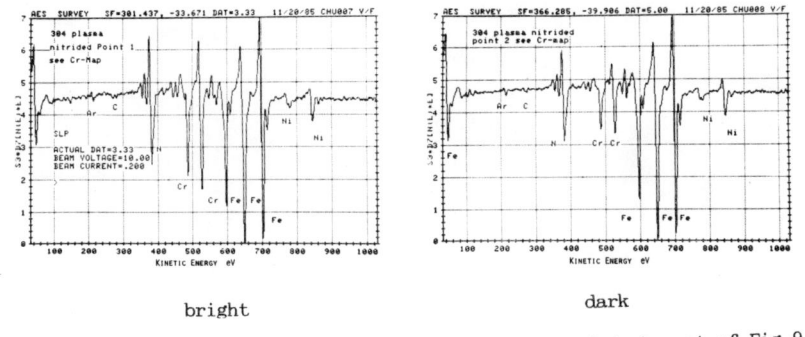

Fig 10 Auger spectra obtained for the bright spot and dark spot of Fig 9.

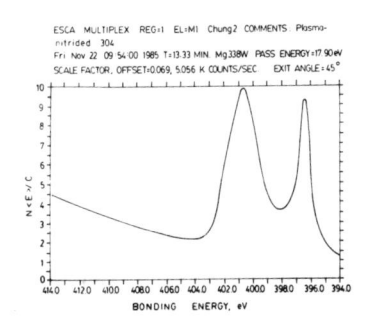

Fig 11 Nitrogen XPS spectrum of plasma nitrided sample.

X-RAY STRUCTURE DETERMINATION OF $Fe_9Mn_6Al_5$ PRECIPITATE IN A Fe-Mn-Al-Cr ALLOY.

Tian-Huey Lu, Tseng-Fong Liu and Chi-Meen Wan
National Tsing Hua University, Hsinchu,
Taiwan 30043, Republic of China

Abstract

The space group of this $Fe_9Mn_6Al_5$ precipitate in a Fe-Mn-Al-Cr alloy is $P4_332$ and is different from that of the typical β-Mn structure which is $P4_132$ (7,8,5). The powder of the precipitate belongs to the cubic crystal system and has the cell dimension of 6.31± 0.01 Å. The residual index is 0.10 in the final least-squares fit for 19 independent reflections. Although the symmetry of typical β-Mn structure can be transformed into another space group which is $P4_332(5)$, yet the atomic coordinates under this transformation does not fit our present data at all. This result is in agreement with the powder diffraction pattern, because the cell dimension of a material determines the positions of the Bragg angles, while the atomic configurations in the unit cell decide the relative intensity of the diffracted peaks. The atoms of the $Fe_9Mn_6Al_5$ precipitate are disordered and are distributed in two sites, 8c and 12d, similar to the structure of typical β-Mn structure, but in different atomic coordinates. There are twelve nearest neighbors surrounding each site. The possible distances, between any two nearest neighbors generated by symmetry operations, range from 2.38 to 2.69 Å which are 0.02 Å longer than those of usual β-Mn structure (5). These increments of 0.02 Å in interatomic distances may explain the brittleness in the alloy formed by various compositions of iron, chromium, manganese and aluminum, when $Fe_9Mn_6Al_5$ precipiate is present.

Introduction

The alloys formed by various compositions of iron, chromium, manganese and aluminum were studied systematically for the different development of a new system of future stainless steel. In order to investigate the crystal structure of the precipitates that are formed under various aging times and aging temperatures of these alloys, X-ray diffraction is carried out. As part of a systematic study of these alloys we have determined the structure of $Fe_9Mn_6Al_5$, which is different from the typical β-Mn structure in space group or atomic coordinates.

Experimental

Testing alloy for the present experiment was prepared in a vacuum induction furnance under a controlled protective argon atmosphere. The chemical composition of this alloy is

Element	Al	Mn	Cr	Fe
wt %	9.1	29.9	2.9	balance

After being homogenized at 1250 °C for 14 hours, the ingot was hot forged and then cold rolled to a final thickness of 2.0 mm. The sheet was subsequently solution heat treated at 980 °C for one hour and rapidly quenched into water at room temperature. Aging processes were performed at 550 °C in a salt bath for time periods ranging from 2 hours to 120 hours.

X-ray diffraction experiment was carried out on a computer-controlled diffractometer using molybdenum as its target. The incident beam was monochromatic (MoK$_\alpha$). Powder specimens were solution treated at 980 °C for one hour in a sealed quartz tube filled with argon, quenched into water, re-heated at 550 °C for 120 hours, and quenched again.

Analysis

Besides the useful pattern of the precipitate, there superimposed the patterns of ferrite and austenite (matrix) on the X-ray diffraction pattern of power alloy. Bragg angle calculations of the known ferrite and austenite reveal the pattern positions of the peaks which are not diffracted by the precipitate. Elimination of these unuseful peaks, 19 nonoverlapped independent reflections of the precipitate were taken for the analysis of the powder crystal structure. Observed structure factor Fo is calculated from the following formula (1)

$$I = |Fo|^2 p \left(\frac{1 + \cos^2 2\theta}{\sin^2 \theta \cos \theta} \right) \qquad [1]$$

where I is the measured diffraction intensity, p the multiplicity factor, and θ the Bragg angle.

On the said powder diffraction pattern, the Bragg angles of the 19 independent reflections are the same as those of typical β-Mn structure within experimental error. But the diffraction peak heights of the present crystal are not proportional to those of β-Mn. Hence the cell dimensions of these two crystals should be the same, with different atomic arrangement in the unit cell. Analysis of the precipitate by electron probe microanalyzer(EPMA) shows that it is containing the composition as $Fe_9Mn_6Al_5$. Least-squares program (2) was used to find the lattice constant to be 6.31 Å for this precipitate. Two possible groups, $P4_132$ and $P4_332$, were presumed to solve the present structure. In order to adopt the N.R.C. VAX Crystal Structure System developed by professor E.J. Gabe et al., (3), input/output conversion program was edited, in addition to the program edition of data reduction in equation [1]. Direct method was applied after an entire failure to solve the crystal structure by Patterson synthesis. Output of direct method gives several possible sets in the phase determination. Each solution of the possible set is tested by Fourier synthesis in accordance with the sequence of high figure of merit and low value of residue. The number of peaks in Fourier map is larger than the number of atom types, Fe, Mn and Al. Restriction of interatomic distance, rough sum of atomic radius (4) of the said atom types, excludes those redundant peaks. Direct method based on the assumed space group of $P4_132$ does not result in reasonable interatomic distances for the present precipitate. But when the space group of $P4_332$ was used for the analysis of direct method, a good solution of reasonable interatomic distances appears. Furthermore, least-squares fit of atomic parameters reveals the following result. The function minimized is $\Sigma w(|Fo| - |Fc|)^2$, where $w = 1.0/Fo^2$, Fo and Fc are the observed and calculated structure factor respectively.

Two most probable peaks on the Fourier map are assumed to be the atom type of Fe and Mn which are located at 12d and 8c respectively, and are used in least-squares fit. Variation of atomic parameters results in a reliability index of 0.16. Third atom, Al, which is selected from each of the several peaks in difference Fourier map, is input in least-squares fit. Interatomic distances among Fe, Mn and Al appears unreasonable. Therefore in the cubic unit cell of lattice constant 6.31 Å, only two atom sites are possible to be squeezed without violating interatomic distance and environmental situaion of each site. This can also be proved by the typical β-Mn structure (5), in which there are also two possible sites located at 8c and 12d for a total number of twenty manganese atoms. The present precipitate also consists of twenty atoms, 9 Fe's, 6 Mn's and 5 Al's, but has different coordinates for the two sites. Variation of occupancies for Fe and Mn atoms in least-squares fit further reduces R factor to 0.10. In the final cycle of least-squares fit, fit of goodness=0.20. The final average and maximum parameter shifts were about 0.2σ and 0.4σ. The maximum electron density in final difference Fourier map is 2.0 ± 0.5 eÅ^{-3}.

Results and Discussion

From the analysis of X-ray diffraction, only two sites are available for these three kinds of atoms, Fe, Mn and Al. Hence the alloy is disordered. Fe and Mn are the most abundant metals at the two sites, and represent the label of the two sites in the listed Tables. Table 1 lists the final atomic coordinates and thermal factors. Observed and calculated structure factors and normalized diffraction intensities are listed in Table 2. A stereoview of the packing for the sites in a unit cell is shown in Fig. 1. A stereoview of the packing of 12 neighbors (6 Fe's and 6 Mn's) surrounding Fe site is shown in Fig. 2, and that of 9 Fe's and 3 Mn's surrounding Mn site is shown in Fig. 3. All figures are plotted by the use of ORTEP(6). There are 24 general equivalent positions for the space group of P4$_3$32. For the disordered Fe$_q$Mn$_6$Al$_5$ precipitate, the possible distances between site Fe and its nearest neighbors generated by symmetry operation are 2.64 and 2.69 Å. One only possible distance between site Mn and its nearest neighbors generated by symmetry operation is 2.38 Å. The possible distances between site Fe and its nearest neighbor sites Mn generated by symmetry operation are 2.55, 2.56 and 2.67 Å. There are 12 nearest neighbors surrounding each site at a variety of distances between 2.38 and 2.69 Å, which are longer than those of typical β-Mn structure (5) whose distances range from 2.36 to 2.67 Å. These increments of 0.02 Å in interatomic distances for Fe$_q$Mn$_6$Al$_5$ may be due to the larger atomic radii of Al and Fe than Mn. The increments of distances between sites might explain the brittleness in the alloy formed by various compositions of iron, chromium, manganese and aluminum, when Fe$_q$Mn$_6$Al$_5$ precipitate is present. One mechanical property of typical β-Mn material is the existence of brittleness that may be due to the large interatomic distances. Increment of the interatomic distances in the present precipitate further weakens the interatomic attractive force, and then shows brittleness in the bulk alloy.

Acknowledgement

The authors thank Professor E.J. Gabe and coworkers in the Chemistry Division, NRC, Canada for their provision of the computing program in VAX crystal structure system. We also thank for the support of this work by the National Science Council in research grant.

References

1. B.D. Cullity. "Elements of X-ray Diffraction" p.132 (1973).
2. J.W. Visser. J. Appl. Cryst., 2, 89 (1969).
3. A.C. Larson, F.L. Lee, Y.Le Page & E.J. Gabe, "The N.R.C. VAX Crystal Structure System", Chemistry Division, NRC, Ottawa, Canada.
4. L.H. Van Vlack, "Elements of Materials Science" 2nd ed., p.142, Addison-Wesley (1967).
5. R.W.G. Wyckoff. Crystal Structure. p.51, John Wiley and Sons, 2nd ed., Vol. 1, (1963).
6. C.K. Johnson, ORTEP II. Report ORNL-5138. (Oak Ridge National Laboratory, Tennessee. 1967.)
7. J.D.H. Donnay & Helen M. Ondik, "Crystal Data, Determinative Tables " c-157, JCPDS, 3rd., Vol. 2, (1973).
8. Nat. Bur. Stand. (U.S.) Monogr. 25, Sec. 18 (1981).

TABLE 1.   ATOMIC PARAMETERS X,Y,Z AND BISO.
          E.S.DS. REFER TO THE LAST DIGIT PRINTED.

	X	Y	Z	BISO
Fe	0.2008(15)	0.04920	5/8	1.00
Mn	0.3094(18)	0.30939	0.30939	1.45(18)

BISO IS THE MEAN OF THE PRINCIPAL AXES OF THE THERMAL ELLIPSOID

TABLE OF U(I,J) OR U VALUES *100.
        E.S.DS. REFER TO THE LAST DIGIT PRINTED

	U11(U)	U22	U33	U12	U13	U23
Fe	0.00(12)	0.00	3.79(16)	0.88(7)	-0.13(7)	-0.13
Mn	1.8 (5)	1.84	1.84	-0.3 (3)	-0.26	-0.26

TABLE 2. 19 reflections: dA denotes the resolution in Å,
          I0/I1 the normalized intensity with a maximum of 1000,
          KFO and FC are 10 times of observed and calculated
          structure factors respectively.

dA	I0/I1	KFO	FC	h k l	dA	I0/I1	KFO	FC	h k l
4.462	235	254	144	1 1 0	3.643	85	230	179	1 1 1
2.822	79	166	172	2 1 0	2.231	71	283	256	2 2 0
1.903	1000	888	901	3 1 1	1.750	55	228	231	3 2 0
1.686	400	451	387	3 2 1	1.345	25	202	221	3 3 2
1.411	170	504	359	4 2 0	1.377	17	115	83	4 2 1
1.115	9	214	241	4 4 0	1.152	188	470	409	5 2 1
1.067	144	449	381	5 3 1	0.962	53	436	281	5 3 3
0.974	11	137	132	5 4 1	1.037	78	481	433	6 1 0
0.998	7	152	149	6 2 0	0.930	13	158	144	6 3 1
0.901	79	404	319	6 3 2					

241

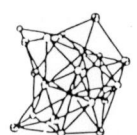

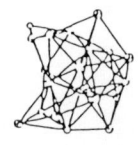

Fig. 1. A stereoview of the packing for the disordered $Fe_9Mn_6Al_5$ at two sites 8c and 12d in a unit cell of space group $p4_332$.

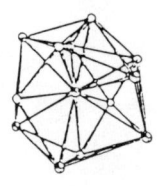

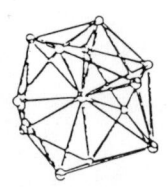

Fig. 2. A stereoview of the packing for 12 neighbors surrounding a site of 12d in the space group of $p4_332$.

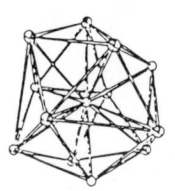

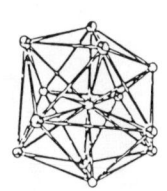

Fig. 3. A stereoview of the packing for 12 neighbors surrounding a site of 8c in the space group of $p4_332$.

THE CREATION AND ELIMINATION OF In ISLANDS ON CLEAN InP SURFACES STUDIED BY ELECTRON ENERGY LOSS SPECTROSCOPY

Xun Wang, Mingren Yu, Xiaoyuan Hou and Xiaofeng Jin.

(Surface Physics Laboratory, Fudan University, Shanghai, P.R.China)

Abstract

Ion sputtering and subsequent annealing is a commonly used method in preparing clean and ordered InP(111) and InP(100) surface under UHV condition. But In islands would always be created on InP surface after Ar ion sputtering. The existance of small and dispersive In islands can be sensitively detected by the electron energy loss spectroscopy (ELS), in which the appearance of surface plasmon loss peak (SP)In at 8.7 eV and bulk plasmon loss peak (BP)In at 11.7 eV is the criterion of In islands formation on InP surfaces. The relative area occupied by In islands can be estimated by the relative intensities of (SP)In and (BP)In as compared with the ELS obtained from clean pure In surface. It has been found that after annealing the In islands become thicker but their relative area decreases. It seems more easy to form In islands on InP(100) surface than on InP(111) surface.

The fomation of In islands probably originates from the preferential sputtering effect of argon ion on InP surface. Because the sputtering yield of phosphorus by argon ion is larger than that of indium, an enrichment of In on InP surface will occur as the result of sputtering. We found increasing the incident angle of sputtering ion beam can raise the relative yield of In, and thus ease the effect of preferential sputtering on InP. The monotonical decreasing of (SP)In intensity and surface In/P ratio as a function of incident angle has been obtained, which verifies the supposed mechanism of In island creation. Although grazing angle sputtering can greatly reduce the amount and area of In islands, but it is hard to completely remove the In islands even following with appropriate annealing. Repeatedly evaporating phosphorus on InP surface accompanying with post heat annealing, In island-free and ordered InP polar surface could be prepared as shown by the absence of (SP)In and (BP)In in ELS. The possible mechanism of In island elimination has been discussed.

I. INTRODUCTION

InP is a hopeful semiconductor material on microwave and optoelectric devices. Usually device characteristics are related to the surface condition of material to a large extent. Based on an atomic clean surface, studying the changes of structure and electron state from the metal covered or gas absorbed surface will fundamentally clarify the interfacial characters of MIS or MES devices, providing an available way for improving the device performance. Among the three crystal planes of low index, the most frequently studied and relatively understood one is the nonpolar face (110), and the polar faces (111) and (100) (commonly adopted in device technique) are less investigated.

In preparing of a clean and ordered InP(111) or (100) surface the cleavage method for (110) face can't be used. A usually employed method is an argon ion sputtering treatment to get rid of the surface contamination and oxidation together with an annealing treatment to restore its ordering. But such a treatment will inevitably lead to the creation of In islands on InP surface[1]. The existance of In islands will largely influence the electric character of the contact between InP surface and metal. How to eliminate the

In islands is of practical meaning to the study on InP(111) and (100) clean surfaces or schottky contact of InP.

Using the molecular beam epitaxy (MBE), an In-island-free InP polar (100) surface with controllable surface compositions can be obtained from depositing several tens or hundreds of InP layer onto the surface[2]. But it is uncertain whether and how the originally existed In islands disappear during the epitaxy process. It was verified in this work that the formation of In island is the result of the preferential sputtering of phosphorous on InP surface by argon ion as well as the gathering of In atoms through surface diffusion during annealing. An ordered In-island-free clean InP(100) surface can be obtained through depositing phosphorous onto InP surface accompanying with annealing. The possible mechanism of In island elimination has been presented which is thought to be helpful in understanding the initial stage of MBE growth.

II. EXPERIMENTAL

The surface treatment and measurement of the samples were carried out on a multifunctional electron spectrometer ESCALAB-5 which has functions of low energy electron diffraction (LEED), X ray photoelectron spectroscopy (XPS) and Auger electron spectroscopy (AES). The samples were chemi-mechanically polished n type InP(100) and InP(111) single crystal wafers. After a conventional treatment employed in device technique the sample was put into the chamber with a pressure lower than 3×10^{-8} Pa after baking . The sample was sputtered by argon ion with certain energy (usually 1 keV) to get rid of C and O contaminations before measurement. The surface cleanliness was monitored by AES and XPS.

Though AES and XPS can determine the richness of In on the surface they can't judge whether these excessive In atoms gather into islands. The most sensitive method for checking the existance of In islands on InP surface is electron energy loss spectroscopy (ELS) [1] . In the presence of In islands, there will appear the bulk plasmon loss peak (BP)In and surface plasmon loss peak (SP)In of In which are located at 11.7 eV and 8.7 eV seperately in the ELS spectrum of the sample, and they are easy to distinguish from the bulk plasmon loss (15.2 eV) and surface plasmon loss (10.2 eV) of InP itself. Comparing the intensity of (SP)In (relative to the scattering elastic peak) with the result from the surface of pure In. We can reckon the percentage of In island area to the total surface area. The approximate thickness of In island can be reckoned through the intensity ratio of (BP)In and (SP)In .

ELS measurements were carried out by the AES faciality of ESCALAB-5. To improve the energy resolution the electron beam energy was turned down to 300 eV, the beam current and the amplitude of modulation voltage were adjusted to around 1.3 uA and 1 V (peak to peak), the electron energy analyser was adjusted to work at CRR mode with a retarding ratio of 10 or at CAE mode with pass energy of 30 eV . From the valuation of full width of half maximum (FWHM) of the elastic electron peak a resolution of about 0.9 eV can be reached. The loss spectra were usually recorded in negative second derivative, i.e the form of $-d^2N/dE^2$, in which the energy position of the peaks are in correspondence with the peak position in original spectra N(E). The reading precision of peak position is about 0.1 eV and its repetition is within 0.3 eV.

III. RESULTS AND DISCUSSION

A typical ELS of InP(100) is shown in Fig.1 where curve (a) was obtained after surface etching by argon ion referring to a disordered surface; curve (b) was obtained after an annealing treatment of 1 hour under 300°C, which gave a weak (4X2) reconstructive LEED pattern, referring to a relatively ordered surface.

The plasmon loss peaks (BP)In and (SP)In of In can be seen in curves (a) and (b). It means the In islands existed on both surfaces. It has been verfied by experiments that argon ion etching will cause the In/P ratio on InP surface to be deviated from its stoicheometric value, there exist some excessive In atoms; and the surface In/P ratio will drop a bit after annealing at 300 C. It can be seen from Fig.1 that the In islands formed after argon ion etching were not eliminated by annealing.

The size of In island does have some change through annealing, which is reflected by the change of peak intensity (relative to elastic peak) of (SP)In and (BP)In. Table.1 shows the typical relative intensity of (SP)In and (BP)In loss peaks of InP(100) and InP(111) surfaces after argon ion etching and annealing taken under same experimental conditions (same incident angle, same scattering angle and same electron beam energy). The relative intensities of (SP)In and (BP)In obtained from pure In surface are also listed for comparison.

Table.1 Intensities of (SP)In and (BP)In (relative to elastic peak)

	InP(111)		InP(100)		pure In
	3keV Ar etching	after annealing	2keV Ar etching	after annealing	
(SP)In	4.4×10^{-3}	9.0×10^{-4}	5.6×10^{-3}	2.8×10^{-3}	2.3×10^{-3}
(BP)In	2.0×10^{-3}	5.5×10^{-4}	3.6×10^{-3}	2.4×10^{-3}	2.0×10^{-3}
(BP)In/(SP)In	0.45	0.61	0.64	0.86	0.87

The relative intensity of (SP)In reflects the size of In island. If the In islands smear on the surface equally and the dimension of each island is far less than the spot size of electron beam (the diameter of the latter is of the order of several tens of micron) then the percentage of islands area to the total surface area can be reckoned by the ratio of surface intensity (SP)In from InP to that from pure In which is in correspondence to a surface of 100% In island. It can be seen from Table.1 that the In island area is about 19% of the total surface area on argon ion etching and shrinks to about 4% after annealing for InP(111) face. For InP(100) face, it is about 24% before annealing and 12% after annealing. That means the (100) face is easier to form In island than (111) face for InP. The ratio of (BP)In/(SP)In reflects the thickness of In island. We assume the intensity of (SP)In to be in proportion to the area of In island and the intensity of (BP)In is in relation to both the area and the thikness of In island. The larger the (BP)In/(SP)In the thicker the In island be. From the data listed in Table.1, (BP)In/(SP)In always increases after annealing, i.e., the thickness of In island increases together with the shrink of In island area. But in comparison with pure In the value of (BP)In/(SP)In is somewhat small. According to the detection depth of electron beam in the sample, the thickness of In island on InP surface is reckoned to be less than 11Å .

The creation of In island is probably due to the preferential sputtering of argon ion on InP surface. Because of the sputtering yield of P by argon ion is higher than that of In, the result of sputtering is an enrichment of In on the InP surface [5]. Taglauer mentioned in his paper, increasing the incident angle of arriving argon ion may increase the forward sputtering probability of heavier elements of the composition [6]. If this conception is available for InP, it means the relative sputtering yield of In may increase through grazing incident of argon ion and the preferential sputtering effect may be decreased, and the surface atomic ratio of In to P may approach its stoicheometric value of bulk, so that the In islands decrease.

Fig.2 shows the ELS spectra of InP(111) at various incident angles of 8 keV argon ion etching. The sample was kept at same position on etching and measuring so that all the results can be compared with each other. Fig.2 shows that the intensity of plasmon loss peaks associated with In island decreases evidently with the increasing of θ_i. Fig.3 is the relationship between relative intensity of (SP)In and θ_i. Using the same method mentioned above to reckon the area percentage of In island to the total surface get a result of 20% for $\theta_i=40°$ argon ion sputtering and it decreases to about 3% while θ_i increases to 89°.

An XPS measurement was taken for the relationship between In/P atomic ratio and argon ion incident angles. The intensity of XPS signal was measured by the area of In 3d5/2 and P 2p peaks. According to our measurement for ESCALAB-5[7], the relative atomic sensitivity factors are: $S_{In}=3.47$ for In 3d5/2 and $S_P=0.38$ for P 2p. The resultant relationship between In/P ratio and θ_i is shown in Fig.4. It is in correspondence with the curve in Fig.3. Thus it is verified that grazing incidence weakens the preferential sputtering effect on InP surface, causing a decreasing of In enrichment so that the forming of In islands is lowered.

The area of In islands from grazing sputtering is indeed very small and further annealing can't decrease the intensity of (SP)In in ELS But annealing makes a diffusion of phosphorous from bulk InP to the surface. It changes the richness of surface In, restores the broken "In-P" bond which is a result of argon ion etching in the surface region and turns the surface to be ordered. During the annealing process because of the surface diffusion of In atoms the numerous seperating In islands are gathered into some bigger in size but smaller in amount ones. Under the situation of nongrazing sputtering, there is a surface with heavily disordered layer and numerous small In islands. After annealing the disorder disappears and the In islands gathered into some smaller amount of bigger ones, their total area decreases and thickness increases. In the situation of grazing sputtering, the surface disorder is shallow and the amount of In islands is small. Also the In islands gathers after annealing but their total area has no apparent change. Fig.5 is the schematic drawings for the process of forming and changing of surface In islands, and it is verified by ELS measurements and microscopic observation of surface morphology.

To eliminate the surface In islands entirely, one has to compensate the surface with phosphorous atoms. A series of ELS spectra have been measured after sequentially evaporating phosphorous on InP surface (which originally contained In islands) and annealing at 300°C for about 1 hour as shown in Fig.6. Two In related loss peaks at 8.7 eV and 11.7 eV decrease gradually and disappear eventually after several cycles of P deposition and annealing, while a loss peak at 10.2 eV which represents the surface plasmon loss of InP increases gradually. Finally, curve (d) is obtained, which involves only the surface and bulk plasmon loss peak of InP and nothing related to plasmon loss of In. This spectrum is quite similar to that on cleaved In-island-free InP(110) surface reported by Tu and Schlier [1].

From the XPS measurements we can find that the surface In/P ratio is 1.8 for a surface with In islands before P evaporating; and it drops down to 0.4 after P evaporating without annealing, which is a P riched but disordered surface. After annealing at 300 C, the In/P ratio reaches 1.4 which is somewhat lower than the original value but is close to the bulk stoicheometric one. There is no decrease of In/P ratio for the further cycles of P evaporating and annealing. This illustrates that there also exist certain amount of P vacancies on an argon ion sputtered InP(100) surface in addition to the In islands. Although the In islands were not eliminated after evaporating certain amount of P and subsequent annealing as shown in ELS, the

P vacancies were basically removed; and this leads to a drop of In/P ratio to a stable value. In islands are removed by repetitive evaporation and annealing; howerer, the area occupied by In islands is only a small fraction of the total surface, so the elimination of In islands will not give rise to further change of In/P ratio.

It was fonnd in XPS measurements that the P deposited on the surface was initially adsorbed atomistically, i.e. atoms did not combine with In atoms to form In-P bonds until the annealing process had been carried out. Fig.7 (a) and (b) are XPS peaks of P 2p recorded before and after annealing. The annealing process changes a broad peak into a narrow one. The latter is the overlapping of an unresolved doublet P $2p_{3/2}$ and P $2p_{1/2}$, with their peak positions at 128.4 and 129.4 eV respectively, as shown by the thin solid lines in Fig.7 (b), which is the result of curve fitting treatment. These correspond obviously to the peaks of P in InP. The curve fitting treatment for Fig.7 (a) leads to two sets of P 2p doublet, as shown by the solid lines and dashed lines. One of them has the same peak energies as in Fig.7 (b), which represents the P signal from InP substrate. Another one corresponds to the signal from absorbed atomic phosphorous with the peak energies at 129.6 and 130.6 eV, which are coincident with the P 2p peak measured from deposited P on GaAs surface.

After annealing, a portion of the deposited P atoms bonded to In atoms, others combined together into phosphorous molecules and evaporated into the vacuum. As a result, an increase in surface In/P ratio from 0.41 before annealing to 1.4 after annealing was obtained. Meanwhile, the InP molevules resulting from P and the uppermost In layer of islands may diffuse or migrate laterally along the surface at the annealing temperature to fill the P vacancies previously existing on surface or form new InP layer. Consequently, the thicknesses of In islands decrease. In Fig.6 the intensity of (BP)In decreases more rapidly than that of (SP)In after P deposition and annealing, this is just an evidence of the reduction of In island thickness. Therefore in order to eliminate the whole In islands, several cycles of P deposition and annealing are necessary. This is in agreement with our experiments.

In conclusion, we propose a schematic process to describe the shrinkage of In islands during P deposition and annealing mentioned above, as shown in Fig.8. The In island elimination process is the repetition of this process.

In the growth of InP by MBE, the substrate is usually maintained at a temperature between 100-405 °C [8]. It is possible that the above-mentioned P deposition and annealing process may happen on the substrate surface. It has been reported also in ref.8, that below 95 °C the InP epitaxial layers grow in a polycrystalline state by MBE. We suggest that this may be partly caused by the existence of In island and surface disorder by the argon ion sputtering treatment before epitaxy.

IV. CONCLUSION

1. ELS is a very powerful tool for observating and checking the existance of In islands on InP surface. According to the intensity of plasmon loss peaks (SP)In and (BP)In of In islands, in combination with the microscpic observation of surface morphology, the change of In islands in the process of various treatments can be determined.

2. The main reasons for the formation of In islands are: the preferential sputtering of argon ion to the InP surface causes an In enrichment on the surface layers and the excessive In atoms may gather into island. After annealing, small and separated In islands can be gathered into some bigger in size and less in amount ones, the total area of In islands decreases.

3. The preferential sputtering of argon ion on InP surface can be lowered by grazing angle incidence, so that a clean surface with less In islands can be obtained. But further annealing treatment after sputtering can't eliminate the In islands.

4. For an argon ion sputtered InP surface through repetitive P evaporating and subsequent annealing, an In-island-free, clean and ordered surface can be obtained eventually. The mechanism for elimination of In islands through such a treatment can be explained as following: the InP molecules formed from the combination of P atoms and the uppermost atoms of the In island may diffuse or migrate laterally along the surface at the annealing temperature, causing thinning and diminishing of the In islands.

REFERENCES

(1) C.U.Tu and A.R.Schlier, Appl. Surface Sci. 11/12, 355(1982)
(2) R.F.C.Farrow, J.Phys. D7, 114(1974)
(3) J.H.McFee and B.I.Miller, J.Electrochem.Soc. 124, 259(1977)
(4) Hou Xiao-yuan, Yu Ming-ren and Wang Xun, Chinese J.Semicond. 5, 171(1984) in Chinese
(5) Yu Ming-ren, Yang Guang and Wang Xun, Chinese Phys. 4, 10(1984)
(6) E.Taglauer, Appl. Surface Sci. 13, 80(1982)
(7) Yu Ming-ren, Pang Yan-wan, Cheng Shi-ning and Wang Xun, J.Applied Sciences, 2, 145(1984) in Chinese
(8) M.T.Morris and C.R.Stanley, Appl. Phys. Lett. 35, 617(1979)

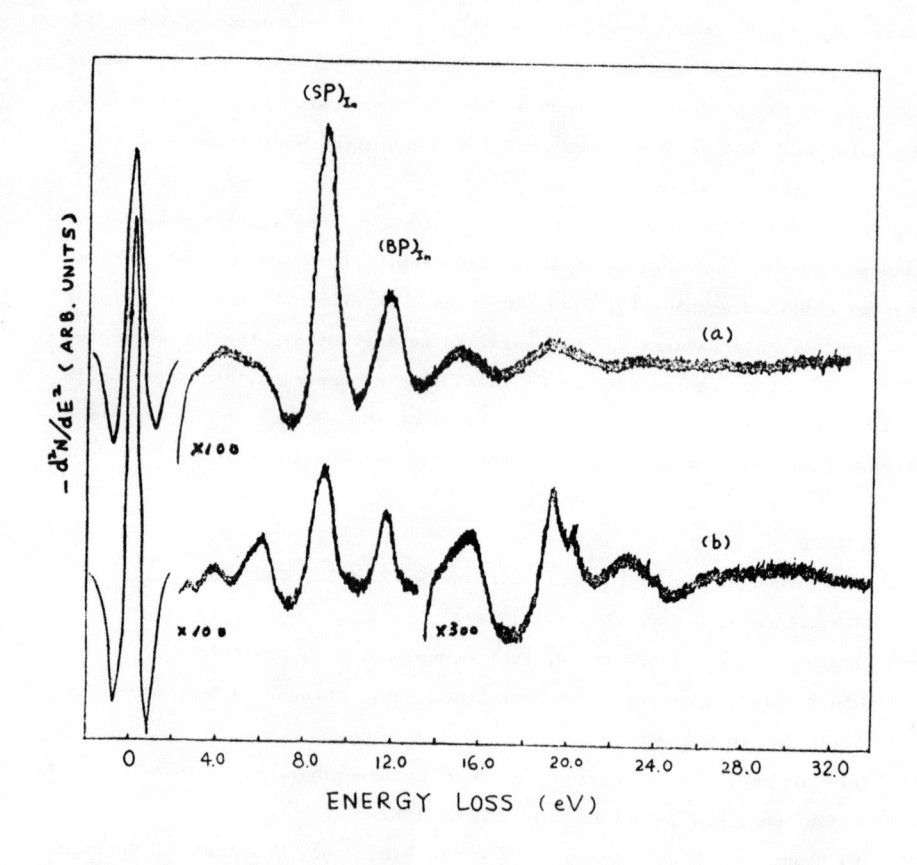

Fig.1 ELS of InP(100) surface

(a) after 5 keV argon ion sputtering;

(b) 1 hour annealing at 300°C.

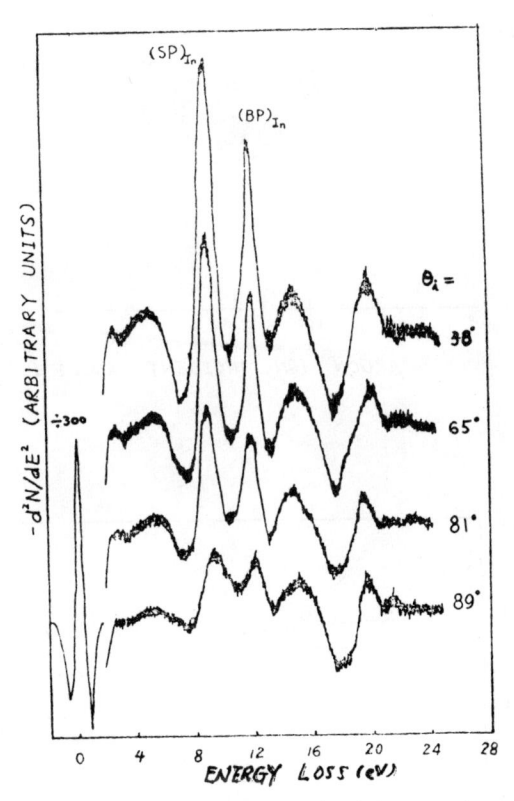

Fig.2 ELS of InP(111) obtained from various sputtering angles θ_i of argon ion.

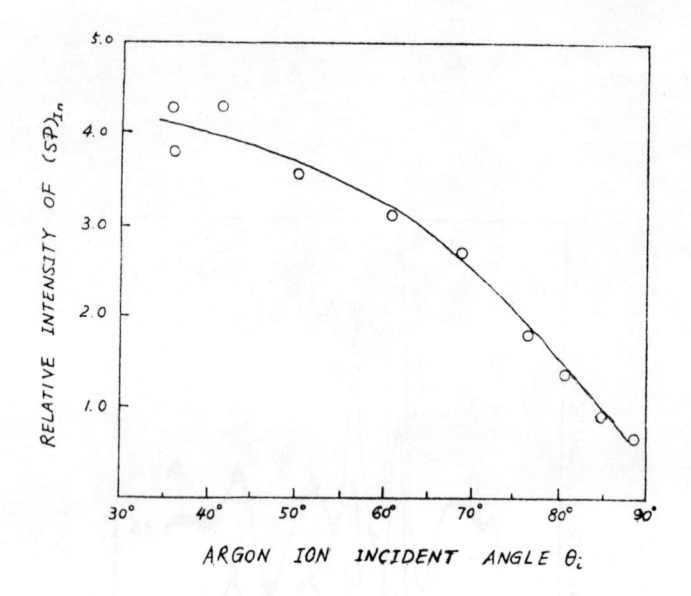

Fig.3 Changes of relative intensity of (SP)In to the incident angles θi of

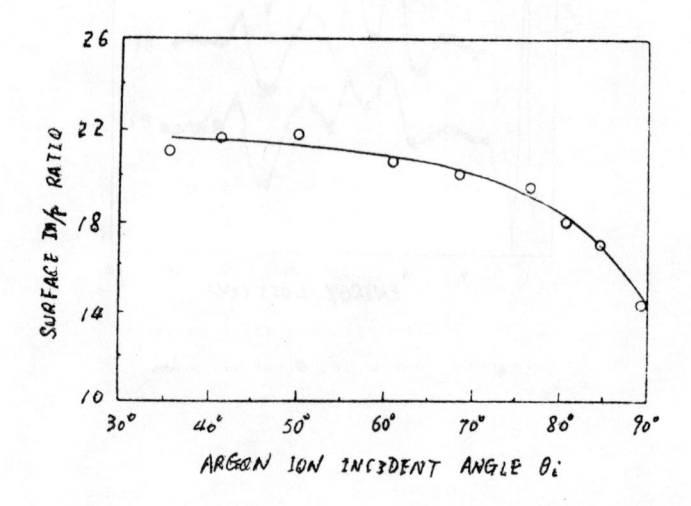

Fig.4 Change of surface In/P ratio to incident angles θi of argon ion.

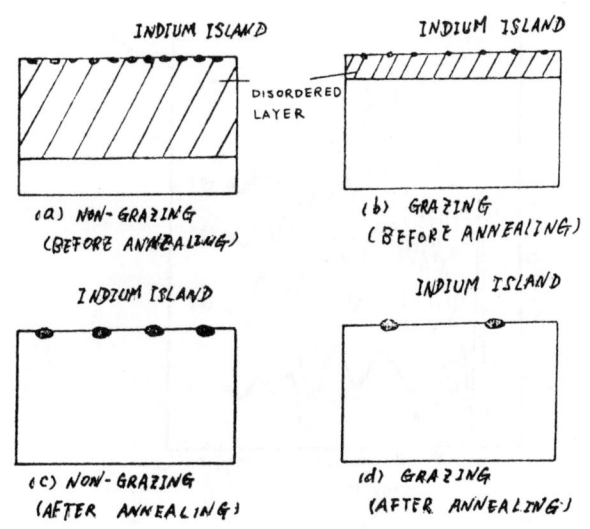

Fig.5 Schematic drawings for the change of In islands to argon ion sputtering and annealing

(a) nongrazing (before annealing);

(b) grazing (before annealing);

(c) nongrazing (after annealing);

(d) grazing (after annealing).

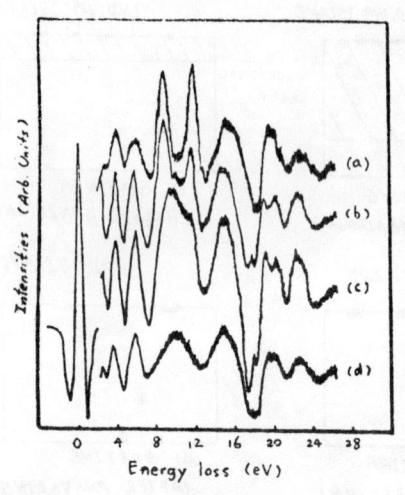

Fig.6 Gradual diminishing of In islands as shown by ELS from P evaporating and after annealing

(a) Befor P evaporating;

(b) After first cycle of evaporating and annealing;

(c) After second cycle of evaporating and annealing;

(d) Atter several cysles of evaporating and annealing.

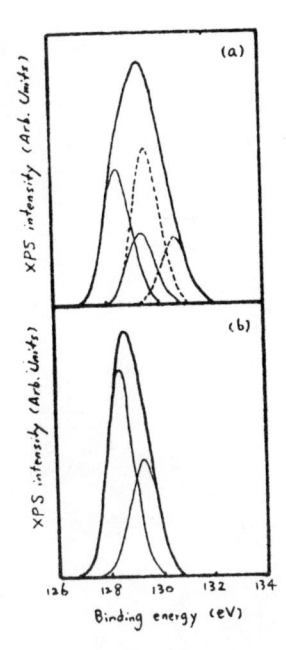

Fig.7 P2p XPS peaks

    (a) Before annealing;

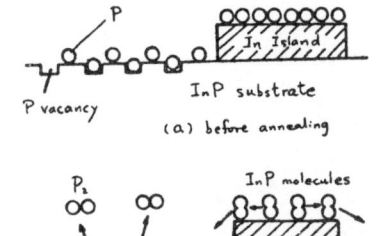

    (b) After annealing.

Fig.8 A schematic process of In island elimination

    (a) Before annealing;

    (b) After annealing.

PERFORMANCE COMPARISON OF CAPACITANCE DISCHARGE PLASMA
(SPARK), INDUCTIVELY COUPLED PLASMA (ICP), AND X-RAY
FLUORESCENCE (XRF) SPECTROMETERS IN THE ANALYSIS OF
AUSTENITIC STAINLESS STEELS

CLAUDE R. MOUNT
Materials Evaluation Laboratory, Inc., 17695 Perkins Road, Baton Rouge, LA
70810

ABSTRACT

It is well known that the compositional analysis of austenitic stainless steels can be difficult for emission spectrometric techniques. How the problems are dealt with by three systems is described while presenting a case history performance comparison. The analysis results reported were collected using routine procedures with no special efforts toward optimization. The systems utilized represent varied computational schemes, as well as, different equipment manufacturers. Included is a discussion of the basic characteristics and merits of each system and its respective methodology.

INTRODUCTION

Materials characterization can invoke many combinations of instrumental techniques and application needs. An interesting case was presented by an industrial client's requirement to monitor the delta ferrite content of austenitic stainless steel weldments. This was of critical importance since excessive "free ferrite" could lead to catastrophic failure during operation. A synopsis of a portion of the analytic process used to resolve this problem is presented.

The initiating request was that the ferrite number, a quantification of the delta ferrite content, be determined for a group of sample weldments. These would be utilized as a set of calibration standards for a commercially available electromagnetic testing device. The primary intent was to evaluate the ferrite content of production weldments in a rapid, portable, nondestructive and economical manner.

TECHNIQUE

An acceptable methodology for determining the ferrite number was established by the American Welding Society. The essential element of that approach is an accurate chemical analysis of the deposited weld metal. To achieve this, it was elected to use the averaged results of the analytic techniques to be discussed.

Three spectrometric systems were available. They were utilized in their normal configuration for routine, bulk chemical analysis. Specifically, they were: Capacitance Discharge Plasma Optical Emission (Spark); Inductively Coupled Argon Plasma Optical Emission (ICAP); and, Energy Dispersive X-ray Fluorescence Emission (EDS/XRF) spectrometers.

The Spark system is a one meter focal curve, vacuum path polychromator built by Labtest Equipment Company. It can analyze conductive solid samples and uses the point-to-plane technique. Either tungsten or carbon counter-electrodes can be selected using argon or nitrogen shielding gas. The plasma is controlled by a dual voltage "Transource" power supply which has tuneable resistance, capacitance and inductance. The system was configured to simultaneously analyze for a maximum of twenty-five elements to characterize a wide variety of engineering materials.

    Sample preparation typically requires nothing more than medium grit sanding to achieve an oxide-free, flat surface. The unknown is then analyzed by a comparison of its response to the response of known materials. A form of "response memory" is achieved through polynomial approximations of the system response to a collection of known materials. Some inter-element, or matrix, effects are compensated for by using known materials which are similar to an anticipated class of unknowns when computing the necessary polynomial coefficients. Day-to-day stability is maintained by relating system response to a single, known reference material.

    The ICAP system is a three-quarter meter focal curve, air path polychromator built by Jarrell-Ash. It can analyze liquid samples with either aqueous or organic solvent bases. The plasma is maintained and controlled by a variable output radio frequency power supply with an auto-tuning feature. The system was configured to simultaneously analyze for a maximum of thirty-five elements.

    The most common sample preparation entails a simple digestion in an inorganic acid. Also being a comparator, unknowns are qualified by referencing the response of known materials. This is achieved by establishing linear relations for elements of interest derived from high and low concentration standard solutions. Some inter-element, or matrix, effects can be compensated for by using standard solutions with compositions similar to the unknown materials. Day-to-day stability can be assured by simply using "fresh" standard solutions.

    The EDS/XRF system was built by EG&G Ortec and represents a different technique than the previous two systems. It operates either at atmospheric pressure or under vacuum and utilizes a dual anode (tungsten/rhodium) X-ray tube. The tube voltage and current are tuneable and post filters can be selected. The silicon, lithium drifted detector is coupled to a 1024 channel multichannel analyzer for spectrum acquisition.

    Some information from virtually any sample form can be obtained. However, the preparation of bulk solid materials usually entails polishing to a 400 grit finish. The actual analysis of routine samples strongly parallels the scheme used by the Spark system.

    The calibration of each of the systems is achieved through the analysis of a known reference material to verify acceptable system response. Normally, for routine bulk analysis response, variation within a 95 percent confidence range is expected. The calibration process is essential to maintain performance traceability to recognized primary reference material sources.

RESULTS

    For the problem being considered, each system was verified to be operating properly. There is no intent to vigorously examine how the results were achieved, rather they will be used to qualitatively discuss performance characteristics. As points of reference, the measured weight percent concentrations of four elements of interest are presented in tabular form.

TABLE 1. CHROMIUM DETERMINATION

SAMPLE	SPARK	ICAP	EDS/XRF	AVERAGE
1A	17.9	18.8	18.5	18.4
2A	19.2	20.0	20.5	19.9
3A	19.2	19.7	20.1	19.7
4A	18.6	19.2	18.7	18.8

TABLE 2. NICKEL DETERMINATION

SAMPLE	SPARK	ICAP	EDS/XRF	AVERAGE
1A	9.81	10.10	9.80	9.90
2A	9.81	9.90	9.12	9.61
3A	10.30	10.60	9.93	10.30
4A	9.81	10.30	9.77	9.96

It can be seen that there is relatively good agreement for the elements chromium and nickel. Generally, all three systems perform quite well for the measurement of these elements over their respective ranges of interest in engineering materials. The Spark system does have a weakness at the upper concentration range where self-absorption tends to limit its sensitivity. This is not a problem with the EDS/XRF system; and, for the ICAP system can be overcome through minor procedural changes.

TABLE 3. MANGANESE DETERMINATION

SAMPLE	SPARK	ICAP	EDS/XRF	AVERAGE
1A	0.84	0.68	0.57	0.70
2A	2.10	2.28	1.97	2.12
3A	1.98	2.01	1.77	1.92
4A	0.83	0.71	0.60	0.71

The measurement of manganese content shows a slightly wider range of variation in the results. In an austenitic stainless steel matrix, manganese has strong spectral interference from other major elements which are present. Although methods are available within each system for coping with interelement interferences, none of the systems were fully utilizing this feature.

TABLE 4. SILICON DETERMINATION

SAMPLE	SPARK	ICAP	EDS/XRF	AVERAGE
1A	1.24	0.88	0.87	1.00
2A	0.88	0.47	0.54	0.63
3A	0.80	0.42	0.47	0.56
4A	1.15	0.90	0.86	0.97

Examining the results for a lower atomic number element, silicon, the range of variation increases even more. In the application discussed, the ICAP system has proven to be more sensitive. Difficulties with matrix effects reduce the capabilities of Spark and EDS/XRF for quantifying silicon content under the given circumstances.

CONCLUSIONS

Each of the systems discussed was considered adequate for applications requiring routine analysis of bulk materials. There are, nonetheless, strengths and weaknesses for the specific techniques. As such, no one system is "the best." In combination, however, they present a powerful tool to support a wide range of analytic questions.

REFERENCES

Materials Characterization, Metals Handbook, Volume 10, Ninth Edition, American Society for Metals, 1986.

Methods for Emission Spectrochemical Analysis, 7th Edition, American Society for Testing and Materials, 1982.

Quantitative X-Ray Spectrometry, Jenkins, Gould & Gedcke, Marcel Dekker, Inc., 1981.

ELECTRICAL AND DIELECTRIC PROPERTIES OF NaYF$_4$ THIN FILMS

NARASIMHA REDDY KATTA
Department of Physics
Osmania University
Hyderabad 500007,INDIA

ABSTRACT:
Thin film capacitors of Sodium yttrium fluoride were fabricated by thermal evaporation.. The current voltage characteristics and dielectric properties of these evaporated films of NaYF$_4$ have been extensively studied at different temperatures and frequencies respectively. It is observed that NaYF$_4$ material exhibits some what unusual electrical and dielectric properties in thin film forms. At low fields the current increases linearly with square root of field and at high electrical fields (above 3x10^2 volts/min^2) the current shoots up suddenly It is also observed that around this transition field the electrical breakdown occurs in the film. The conduction behaviour was found to be predominently because of pool-frenkel. The capacitance of these films was found to depend upon both frequency and temperatures. The plot of Tan vs Temperature exhibited pronounced maxima at 260,290,340K. The dielectric relaxation proccess in these films was explained on the basi of dipolar orientations.

1.INTRODUCTION:

Sodium yttrium fluoride is a material that exhibits some what unusual electrical and dielectric properties both in crystalline and Thin film forms. Fluorite structured compounds like NaYF$_4$ forms a rare example containing vacancies and interstitials in very high concentrations as described in our earlier paper (1). By systematically monitering the formation of substitutional and interstitial defects in fluorite structure compounds, there exists a probability of controlling a whole series of properties for applications such as solid state laseres and electrolytic Batteries. Therefore in search of similar compounds some attention has been given to NaYF$_4$ compound. The knowledge of the effec of the nature of conducting species in thèss system is of vital signifi cance. This paper deals with the steady state electrical conductivity and dielectric nature of NaYF$_4$ thin films.

2.EXPERIMENTAL

NaYF$_4$ compound was prepared in our laboratory as described earlier

(2). Vacuum deposited layers of NaYF$_4$ were prepared by conventional thermal evaporation technique, on glass substrates. Thick vacuum deposited films of Aluminium electrodes served as ohmic contacts with NaYF$_4$ layers. The conduction current with temperature was measured as a function of temperature. The change of current with temperature was taken as the temperature dependence of conductivity. Dielectric measurements have been made using a LCR bridge. These measurements were carried out in the frequency range 100Hz to 16 KHz and in the temperature range from LNT to above room temperature.

3. RESULTS AND DISCUSSION:

3.1. Electrical conductivity:

The current-voltage characteristics of NaYF$_4$ film of 423nm thickness is shown in Figure 1.at various temperatures. The I-V characteristics curve shows two distinct regions as discussed in our earlier paper (1). At low fields the slopes of I-V curve exhibits ohmic behaviour. However at high fields these curves indicate non ohmic behaviour

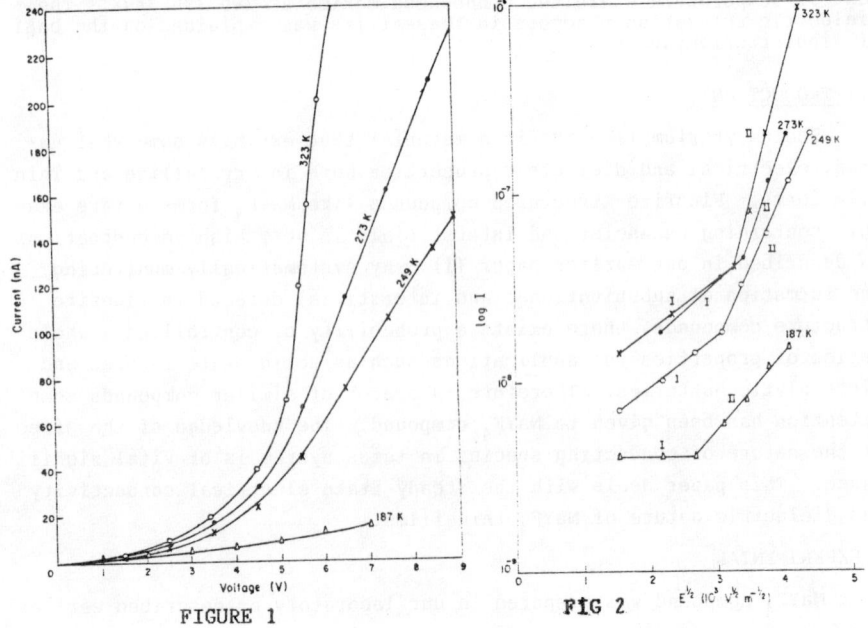

FIGURE 1

FIG 2

Electrical tunneling in these films has been ruled out in view of the film thickness (423nm). The current density in the spacecharge limited conduction region is not followed (3). Hence the posibility of space charge limited conduction is also ruled out for $NaYF_4$ films.

To explain the conduction mechanism in these films variation of logarithmic current with sqare root of field has been ploted in Fig 2. The slopes of the straight lines in Figure 2 (Region II) are estimated and plotted against e/kT in Figure 3. The curves shown in the Figure 2 exhibit two distinct regions. The observed high field region(II) behaviour may be undestood interms of either schottky or Pool-Frenkel mechanisms. At high electric field $(E > 10^6 \, VM^{-1})$ several dielectrical and semiconducting thin films (4-6) exhibit current-voltage relation ship characteristic of the form (7,8)

$$I = I_0 \exp(e\beta E^{\frac{1}{2}}/KT) \quad \ldots 1$$

where E=V/d and β is a constant given by

$$\beta = (e/a\pi\varepsilon_0\varepsilon')^{\frac{1}{2}} \quad \ldots 2$$

where e is the electronic charge, ε_0 is the permitivity of the free space and ε' is high frequency dielectric constant.

The value of a=1 is adopted for Pool-Frenkel mechanism and the value of a=4 is adopted for schottkey mechanism. The Theoritical values of 'β' obtained from equation 2 are

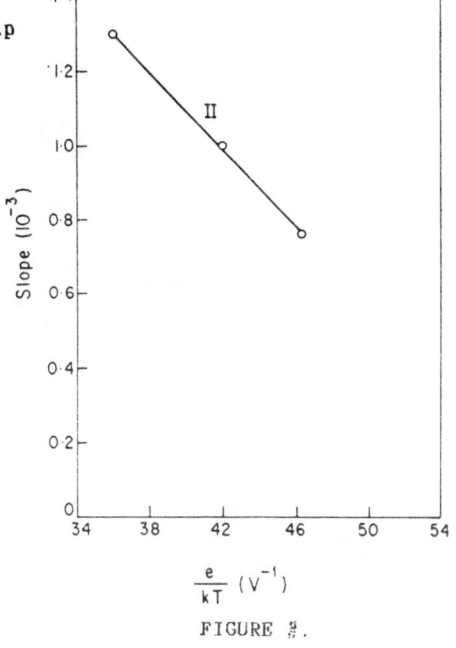

$$\beta_S = 1.80 \times 10^{-5} (mV)^{\frac{1}{2}}$$

$$\beta_{pF} = 3.60 \times 10^{-5} (mV)^{\frac{1}{2}}$$

FIGURE 3.

From the slope of the straight line in Figure 3 the experimental values of β (7) can be obtained using the relation

$$\beta_{exp} = (d\log I/dE^{\frac{1}{2}})(KT/e^{0.478}) \quad \ldots 3$$

The experimental value of ß evaluated from equation 3 is

$$\beta_{exp} = 5.19 \times 10^{-15} (mV)^{\frac{1}{2}}$$

since the value of β_{exp} is close to that of β_{PF} than β_{S}, the conduction mechnism seems to be predominently that of Pool-Frenkel emission in these films.

3.2. Dielectric properties:

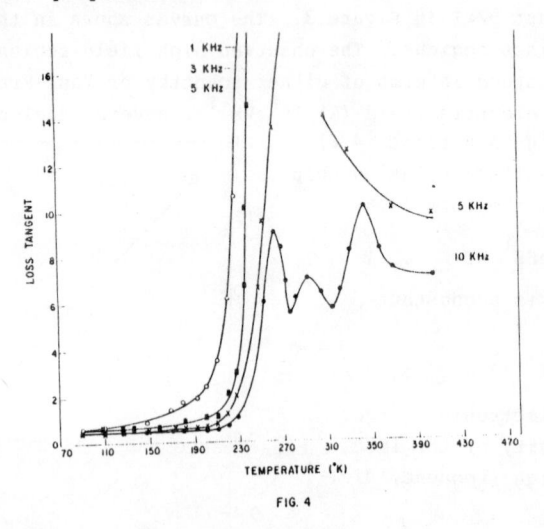

FIG. 4

Figure 4 shows the variation of Tanδ with temperature at different frequencies for the film of thickness 423nm. The figure shows that the loss factor is independent of temperature upto 200k. Above 200K the loss factor increases with increase in temperature of the structure. The loss factor excceeds the instrument limit of observation from the temperature 230K for 1KHz, 240K for 3KHz and over a range of 270K to 310K for a frequency of 10KHz. The loss factor with temperature plot shows three peaks at 260K, 290K and 342K which may be due to some type of relaxation proccess. Similar behaviour was obserbed in various other rare earth fluoride films (9-11). These peaks may be due to the dielectric relaxation phenomenon arising from dipoles. In the present study the dipoles are probably formed at high temperatures owing to the concentration of defects in these films. The high temperature region given in Figure 4 may be related the formation and decay of dipoles. Figures 5and 6 show that the variation of capacitance and loss factor with frequency at different temperatures in the temperature range 77-400K. From these figure it is found that at any temperature, variation of capacitance and loss factor

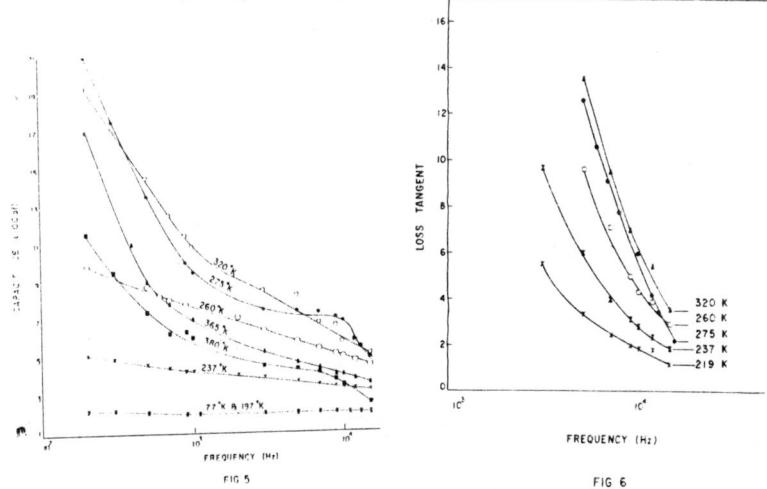

FIG 5

FIG 6

with frequency is same that is the capacitance and loss factor increases with decrease in frequency.

ACKNOLEDGEMENTS:

The author thanks Mr.M.V.Ramana Reddy research scholar for taking some experimental data.

REFERENCES:

1. M.V.Ramana Reddy and K.Narasimha Reddy Phys.Stat.Sol(a) 95,1986,K193
2. K.Narasimha Reddy,M.A.H.Shareef and N.Pandaraiah J.Mat.Sci.Lett 2,1983,83.
3. A.K.Jonscher, Thin solid films 14,1980,189.
4. M.Sheart Phys.stat.sol 23,1967,595.
5. S.M.Sze   J.Appl.Phys 38,1967,2951.
6. P.A.Walley Thin solid films 2,1968, 327
7. A.K.Jonscher Thin solid films 1,1967,213
8. A.Servini and A.K.Jonscher Thin solid films 3,1969,341.
9. M.C.Lancaster J.Phys.D 5,1972,1133.
10. T.M.Lingam,M.R.Krishnan and C.B.Subramanyam Thin solid films 59,1979 221
11. A.Goswamy and P.Amit Goswamy Thin solid films 16,1973,195.

DETERMINATION OF TRACES OF SULPHATE INDIRECTLY
BY ATOMIC ABSORPTION SPECTROPHOTOMETER

D.C. Parashar & A.K. Sarkar,
Analytical Chemistry Division
National Physical Laboratory
New Delhi - 110012 (India).

SUMMARY

A sensitive method has been developed for the determination of sulphate in parts per million level indirectly by atomic absorption spectrophotometer (A.A.S.). The method is based on precipitation of sulphate with a known amount of lead as lead sulphate in ethanol and subsequent determination of the remaining lead content in the supernatant liquid, after centrifuging the precipitated lead sulphate by A.A.S. The concentration of sulphate is calculated by determining the concentration of lead as the ratio of sulphate to lead is found to be linear. The method is successfully employed in estimating sulphate in different samples of water including those from power plants and antartica glaciar. The results are discussed in this paper.

INTRODUCTION

A systematic study of determining non-metals by atomic absorption spectrophotometer (AAS) is in progress in our laboratory [1,2,3,4]. Method for the determination of sulphate in biological samples has been reported earlier[5] where, after converting sulphur into sulphate, it is precipitated with barium as barium sulphate and measuring the barium content by AAS. The indirect colourimetric methods for sulphate with barium chloroanilate are having their own limitations viz.,the methods are laborious and subject to interference by many ions [6,7,8] Ion selective electrode method [9] has been used for the determination of sulphate in antartic snow melts. The method cannot be applied without preconcentration and suffers precision.

In the present investigation, an indirect atomic absorption spectrophotometric method has been developed in which sulphate is precipitated as lead sulphate with excess of lead and remaining lead in the solution is measured by AAS after removal of lead sulphate by centrifugation. This method has been applied by the authors for the determination of sulphate in high purity water used in power plants like D.M. water, boiler water, feed water, saturated steam etc. and in antartica glaciar melts.

EXPERIMENTAL

Apparatus

All the experiments are performed on a Pye Unicam SP 1900 double beam atomic absorption spectrophotometer equipped with digital read out indicator system. Experimental parameters selected are: wave length 217.00 nm; lamp current 5 mA; slit width 0.15 - 0.20 nm; air acetylene burner (10 cm width); oxidant flow rate 5.0 litres/minute; acetylene flow rate 1.0 litre/minute.

Reagents

All the reagents and chemicals used are of A.R. or G.R. grade until mentioned otherwise.

Standard lead (II) solution: Dissolve 1.0 gram of lead metal in 50 ml of 2M nitric acid (E.L. grade). Dilute to 1 litre in a volumetric flask with double distilled water to get the final concentration of 1 mg/ml of lead. Stored in a polythene bottle and further diluted to get the working solution for atomic absorption spectrophotometry.

Standard sulphate solution: Dissolve anhydrous sodium sulphate (1.48 gram) in double distilled water and make the volume to 1000 ml to get the final concentration of 1 mg/ml of sulphate.

PROCEDURE

Known quantity of sulphate solution (1 to 10 ppm) is taken in 10 ml of volumetric flask. To this solution, known quantity of excess lead solution (10 ppm), 1 drop of nitric acid, 2 to 5 ml of absolute ethanol are added in each flask and volume made to 10 ml with water and allowed to stand for half an hour. The contents of the flask are transferred in a centrifuge tube and centrifuged for 5 minutes. Standard lead solutions are aspirated to get standard graph and then the supernatant liquid of the sulphate precipitate aspirated to get absorbance values and concentration of sulphate is evaluated. The decrease in the absorbance value of known amount of lead gives the amount of sulphate present in the solution. The decrease in the absorbance value is found to be linear and Beer's law is obeyed upto 10 ppm of lead/sulphate.

Interfering ions

Interference of a number of ions has been studied with 5 ppm of sulphate. It has been found that 100 ppm of chloride, nitrate, nitrite, perchlorate, carbonate, sodium, potassium, silver, ammonium, calcium, magnesium, copper, nickel, manganese, iron, aluminium do not interfere. The permissible limits of other interfering ions are: chromate: 5 ppm, bromide: 10 ppm, iodide: 10 ppm, barium: 10 ppm, strontium: 10 ppm.

Determination of sulphate in Power Plant Water

Definite amount (100 ml and 200 ml of different water samples are taken, 1 to 2 drops of nitric acid (E.L. grade) are added in each solution and evaporated in a pyrex beaker till it reduces 2-3 ml. The solutions are transferred in 10 ml volumetric flask and 1 ml of standard lead solution (10 ppm) is added in each flask. 2 to 5 ml of ethanol are added in each flask. The volume of the flask is made upto the mark with double distilled water and thoroughly shaked. It is allowed to stand for half an hour, centrifuged and the lead content is determined in the supernatant liquid by atomic absorption spectrophotometer. The results are given in table I.

Determination of sulphate in Antartica Glaciar Melts

1 ml of the glaciar melt is taken in a 10 ml volumetric flask. One drop of nitric acid (E.L. grade) and 2 ml of standard lead solution (10 ppm) are added in each sample with 2 to 5 ml of ethanol and volume made to 10 ml with double distilled water. The lead content is determined in the supernatant liquid after following the procedure given earlier. The results are given in table I.

Table I

Sample No.	Sample	Sulphate added (ppm)	Sulphate Expected (ppm)	Sulphate Obtained (ppm)	Difference (ppm)
1.	Standard Sulphate	0.1	0.1	0.10 / 0.11	0 / +0.01
2.	Standard Sulphate	0.5	0.5	0.51 / 0.50	+0.01 / 0
3.	Standard Sulphate	1.0	1.0	1.0 / 1.0	0 / 0
4.	D.M. Water	1.0	1.07	0.07 / 1.07	0
5.	Feed Water	1.0	1.08	0.08 / 1.07	−0.01
6.	Condensate	1.0	1.03	0.03 / 1.03	0
7.	Antartica Glaciar melt*	1.0	1.5	0.5 / 1.55	+0.05
8.	Antartica Glaciar melt*	1.0	1.6	0.6 / 1.6	0

*Samples received from the Department of Oceanography, Government of India.

Result & Discussion

The insolubility of lead sulphate in ethanol is well established by the earlier author [10]. The results given in table I shows that the sulphate content is determined in different types of water samples and the relative mean deviation is 0.8%. Different water samples with standard addition of sulphate are analysed. The sensitivity of this method is 0.1 ppm based on 1% adsorption and Beer's law is obeyed upto 10 ppm of lead sulphate. The method is specific and rapid. Large number of samples from different power plants have been analysed for their sulphate content as it is one of the principal scale forming ingredient and monitoring of this anion is of extreme importance.

Impurities including 5 to 10 ppm of chromate, bromide and iodide and 100 ppm of chloride, nitrate, nitrite, perchlorate, carbonate, sodium, potassium, silver, ammonium, calcium, magnesium, copper, nickel, manganese, iron, aluminium do not interfere.

Acknowledgement

The authors are grateful to Dr. S.K.Joshi Director National Physical Laboratory for his encouragements and Dr. K.Lal, Deputy Director and Head Material Characterization Division, National Physical Laboratory for his keen interest in this work.

REFERENCES

1. P.K. Gupta and A.K. Sarkar, Indian J.Chem. <u>17A</u>, (1979) 317.

2. A.K.Sarkar and D.C.Parashar, Recent Trend on Microanalytical Chemistry, 50, (1982).

3. A.K. Sarkar, Proceedings of the Golden Jubilee Celebration of Indian Chemical Soc., Anal-1, (1984).

4. D.C. Parashar and A.K. Sarkar, Analytical Letters, <u>17(A11)</u>, (1984) 1269.

5. D.A. Roe, P.S. Miller and L.Lutwar, Anal.Biochem, <u>15</u>(1966) 313.

6. R.J.Bertolacini and J.E.Barney, Anal.Chem <u>29</u>, (1957) 281
   <u>30</u>, (1958) 202

7. R.M.Carlson, R.A. Rosell and W.Vallegos, Anal.Chem., <u>39</u>, (1967) 688.

8. H.N.S. Schafer, Anal.Chem. <u>39</u>, (1967) 1719.

9. R.Delmas and C. Bourton, Atmospheric Environment, <u>12</u>, (1978) 723.

10. T.W.Guilbert, I.M. Kolthoff and P.J. Elving, Treatise on Analytical Chemistry Part II, Vol. 6, Interscience, N.York, 1964, p.111.

EXPERIMENTAL TECHNIQUES FOR CHARACTERISATION OF FAST ION CONDUCTING MATERIALS

P. SATHYA SAINATH PRASAD and S. RADHAKRISHNA
Department of Physics
Indian Institute of Technology
MADRAS - 600 036 - INDIA

ABSTRACT

Silver based quarternary fast ion conducting polycrystalline and glasses have been prepared by the open air crucible melting method. A preliminary investigation of the quarternary system revealed that the electrolyte material when used in a electrochemical cell has a high capacity of 15 mAH for the 0.66 AgI - 0.22 Ag_2O - 0.11 (0.8 V_2O_5 + 0.2 P_2O_5) composition. On a further study to establish the highest conducting composition, the composition with 0.70 AgI - 0.20 Ag_2O - 0.08 V_2O_5 - 0.02 P_2O_5 was observed to have an ionic conductivity of 0.082 $Ohm^{-1}cm^{-1}$ and an electronic conductivity of 2.6×10^{-8} $Ohm^{-1}cm^{-1}$. The glass forming region has been established and the glass transition temperature was determined from the DTA. The infrared, Raman and electron paramagnetic resonance spectroscopic techniques were used to analyse and confirm the presence of ionic clusters in the glassy electrolyte thus highlighting the structural properties of the solid electrolyte, The existence of ionic clusters in the glass was assumed to be responsible for enhanced ionic conduction and the thermal and electrochemical properties of the system were studied to explore the effect of mixing two different glass formers. The T_g obtained from DTA technique was well correlated with the obtained glass transition temperature from the $\log \sigma T$ vs $10^3/T$ plot. The resistance-time and the resistance-frequency characteristics of the pellets were investigated to study the behaviour of the material as an electrolyte in the solid state batteries.

1 INTRODUCTION

In a search for silver based superionic conducting materials to be used as electrolytes in Solid State Batteries and analog memory device efforts are being channelised to develop vitreous Electrolytes with an ionic conductivity of 0.1 $Ohm^{-1}Cm^{-1}$ at room temperature. The silver based ternary systems with Ag_2O as a glass modifier and V_2O_5 & P_2O_5 as glass formers were already well studied and reported (1-3). On the assumption that a mixture of two glass formers will be a better glass fromer with an enhancement in the Ionic Conduction due to the increased random glassy matrix and the formation of ionic clusters, a new quarternary system with V_2O_5 & P_2O_5 as glass formers was studied and a preliminary investigative report is communicated (4). It was found that the glass with 66% AgI - 22% Ag_2O - 11% (.8 V_2O_5 + 0.2 V_2O_5) composition cooded as 66VP82G has an ionic conductivity of 4.2×10^{-2} $(Ohm\ Cm)^{-1}$ and an Electronic Conductivity 1.8×10^{-10} $Ohm^{-1}Cm^{-1}$ at room temperature. In the present work with an aim to enhance the ionic conductivity further the concentration of AgI was changed from 40% to 85% in steps of 5% and the glass forming region for the glass modifier to glass former ratio of 2 : 1 keeping the V_2O_5 - P_2O_5 ratio as 0.8 : 0.2 was studied. It was observed that glasses are formed for the range $45 \leqslant x \leqslant 85$ where x = AgI%. From the above studies the glass with the composition of 70% AgI - 20% Ag_2O - 10% (0.8 V_2O_5 + 0.2 P_2O_5) cooded as

70VP 82G was found to have maximum ionic conductivity $.82 \times 10^{-1}$ ohm^{-1}cm^{-1} which is very near to the required value of 0.1 ohm^{-1}cm^{-1}. To use the material as an electrolyte in a solid state battery, it was characterised by optical, thermal and electrical analytical techniques that are frequently used for material characterisation.

2. METHOD OF PREPARATION

The three methods of preparation of glasses were explained in detail in a previous communication (4). It was shown that the open air crucible method had an advantage to offer than the other two methods of preparation as described therein. In this method of preparation the bulk glass formation is easy which has an advantage over the pulverised glass, like less variation of resistance with time and frequency. Moreover the activation energy was comparitively lower than that of the pulverised glass, with the glass transition temperature remaining the same. Hence the open air crucible method was adopted in the preparation of polycrystalline and vitreous samples 70VP82 from the same melt. The appropriate amounts of respective compounds AgI, Ag_2O, V_2O_5 & P_2O_5 corresponding to 70% AgI, 20% Ag_2O, 10% (0.8 V_2O_5 + 0.2 P_2O_5) were taken in a quartz crucible and heated in an electric muffle furnace at a temperature of 600°C for 3 hours. One half of the melt is quenched in a liquid nitrogen cooled stainless steel plate, and the other half was allowed to cool to room temperature. Since the vitreous and polycrystalline samples were taken from the same melt with the same composition, a comparitive study could be undertaken. The formation of vitreous and polycrystalline compounds was confirmed by the x-ray diffractogram shown in Fig.(1). In Fig (1) the diffractogram for AgI, Ag_2O, V_2O_5 and P_2O_5 was shown along with the polycrystalline 70VP82 to confirm that a completely new compound was formed with slight traces of pure AgI.

The glassy and polycrystalline materials were pulverised separately and weights of 3gms and 2gms were pressed into pellets of 1.3cms diameter for the ionic conductivity and electronic conductivity experiments respectively. A uniform pressure was applied and for a better electrode-electrolyte contact, silver powder mixed with electrolyte (2:1 ratio) was used as electrode material. The samples 70VP82G and 70VP82P were dispersed in KBr and polyethylene. Thin pellet samples were made for the IR and far IR spectral measurements respectively. A 1.000 gm weight of the sample was taken for EPR measurements. A small glass piece was taken from the bulk glass was used for recording the Raman Spectrum at room temperature.

3. EXPERIMENTAL TECHNIQUE DETAILS

In the process of characterising Fast ionic conducting Silver based Vanadium phosphate glassy electrolyte for the application in solid state battery, the diffractogram, IR , Far-IR, Raman, E.P.R., D.T.A., Ionic conductivity and electronic conductivity studies were undertaken. A brief description of all the instruments in the above mentioned techniques is presented below.

X-ray diffractograms were recorded in a Philips pw-1130 x-ray generator with a vertical diffractometer model pw-1050 and a diffractometer control unit pw-1710. The radiation used was CuK_α with a nickel filter for 35KV and 25mA. The d-values along with the 2θ and intensities were printed out from the data station. Hence it had been very easy to compare the polycrystalline material formed and the individual AgI, Ag_2O, V_2O_5, P_2O_5 compounds so as to confirm the new compound formation. The differential thermal analysis were made using a Du point series 900 differential scanning calorimeter. The optical spectra in the infrared region 4000 cm^{-1} - 400 cm^{-1} for both the polycrystalline and amorphous material were recorded on a perkin Elmer 839 I.R. spectrophotometer with a perkin Elmer 810 data station to store the data. The Far-IR transmission spectra were recorded on a polytec 30 Fourier Spectrometer with a Far I.R. Michelson interferometer in the spectral region 500 cm^{-1} - 50 cm^{-1}. The Far I.R. spectrometer works on line with a small (4k memory) digital computer to provide an instantaneous 'real' time spectral display.

Glassy samples were cut into small parellopipeds of about 5x5x3 mm cube size and their faces carefully polished. The Raman spectrum was recorded at room temperature with a CARY 82 spectrometer coupled with a spectra physics Krypton Ion Laser. The 6471 A° line was used at an instantaneous power of 0.2 w. The spectra were recorded in the region 1200 cm^{-1} - 20cm^{-1} with a resolution of 5 cm^{-1}.

In the present study a VARION E4 x-band spectrometer operating at 9 GHz with a variable temperature accessory attached to it for recording the spectra in the temperature range 90k - 500k was used. The basic principles of the spectrometer have been described in detailed by INGRAM and POOLE (5-6).

The ionic conductivities of 70VP82P and 70VP82G pellets were measured using General Radio 1650 B, impedance bridge with an internal oscillator at 1KHz and a null detector. The cell configuration for the ionic conductivity measurements was as shown below

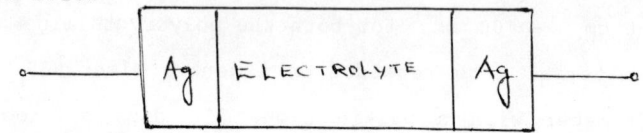

o——————[Ag | ELECTROLYTE | Ag]——————o

The pressed pellets were mounted in a glass cryostat, kept at a pressure of 10^{-2} torr. The temperature at the sample was measured with an Iron - constantan thermocouple using a d.c. millivoltmeter that directly reads the temperature in degrees centigrade. The variation of temperature is achieved with a furnace type heater mounted around the outer jacket of the cryostat.

To measure the Electronic conductivity for both 70VP82G and 70VP82P pellets, various stable d.c. potentials (ranging from 10 mv to 120 mv) were applied with the pellet loaded in a glass cryostat at 10^{-2} torr pressure. The configuration of the cell for the d.c. polarisation technique measurements was as shown below.

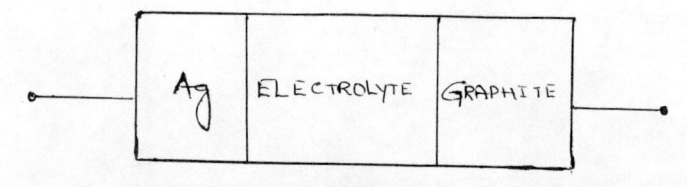

o——————[Ag | ELECTROLYTE | GRAPHITE]——————o

For the various applied potentials, the currents corresponding to the steady state were recorded using a Kiethley 610C Electrometer in series with the specimen. All the measurements were carried out only at room temperature $32^{\circ}C$.

4. SPECTROSCOPIC CHARACTERISATION

Investigations of the interaction of light with glasses are closely connected with the discovery and development of spectroscopic methods. Spectroscopic techniques were used in glasses to identify the structural units and their variations with composition and thermal history. Since diffraction methods are at a disadvantage in glasses due to the absence of long range order, the elucidation of the structural elements and symmetry in glasses are a result of the experiments involving the transmission, reflection, refraction and scattering of light in the $50 cm^{-1} - 50000 cm^{-1}$ range. The experiments namely UV-Visible-NIR, IR-FIR, Raman and EPR constitute the spectroscopic techniques available for the analysis of the amorphous materials. Here the vibrational frequencies of the constituents of the glasses can be observed using IR-FIR and Raman, whereas the UV-VIS-NIR techniques are useful in the band gap determination of the semiconducting materials. Hence spectroscopic techniques can be used as analytical tools to study the atomic and molecular arrangement, electronic structures, defect states and the relaxation processes in the glasses.

(A) Infrared Studies: To investigate the ionic clusters

and the condensed macro molecular structure units which are assumed to be responsible for the enhanced ionic conduction in fast ionic conducting glassy electrolyte, the vibrational spectra of polycrystalline and vitreous 70VP82 compound were recorded in the range 4000 cm^{-1} - 400 cm^{-1}. The I.R. spectra of both 70VP82G and 70VP82P are shown in Fig (2). The I.R. spectra of polycrystalline 70VP82 in Fig.(2) contains well defined bands with peaks at 1600 cm^{-1}, 1380 cm^{-1}, 1350 cm^{-1}, 1080 cm^{-1}, 990 cm^{-1}, 850 cm^{-1}, 540 cm^{-1} and 420 cm^{-1}. The bands are assigned to particular vibrations by comparing the polycrystalline 70VP82 spectra with those of the features observed in the putative model compounds. The glassy 70VP82 spectra was compared with the above polycrystalline spectra to identify the ionic clusters existing in the glassy material.

The two spectra corresponding to the vitreous 70VP82 and polycrystalline 70VP82 appear to be difficult to interpret. Significant changes can be seen in both the spectra. In the spectra of the polycrystalline material, very clear and broad peaks were observed at 1600 cm^{-1}, 1080 cm^{-1}, 850 cm^{-1} and 540 cm^{-1}, small and sharp peaks were observed at 1380 cm^{-1}, 1350 cm^{-1}, 990 cm^{-1}, 930 cm^{-1} and 420 cm^{-1}. In the glassy material the spectra as can be seen in Fig. (2) are not clear and the peaks are very sharp. This is in confirmity with vitreous nature of the material. A broad band was observed at 1340 cm^{-1} and another at 1050 cm^{-1}. The other bands observed at 890 cm^{-1}, 860 cm^{-1}, 775 cm^{-1} and 500 cm^{-1} were weak and sharp. Several peaks that appear on the high frequency side of polycrystalline spectra were due to the splitting of V_2O_5 ion frequencies and the appearance of other resolved frequencies were due to the phosphate like units. The absorption peaks on the high frequency side of the 850 cm^{-1} peak are attributed to the vibration arising from phosphate related groups. The absorption peaks in the range lower than 540 cm^{-1} were due to the deformation of V-O-V linkages and in the region 850 cm^{-1} due to the deformation of P = O bands and VO_4 like units. By comparing the I.R. spectra of the glassy material with that of the polycrystalline material and considering the vibrational shifts due to the vitreous nature, it can be concluded that the glass consists of ionic clusters of PO_4^{3-}, VO_4^{3-} and their combinations. Hence, from the I.R. spectral studies, the glass under investigation can be classified as an ionic glass rather than a condensed glass.

(B) <u>Far I.R.</u> : The solid electrolytes in the polycrystalline form have higher ionic conductivities when compared to that of single crystals because of the free motion of the ions in the lattice. This is attributed to the diffusion mechanism which is neither continuous as the motion of atoms in liquids nor discontinuous as Hydrogen ion motion in Hydrogen metal system (7-8).. The observed very high ionic conductivity of the order of 10^{-2} in glasses is assumed to be due to the freezing of liquid like ionic motion in the electrolyte. It is assumed at this point that ionic clusters form due to the presence of different polarising species namely V_2O_5 and P_2O_5 which is confirmed from the I.R. studies. The absence of long range order and great randomness in the skeleton glass structure, contribute to the mobile ions greater mobility due to the formation of open channels for ion migration. Because of this local diffusive motion the glassy network provides open channel like pathways for the mobileions. The mobile ions which have translational motion interact with the lattice vibrations of the frozen glassy structure which is a characteristic feature of the cations in the vitreous electrolyte materials. Since the lattice vibrations of the mobile Ag^+ ions are of lower frequencies when compared to the electron transitions, FIR and Raman studies can be used as analytical techniques for understanding the presence of AgI, I_2, Ag - I and I-I lattice vibrations in superionic solids (9-10). Hence the FIR and Raman studies were used in the vitreous electrolytes to obtain information on the lattice dynamics associated with the ion transport of silver in the glasses.

The FIR spectra of the polycrystalline and the glassy 70VP82 material were shown in Fig. (3) respectively. From figure 4 it was observed that the 3 bands at 195 cm^{-1}, 85cm^{-1} and 65 cm^{-1} were due to the AgI lattice vibrations. The bands are very clear when compared to the bands of glassy material as shown in Fig.(3). The bands at 190 cm^{-1}, 87 cm^{-1} & 65 cm^{-1} for the glassy material as shown in Fig.(3) are very sharp and are not clear. This nature of the bands reflect the state of the disorder of AgI in the vitreous form. Hence it can be confirmed that AgI exists in the glassy matrix in a highly disordered form. The band at 65 cm^{-1} is assigned to the E-symmetry of the Ag$^+$ ion translation mode. It is attributed to the characteristic nature of the Ag$^+$ ion, attempt frequency from the diffusive behaviour to the oscillatory behaviour. The band at 87 cm^{-1} corresponds to the vibration of Ag$^+$ ions against the Iodine ions. As the silver content increases in the composition of a glass the intensity of the band increases which also enhances the ionic conductivity until 70% AgI (11). This proves that the Ag$^+$ ion contributes to the overall conductivity of the material.

(C) RAMAN STUDIES

Raman spectroscopy has become an effective probe to study the vibrational excitations in ordered solids and considerable progress has been made in using this technique to study the vibrational excitations in vitreous materials. This technique has become an effective one due to its experimental simplicity, high resolution and availability of high power quality laser sources at different wave lengths. The motion of the mobile ions in superionic conductors is associated with the stray electric moments and hence can directly interact with the electric field of the incident electromagnetic radiation of proper polarisation.

In the silver based oxysalt superionic conducting glasses the ionic conductivity tends to increase with a decrease in the activation energy. This was assumed to be due to the structural breakdown of the glass as the glass approaches the glass transition temperature. It is also possible for the cation to be highly mobile in a random glassy structure with the anion not involving itself in the motion. Thus with an assumption that the anion is not involved in the macromolecular chain and assuming that the chain preserves the characteristics of the original glass, the Raman spectroscopic technique was used as an analytical technique to confirm the structural hypothesis. If a change in the Raman shift is observed for the pure AgI and the glass under investigation, then it can be confirmed that the anion enters into same of the vacant sites of the macromolecular structure. If the Raman spectrum does not show any modification corresponding to P-O or V-O vibrational mode upon addition of large quantities of silver halide, then the assumption that 'free' anions and cations exist can be fit into a free ion model introduced by Rice and Roth (12).

The Raman spectrum at room temperature for the vitreous electrolyte 70VP82 in the form of a small cube is shown in Fig.(4). It can be seen from Fig.(6) that six bands exist at 87 cm^{-1}, 120 cm^{-1}, 250 cm^{-1}, 340 cm^{-1} and 420 cm^{-1} in the higher region of 20 - 500 cm^{-1}, and four bands exist at 500 cm^{-1} 540 cm^{-1}, 850 cm^{-1} and 890 cm^{-1} in the lower region 500-1200 cm^{-1}. Comparing the Far IR & IR Spectra of the vitreous material 70VP82 as shown in Figs. (2) and (3) respectively to that of the Raman spectrum as shown in Fig.(4), it can be concluded that the band at 87 cm^{-1} is both IR and Raman active whereas the band at 120 cm^{-1} is only Raman active. The other two bands observed in the FIR at 190 cm^{-1} and 65 cm^{-1} are not observed in Raman, hence these bands are IR active. The Raman bands at 500 cm^{-1} and 890 cm^{-1} are due to the $(PO_4)^{3-}$ ionic clusters which are IR active also, whereas the bands at 540 cm^{-1}, 850 cm^{-1} are Raman active only. Hence additional information on the vibrational frequencies of the ionic clusters was obtained from the Raman spectra. Though IR spectra itself is rich in giving information about the structure, the Raman spectrum contains data that can supplement IR data. The observed difference in the vibrational frequencies of the polycrystalline and amorphous material for the same composition is due to the difference in selection rules and the different mechanisms that exist in the polycrystalline material for the coupling of vibrations that constitute a unit and the rest of the material. Whereas in the glass the difference is due to the structural variation which is intrinsic to the vitreous nature.

Hence, from the FIR, IR and the RamanSpectra it can be concluded that the ionic clusters of VO_4^{3-}, PO_4^{3-} and a combination of these exists in the present 70VP82 glass. The IR spectra in the region 4000 cm^{-1} - 400 cm^{-1} confirmed the presence of ionic clusters whereas the Far IR spectra confirmed the presence of disordered AgI which is responsible for the high ionic conductivity in the vitreous state. The Raman spectra confirmed the formation of the ionic clusters by supplementing the unobserved bands of the ionic clusters from the IR spectra. Also the lattice vibrations of the disordered AgI were obtained from both FIR and Raman Spectra. Hence the glass is calssified and characterised as an ionic glass.

(D) E.P.R. STUDIES

The Electron paramagnetic resonance technique is used in the present study of 70VP82 material to obtain the concentration in the glassy and polycrystalline samples. The d.c. conductivity measurements indicates that the Electronic conductivity decreases as the concentration of V_2O_5 decreases. It was proved in the case of GeO_2 - V_2O_5 - P_4O_{10} semiconducting glass that the basic unit responsible for the electronic conduction in glasses is the presence of VO_4 octahedra. The existence of vanadium ion in different valence states in equivalent positinns in the VO_6 octahdera linked as chains or sheets was beleived to provide a continuous path for the Electronic conduction through the glass (13). VERWEY etal (14) suggested the above assumed 'partial valence' mechanism for the Electronic conduction which was supported by IOFFE etal (15) and REUTER (16) in the case of V_2O_5 - P_2O_5 semiconducting oxide glasses. From the I.R. spectra as shown in Fig.(2) it was confirmed that the VO_6 octahdera retains its identity in the glassy state. The EPR gives an indication of the presence and valence state of a paramagnetic species and the NMR can be used as a microscopic probe to study the electric and magnetic fields present in the material. A typical EPR spectrum of AgI - Ag_2O - P_2O_5 glass is shown in Fig.(7).

The spectrum resolves into five perpendicular components and six parallel components for different values of m_I (-7/2 to +7/2) of Vanadyl ion in the form of $(VO_4)^{4-}$ tetrahedra or $(V_2O_6)^{4-}$ octahedra. The calculated spin Hammittonian parameters for the V_2O_5 - P_2O_5 semiconducting glass and the AgI - Ag_2O - V_2O_5 - P_2O_5 glass are given in table 1. From table 1, it is evident that the A values are approximately the same, which clearly indicates that the species responsible for the electronic conductivity are the same in both V_2O_5 - P_2O_5 and AgI - Ag_2O - V_2O_5 - P_2O_5 systems. However the difference in the g values indicate a change in the "order" of the system. The near isotropic value of g (g_1 = 1.986 & g_{11}= 1.978) in the present 70VP82 system predicts a high ordering of the conducting species when compared to the conducting species of the semiconductor glasses (g_1 = 2.00 & g_{11} = 1.95) V_2O_5 - P_2O_5.

Fig.(5) shows the EPR spectra of 70VP82 polycrystalline material. On comparison of the intensities of the EPR spectra of 70VP82G and 70VP82P sample, we could conclude that the existence of vanadium in 4^+ state is more in the case of polycrystalline than glassy sample. Hence a high electronic conductivity is observed in the polycrystalline, since V^{4+}/V^{5+} partial valence mechanism is responsible for the electronic conductivity. Thus it can be concluded that V^{4+}/V^{5+} concentration is more in polycrystalline than in the glassy sample. From the isotropic nature of the g values for 70VP82G, it is evident that Ag_2O (glass modifier) brings order to the glassy matrix containing V_2O_5 and P_2O_5 structural units thus aiding ionic conductivity and reducing the electronic conductivity drastically.

5. THERMAL AND XRD CHARACTERISATION

The nature of the sample is determined from the x-ray diffractogram. The amorphous nature is characterised by a diffractogram as shown in Fig.1(a) and the polycrystalline nature by 1(b). The XRD patterns for AgI, Ag_2O, V_2O_5 & P_2O_5 are shown in Figs. 1(c), 1(d), 1(e) & 1(f) respectively. The XRD patterns of the individual compounds are shown here to confirm the formation of a new polycrystalline compound. New 'd' spacings are observed for the polycrystalline 70VP82 which are different from that of the individual compounds, thus confirming the formation of a new compound.

From the XRD, only the vitreous nature can be predicted where as to compare them out of thermodynamic equilibrium with the same compound in crystallized form in thermodynamic equilibrium, a D.T.A. has to be performed to establish the glass transition temperature (T_g) and crystallization temperature (T_c). T_g seperates the high temperature domain where the vitreous material possess thermodynamic and mechanical properties of a liquid from the low temperature domain where the material possess the properties of a solid. The significance of T_g is due to its relation to the kinetic parameters and the duration of experiment $\log \sigma T$ Vs $10^3/T$. On a microscopic level, T_g corresponds to the time required bythe elements of the

macromolecular chains consituting a glass to move a distance comparable to their size. Hence thermal characterisation of a material is very important to study T_g, which reflects the degree of freedom of the macromolecules in the glassy matrix. The analysis developed by GIBBS (17) and DIMARZIO (18) and then by ADAM and GIBBS (19), associates the ideal glass transition temperature with the structural relaxation time of the macromolecular chains porportional to $\exp\left(\frac{-W}{R(T-T_o)}\right)$
Where w is the activation energy, T_o the temperature below which the properties of liquid can't be extrapolated.

At low temperatures ie below T_g, cationic displacement is decoupled with the slow motion of the elemental structural units (PO_4, VO_4, PO_6 etc.), ofthe macromolecular chains and thus conductivity is an Arrehenius plot. At high temperatures i.e. above T_g, the cationic relaxation time is less than the relaxation of the macromolecular chain, thus the ion transport is cooperative with the movement of the macromolecular chain elements. Hence from DTA and TGA, the thermodynamic information that is relevant to T_g and the relaxation time of the basic structural units can be obtained. Also the low temperature thermodynamic properties of the liquid like nature of vitreous samples can be analysed through the DTA and DSC curves. This can be correlated to the Ionic conductivity data by the Arrehewius plot and the quenching rate 'q' by the equation given by J. ZARZYCHI (20) as

$$q = q_o \exp\left(\frac{-Ea}{RT_g}\right)$$

The DTA experiments were extended from room temperature to about 400°C to oberve the absence of thermal arrests upto the glass transition temperature and the presence of Endothermic inflection at the T_g which is characteristic of a glass. Since DTA is generally considered to be a sensitive indicator of a structural change it was thought worthwhile to examine the stability of polycrystalline and glassy material below the glass transition temperature so as to study the thermal stability of the electrolyte. Fig.(6) shows the typical DTA trace for polycrystalline and glassy 70VP82 in the temperature region 30°C - 400°C. Endothermic peak was observed at 72°C which is attributed to the glass transition temperature, from the TGA at the same temperature a weight loss of 0.012 mg was observed which is due to the crystallization of the glassy to the polycrystalline state. The Endothermic inflection at 380 C is due to the absorption of thermal energy for the polycrystalline compound to melt into a liquid state. Qualitatively the glassy compound appears to possess greater stability with respect to moist air than the polycrystalline material. The DTA measure-show that after the exposure of glassy material to the laboratory atmosphere for 60 days, additonal endothermic peaks do not develop in the region of 30 - 400°C.

6. ELECTRICAL CHARACTERISATION

Apart from the complex Impedance Analysis and the Electrical Modulus Analysis, electrical conductivity measurements are widely used for the material characterisation of solid electrolytes because of thesimplicity in the experimental technique and the versatile nature of the obtained data. The ionic conductivity of the electrolyte projects the ability of the mobile ion as a charge carrier in the electrolyte. The d.c. conductivity of the sample gives the contribution of electronic conductivity to the total conduction. The lower the electronic conductivity, the less will be the self-discharge of the battery. Hence a long-shelf life can be expected. Thus from the ionic conductivity, the ohmic resistance of the electrolyte can be predicted. Thus the 2R polarisation, ohmic and non-ohmic leakage currents can be studied to observe the mechanical relaxation of the charge carriers due to the mechanical stress applied due to the pressing of the electrolyte into pellets. Then the frequency dependance of conductivity is studied to obtain the electrolyte-electrode interface performance in the solid electrolyte cells. Thus from the electrical characterisation of the electrolyte, the performance and the shelf-life of the solid state battery can be predicted.

A. IONIC CONDUCTIVITY

After the preparation of the polycrystalline and glassy6 samples 70VP82 from the same melt, pellets of 1.3 cms in diameter are pressed at a pressure of 1000 kg/cm^2. Then the time dependance of the resistance is studied. For the polycrystalline material it was observed to be nearly constant, whereas for the glassy material a very large variation was observed as shown in Fig.(7). This is attributed to the relaxation and homogenisation of the charge carriers to a stable state.

The ionic conductivity of the polycrystalline and glassy 70VP82 material was measured at 1KHz, in the temperature range 30 - 100°C. The logarithm of σT product Vs reciprocal absolute temperature for the polycrystalline and glassy 70VP82 were shown in Fig.8 and 9 respectively. In the temperature range 30 - 70°C, the glassy sample obeys the Arrehenius relation and the calculated activation energy is .264 ev.

In Fig.(9), the observed ionic conductivity in the polycrystalline form of the material 70VP82 is shown. The plot of $\log \sigma T$ Vs $10^3/T$ shows the linear regions with the activation energies of 0.289 and 0.221 . The region with low activation energy obeys the Arrehenius relation whereas the region with high activation energy deviates from the above mentioned relation, thus proving the point that in a polycrystalline material, the existence of grain boundaries limit the conduction or migration of ions at low temperatures. At high temperatures, the tendancy of the polycrystalline material to become a random network structure will enhance the ionic conductivity. This is in true confirmity with the proposal that, because of high randomness in the liquid like state of the polycrystalline material, the observation of high ionic conductivities in glasses is due to the freezing of the liquid state without any grain boundaries. The overall activation energy was obtained as 0.264 ev, whereas for the glassy 70VP82 it was 1.305ev which is very high when compared to the other compositions. Even for a pure V_2O_5 - P_2O_5 glass with a ratio of 8:2, a high activation energy was observed (21).

The frequency dependent conductivity in the glassy state is shown in Fig.(10) which is somewhat different from the other superionic glasses (22,23). At 100 Hz, the conductivity is minimum and from 10^3 Hz to 10^6 Hz a linear increase was observed. The frequency dependance in the low frequency region is due to the space charge polarisation or an interfacial effect. No frequency dependant conductivity was observed for the polycrystalline material.

ELECTRONIC CONDUCTIVITY

The criterion for a good electrolyte is its ability to discharge currents for longer periods when connected to an external load. To achieve high shelf-life, a solid electrolyte should necessarily have a very low electronic conductivity. Hence the lower the electronic conductivity, the lesser will be the self-discharge and hence a long shelf-life. Hence the study of the contribution of electronic conductivity to the total conductivity reflects on the life of the battery and the extent of self-polarisation.

For the d.c. polarisation cell, Wagner (24) has derived the relationship between electronic current and applied voltage. When only n-type conduction is significant, the steady state electronic current is given by

$$I_e = \frac{RTA}{LF} \sigma_e \left[1 - \exp\left(\frac{-EF}{RT}\right) \right] \quad (1)$$

Where A is the area; L, the thickness of the pellet; and σ_e, the electronic conductivity, For an applied voltage EF $\gg$ RT equation (1) reduces to

$$I_e = \frac{RTA}{LF} \sigma_e \qquad (2)$$

According to equation (2), the steady state current becomes independant of applied voltage, with increasing voltage. This voltage is known as the saturated potential. The current in the present system 70VP82 reached steady state values in about 3 hours. As can be seen in Figs (11) and (12) which illustrates the I-V characteristics of the polycrystalline and glassy 70VP82 material, the current is nearly constant in the applied voltage range 80-200mv in accordance with equation (2). At voltages less than 80mv, current varies exponentially satisfying equation (1).

CONCLUSIONS

From the x-ray diffractogram, the formation of the vitreous material is confirmed with an observation of no peaks whereas for a polycrystalline peaks were observed with definite 'd' spacings. The spectrosocpic studies proved the existence of ionic clusters as can be inferred from the IR spectra. The FIR spectra indicates that AgI exists as Ag-I, which may not paritcipate in the transport mechanism. The ionic conductivity measurements indicate that the mobile species are Ag^+ ions, and are responsible for the ionic conduction. From the electronic conductivity, it was found that the electronic contribution to conductivity is almost negligible.

On the overall studies, it can be observed that a 0.2% addition of P_2O_5 to 0.8 V_2O_5 as a glass former, increases the number of oxygen ions per cation, which resulted in a good quality glass with a large improvement in the ionic conductivity.

The observation of high ionic conductivity in rapidly quenched V_2O_5 - P_2O_5 oxides opens up a new type of oxides, and a combination of them with As_2O_5 as material for applications as vitreous electrolytes. From the data obtained in this work, there is no clear-cut direction to follow, to optimise the composition for high ionic conductivity except to study all the compositions and establish the highest conducting composition. The stability of the glass in the temperature region 30 - 300°C is very high and thus the material can be a very good electrolyte in a solid state battery.

REFERENCES

(1) T. MINAMI, Y. TAKUMA and M. TANAKA,
J. Electrochem. Soc. 124, 1659 (1977).

(2) T. MINAMI, K. IMAZAWA and M. TANAKA,
J. Non Cryst, Solids 42, 469 (1980).

(3) K. HARIHARAN, RAJIV KAUSHIK and S. RADHAKRISHNA,
Materials forSolid State Batteries, 325 p.No.,
Ed. by B.V.R. CHOWDARI and S. RADHAKRISHNA.

(4) P. SATHYA SAINATH PRASAD and S. RADHAKRISHNA,
J. of Mat. Sci. (Communicated).

(5) D.J.E. INGRAM,
Spectroscopy at the Radio and Microwave Frequencies,
Butterworth, London (1967).

(6) C.P. POOLE Jr.,
Electron Spin Resonance - A Comperhensive treatise on
Experimental Techniques, Inter Science, New York (1967).

(7) P.A. EGELSTAFF,
An Introduction to the Liquid State,
Acad. Press,London (1967).

(8) T. SPRINGER,
Springer Tracks in Modern Physics 64, 50 (1972).

(9) S.S. MITRA,
Solid State Physics 13, 1 (1962).

(10) W. VEDDAR and D.F. HORNING,
Advan. Spectroscopy 2, 189 (1961).

(11) P. SATHYA SAINATH PRASAD and S. RADHAKRISHNA,
J. Phys. Chem. Solids. (to be communicated).

(12) M.J. RICE and W.C. ROTH,
Solid State Chemistry 4, 294 (1972).

(13) B.H.V. JANKIRAMA RAO,
J. Am. Ceram. Soc. 48, 311 (1965).

(14) E.J.W. VERWEY, P.W. HAAIJMAN, F.C. ROMEIJN and G.W. VAN etal,
Philips Res. Rept. 5, 173 (1950).

(15) V.A. IOFFE, I.V. PATRINA and S.V. POBEROVSKAYA,
Sov. Phys. Solid State 2, 609 (1960).

(16) BERTOL REVTER, JORG JASKOWSKY and ERWIN RIEDEL,
Z. Electrochem. 63, 937 (1959).

(17) J.H. GIBBS etal.,
J. Chem. Phys. 28, 373 (1958).

(18) E.A. DI MARZIO and J.H. GIBBS,
J. Chem. Phys. 28, 807 (1958).

(19) G. ADAM and J.H. GIBBS,
J. Chem. Phys. 43(1965) 139.

(20) J. ZARZYCHI, LES VERRES etal l' ETUT VITREUX,
Masson Paris (1982).

(21) A.P. SCHMID,
J. Appl. Phys. 39, 3140 (1968).

(22) R.J. GRANT, M.D. INGRAM et al.,
J. Phys. Chem. 82, 2838 (1978).

(23) J. KAWAMURA and M. SHIMOJI,
J. Non-Cryst. Solids 79, 367 (1986).

Table 1 : The calculated Spin Hamiltonian Parameters from the recorded EPR spectra of semiconducting 80% V_2O_5 - 20% P_2O_5 glass and the superion conducting 70VP82 Glass.

SPIN HAMILTONIAN PARAMETER	80% V_2O_5-20% P_2O_5	70VP82 GLASS
$g_{\parallel}$	1.950	1.978
$g_{\perp}$	2.000	1.986
$A_{\parallel}$	160 ± 2	160 ± 5
$A_{\perp}$	72 ± 2	70 ± 2

305

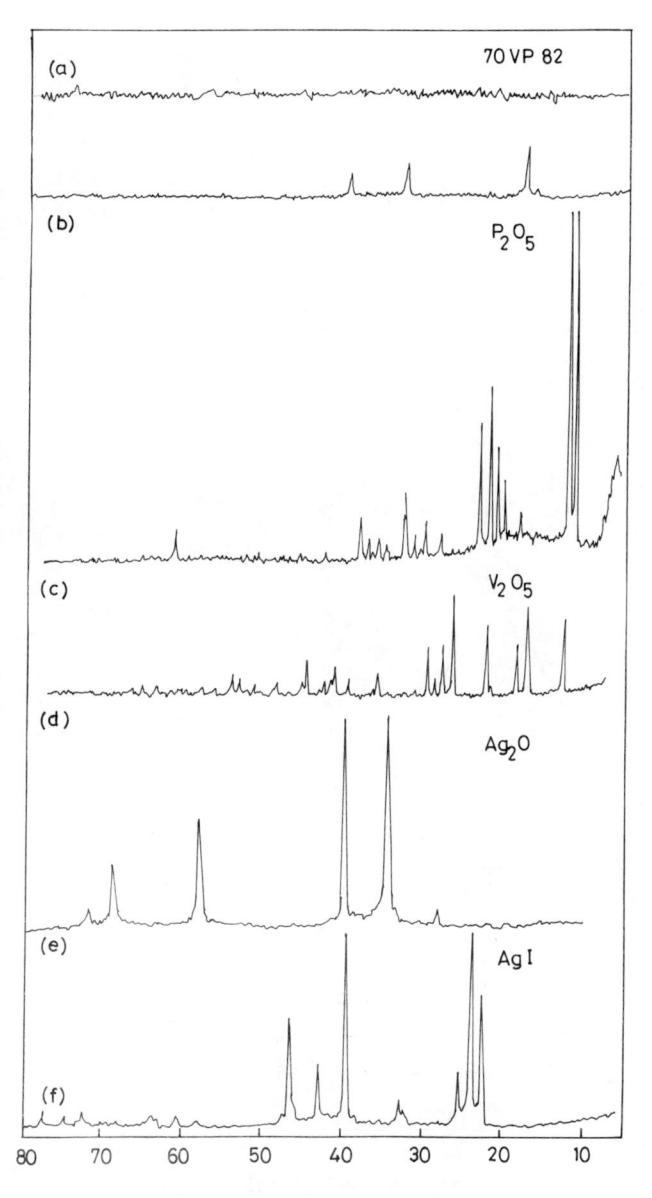

Fig.1 X-ray diffractogram of AgI, Ag_2O, V_2O_5, P_2O_5, 70VP82P and 70VP82G.

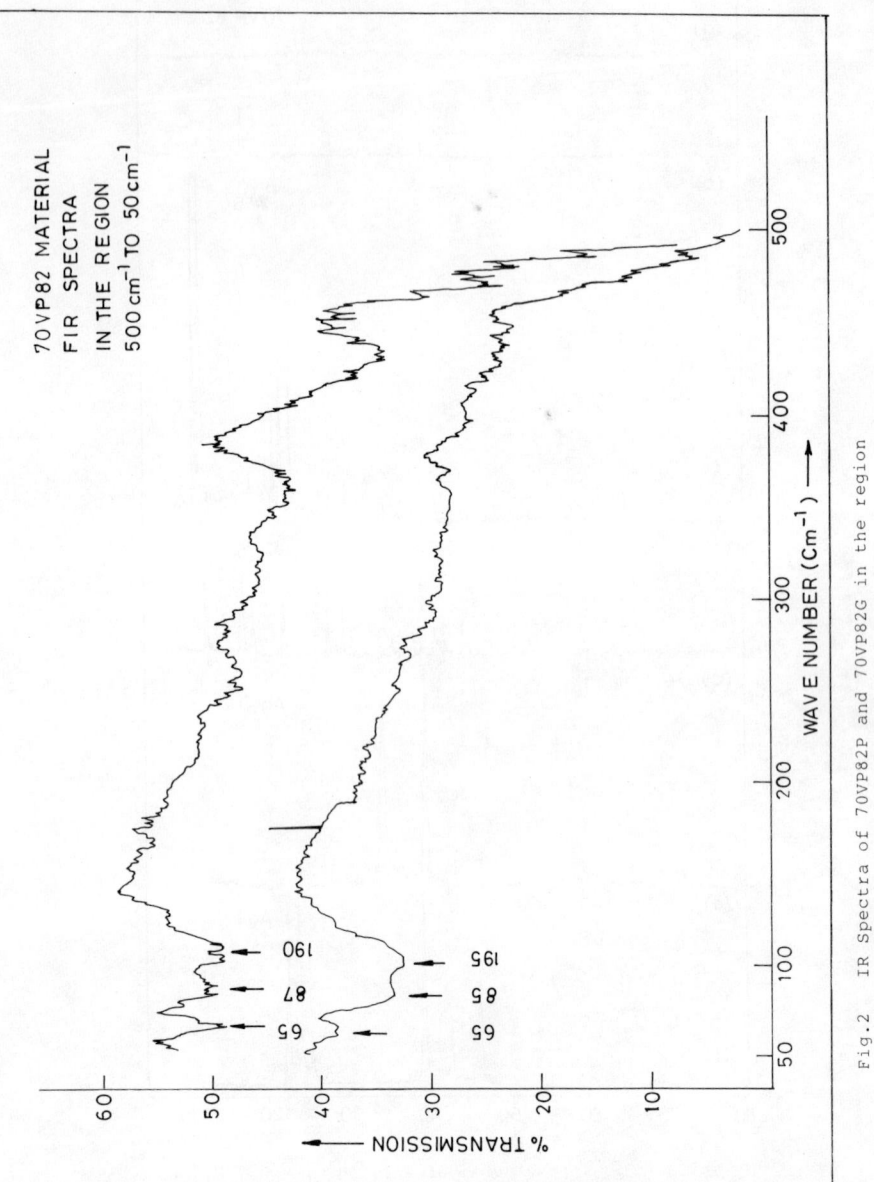

Fig.2 IR Spectra of 70VP82P and 70VP82G in the region 1600 – 200 cm^{-1} in KBr Matrix.

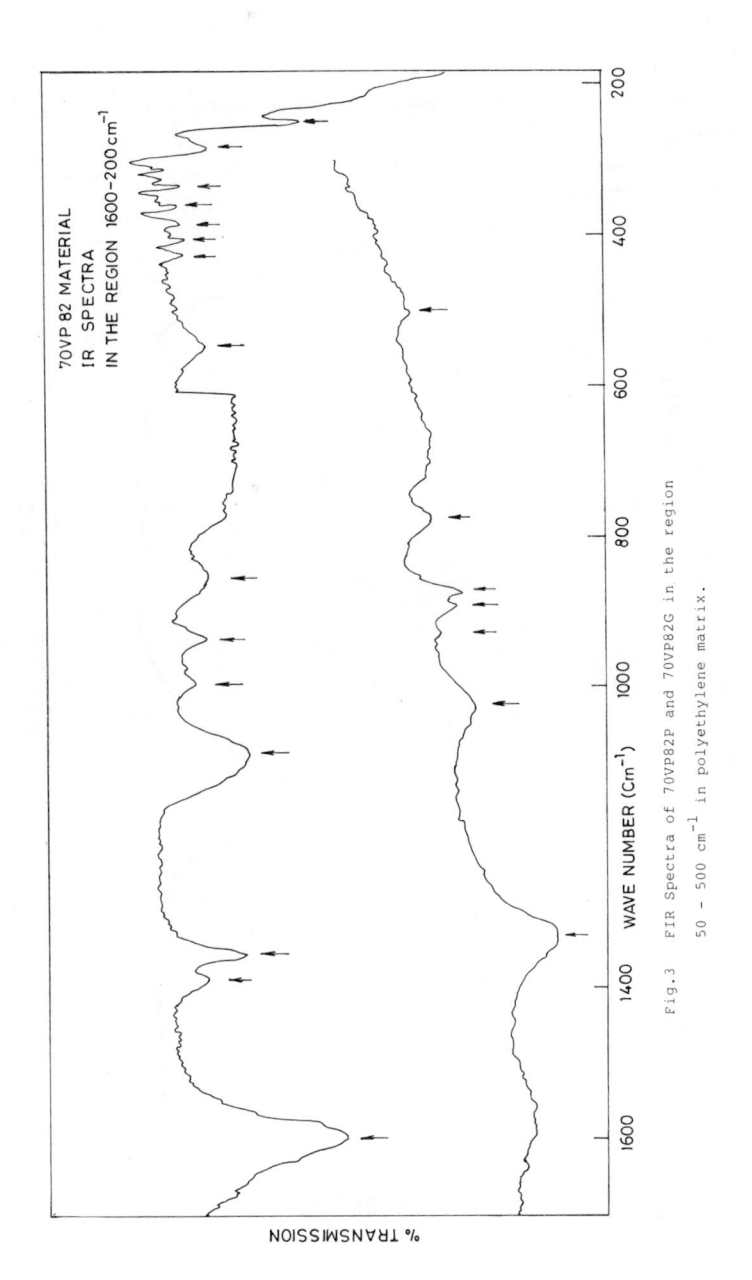

Fig.3 FIR Spectra of 70VP82P and 70VP82G in the region
50 – 500 cm^{-1} in polyethylene matrix.

308

RAMAN SPECTRA
OF 70 VP 82G
IN THE REGION
1250 - 20 cm^{-1}

WAVE NUMBER cm^{-1}

INTENSITY

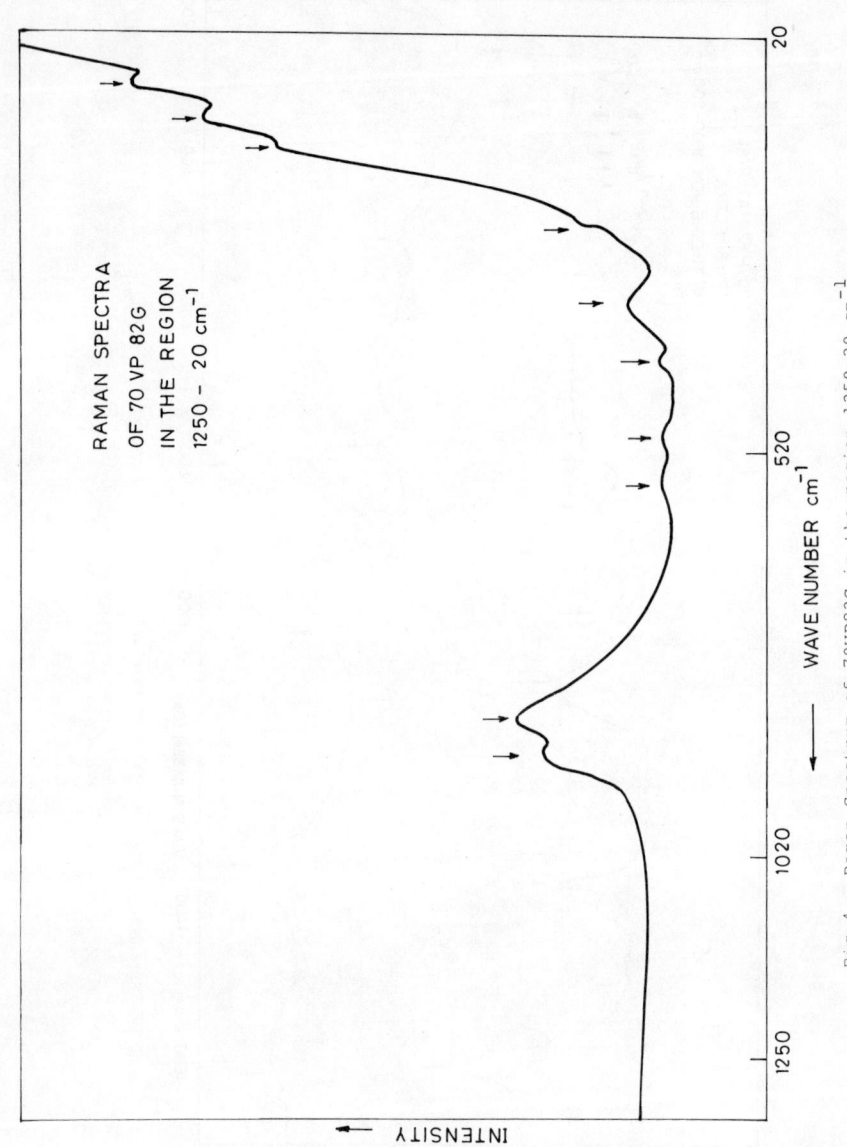

Fig.4 Raman Spectrum of 70VP82G in the region 1250-20 cm^{-1}.

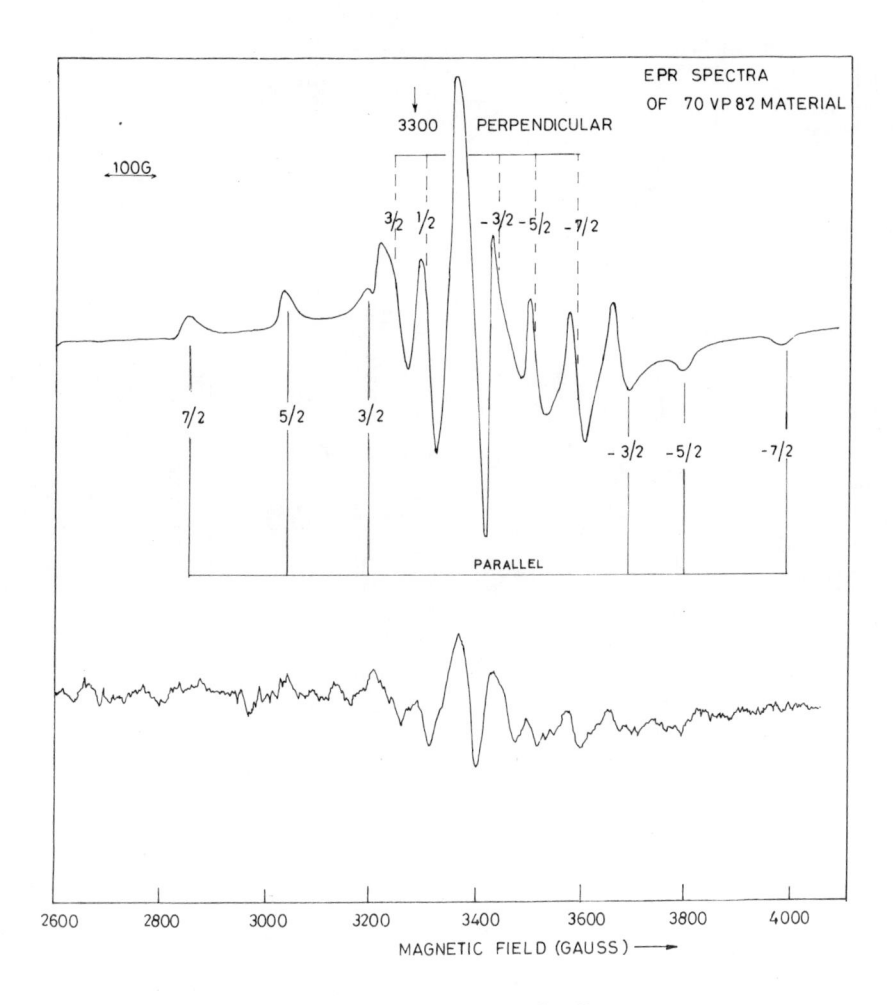

Fig.5 EPR Spectra of 70VP82G and 70VP82P.

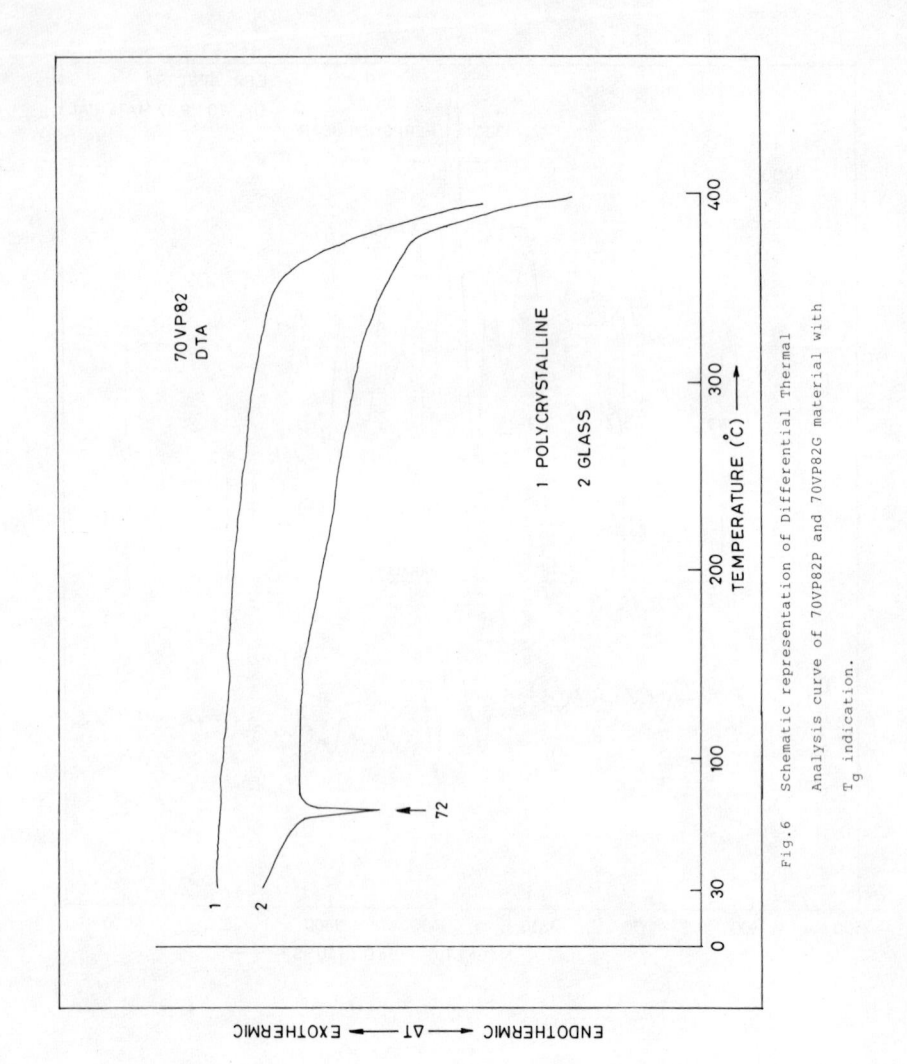

Fig.6 Schematic representation of Differential Thermal Analysis curve of 70VP82P and 70VP82G material with T_g indication.

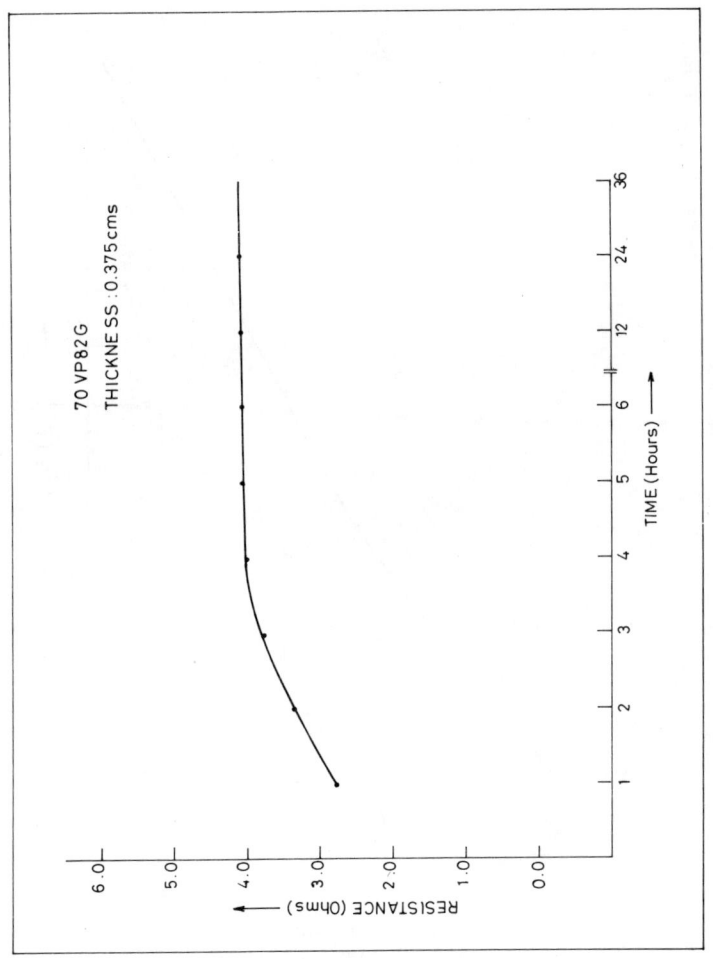

Fig.7 Resistance Vs Time curve for the 70VP82G pellet.

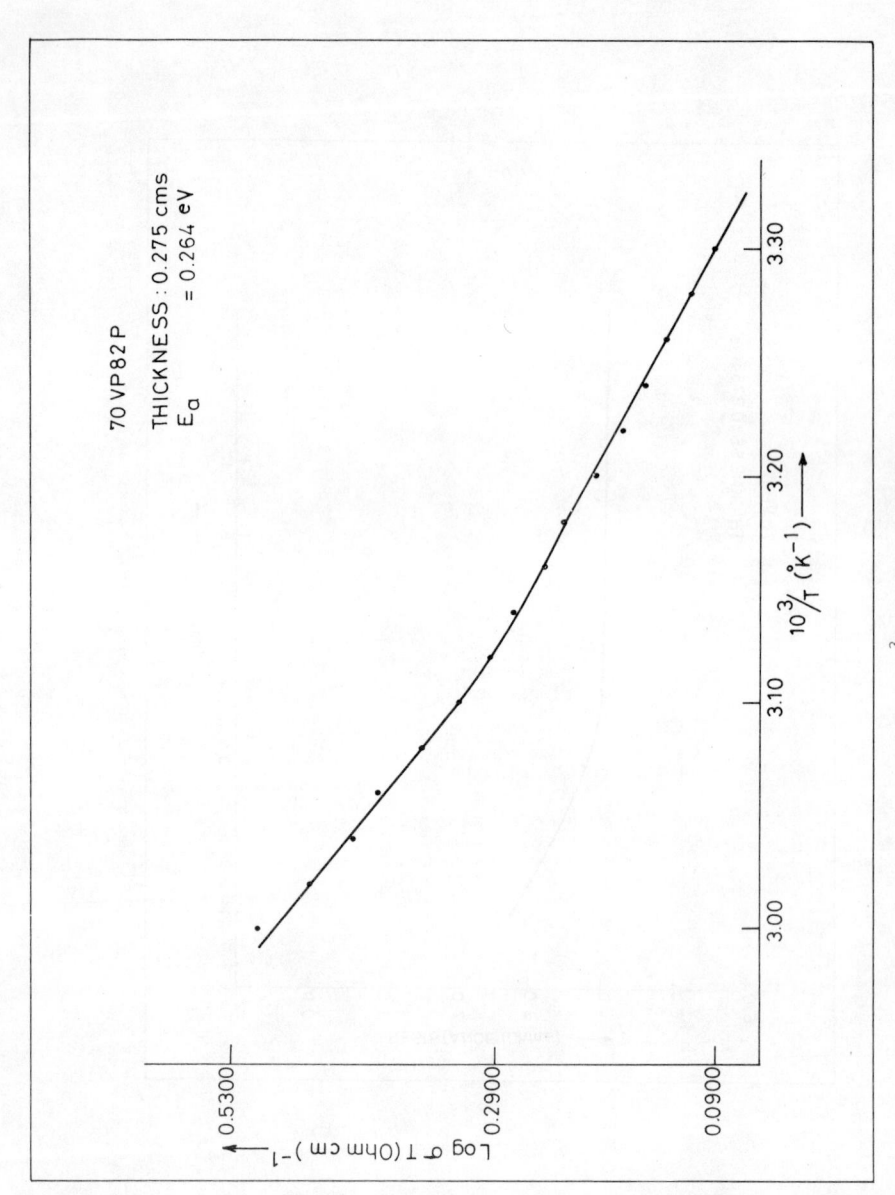

Fig.8 Plot of log σT vs $10^3/T$ for the 70VP82P material.

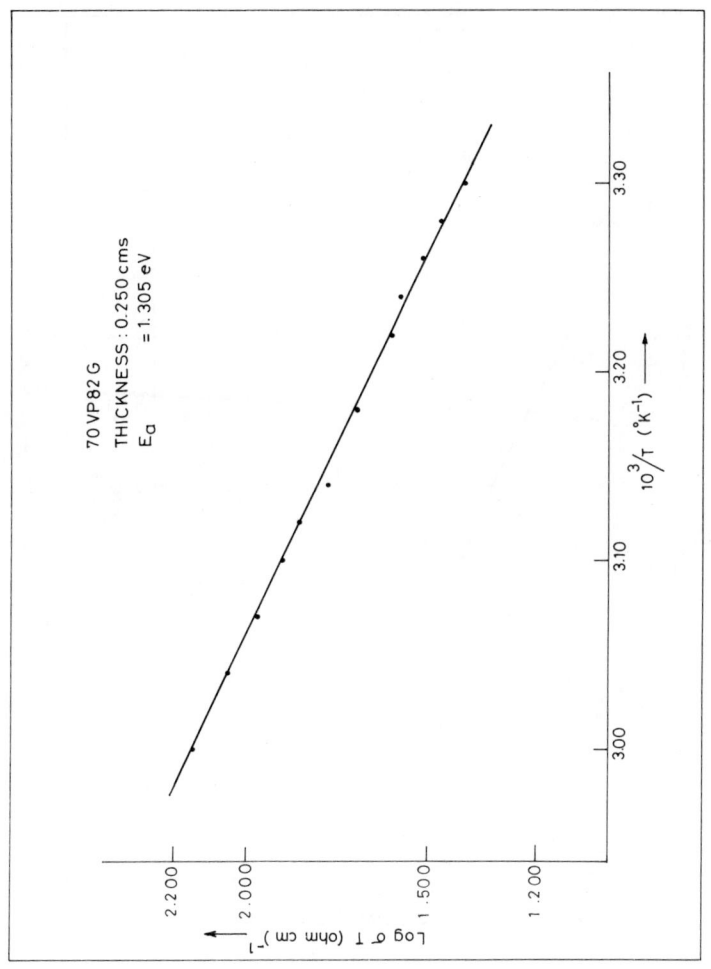

Fig.9 Plot of log σT vs 10³/T for the 70VP82G material.

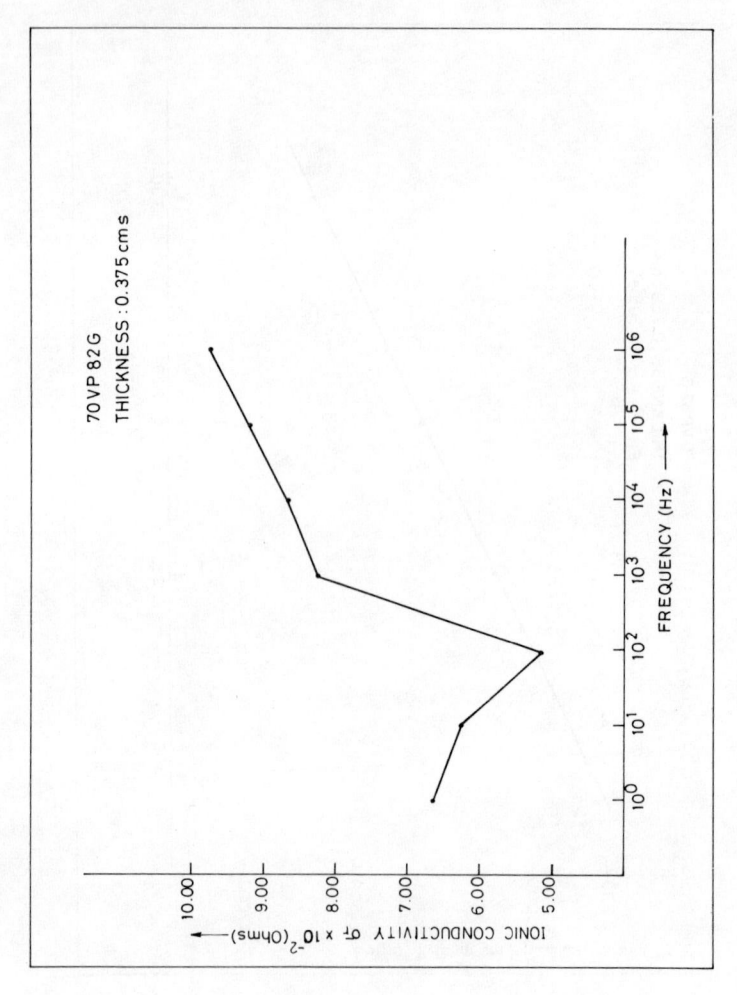

Fig.10 Frequency dependance of conductivity curve for 70VP82G pellet with silver electrodes.

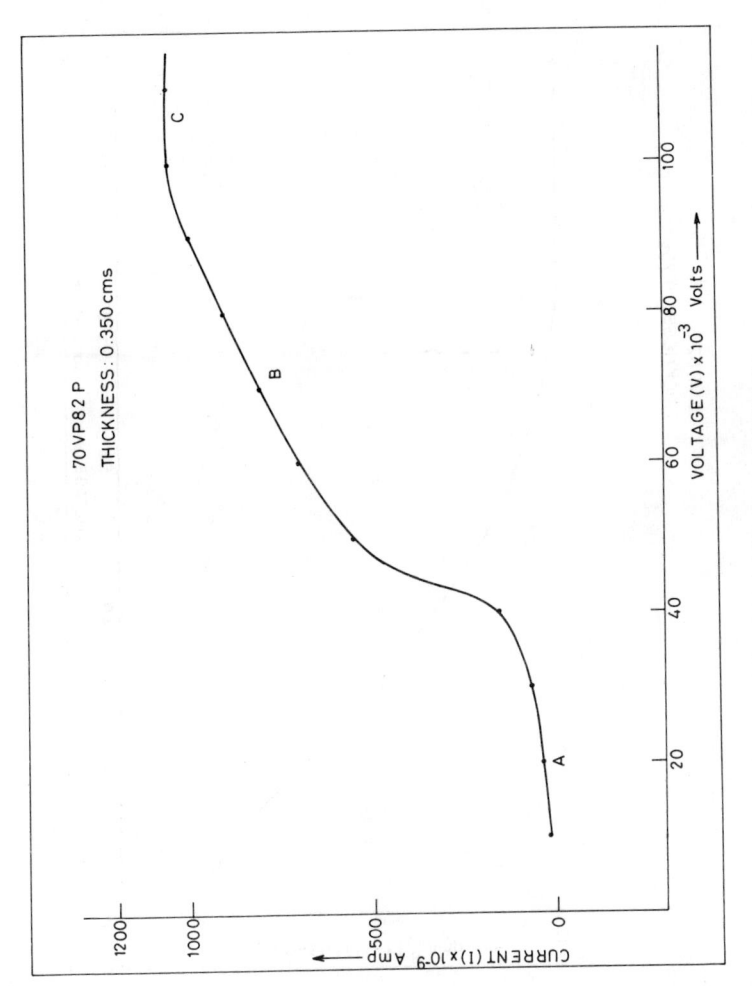

Fig.11 I - V Characteristic curve of 70VP82P.

Fig.12 I - V Characteristic curve of 70VP82G.

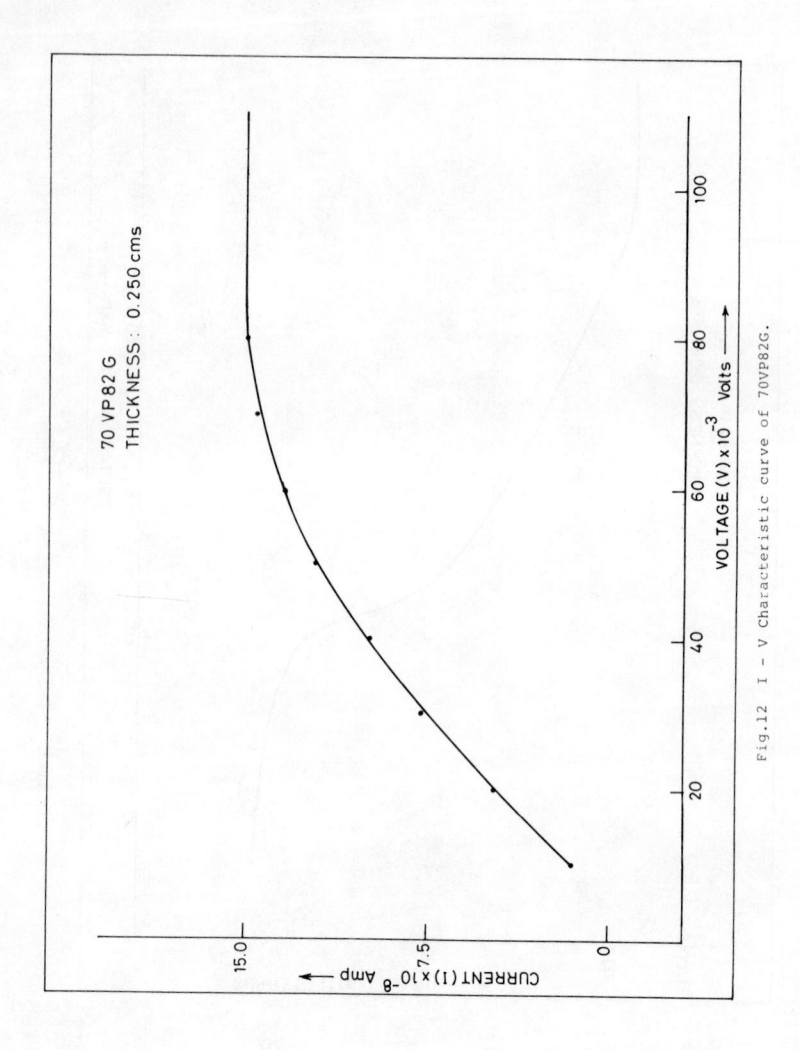

MATERIAL CHARACTERISATION OF $Ag_{1-x}Pb_xI$ SOLID SOLUTION IONIC CONDUCTOR

RVGK SARMA, P SATHYA SAINATH PRASAD
AND S RADHAKRISHNA
Department of Physics
Indian Institute of Technology
MADRAS - 600 036 - INDIA

ABSTRACT

The $Ag_{1-x}Pb_xI$ solid solution is prepared by the co-precipitation method, so as to obtain $Ag_{1-x}Pb_xI$ for different stochiometric ratios of x ($0.1 \leq x \leq 0.9$). From the ionic conductivity studies the best conducting ratio was established as $Ag_{0.3}Pb_{0.7}I$. The polycrystalline nature of the prepared solid solution is characterised by X-ray diffraction pattern. The absence of agglomarated AgI and PbI_2 in the formed poly-crystalline $Ag_{1-x}Pb_xI$ was confirmed using FIR Spectroscopic technique. The phase transition temperature was determined from DTA technique. The observed phase transition temperature was well correlated with that from ionic conductivity measurements. The dielectric behaviour of $Ag_{0.3}Pb_{0.7}I$ was studied to establish the nature of the defects and the disordered state of Ag^+ ions in the solid solution.

INTRODUCTION

In a mixed system the ionic conductivity increases either due to the formation of open channels through which ions can migrate freely [1,2] or due to the generation of strains arising from the displacement of ions of dissimilar sizes, charges and polarizabilities. In the AgI, AgBr system it was shown that the strain developed due to the presence of I^- and Br^- with dissimilar ionic sizes and same charge lossens the lattice structure which causes an enhancement in ionic conduction and lowers the superionic transition temperature [3].

In a recent study on the 2:1 composition of AgI : PbI_2 it was observed that the ionic conductivity increases, and this was attributed to the presence of PbI_2 in the matrix [4]. This large change in the conductivity is due to the set of seggregated PbI_2 layered structures which was also observed in AgI-CdI_2 [5]. The addition of dissimilar ions, generally results in lattice distortion, leading to more thermal disorder due to the expansion or contraction of lattice points, which lowers the transition temperature value. In the case of mixed conductor of the type AgI - PbI_4^{2-} where PbI_4^{2-} and I^- with different ionic sizes effect the superionic transition properties, it was observed that the enhancement in conductivity is due to the formation of an aggregate in the form of Ag $[PbI_4]$ at the molecular level and the loosening of lattice due to the different ionic sizes. With the above conclusions, for the AgI - PbI_2 mixed system it can be assumed that a higher enhancement in ionic conduction can be achieved, if the two compounds AgI - PbI_2 are taken in the form of a solid solution with different ratios of PbI_2, inorder to see the effect of univalent, divalent interaction of Ag^+ - Pb^{2+} on the creation of defects and vacant sites. Because of the creation of vacant sites AgI undergoes a thermal disorder at higher temperatures which will migrate more rapidly than in pure AgI. Thus $Ag_{1-x}Pb_xI$ solid solutions have been prepared by the co-precipitation method for $0.1 \leqslant x \leqslant 0.9$.

2 PREPARATION AND CHARACTERISATION

The mixed coionic conducting $Ag_{1-x}Pb_xI$ was prepared by the co-precipitation method for different values of 'x' ranging from 0.1 to 0.9 in steps of 0.1. The corresponding solutions of silver Nitrate and Lead Nitrate were taken and mixed in slight excess of Potassium Iodide solution. The solutions were kept as concentrated as possible, inorder to minimise the loss of lead Iodide. The solutions were heated inorder to obtain a homogeneous solution with complete chemical reaction. The resultant yellow to orange precipitate was filtered after the solution was washed with cold water and then with Acetone to remove excess of Iodine. The precipitate is dried at 120^oC for several hours. The poly-crystalline thus formed is characterised by systamatic Experimental Techniques.

3 XRD and DTA

The X-ray diffraction pattern was studied using a Philips instrument. XRD of the samples of various concentrations showed qualitatively the presence of only one compound which is different from AgI and PbI, conforming the formation of solid solution. The X-ray diffraction pattern is recorded under identical conditions for all the samples. The obtained X-ray diffractogram intensity of Pb^{2+} and I^- were qualitative indication of the amount of Pb^{2+} present in the total compound. As expected, the peaks were very sharp indicating that the disorder in the system is very less. The XRD pattern of the compound $Ag_{0.3}Pb_{0.7}I$ shows great similarity with rest of the compounds with the general formula $Ag_{1-x}Pb_xI$.

Although certain lines are found in common in all patterns as seen from Fig.1, the occurance of strong lines at 6.4929, 2.6105, 2.3270, 1.7463 and 1.3863 suggests that the compound $Ag_{0.3}Pb_{0.7}I$ solid solution is not simply disorder or contained $AgI - PbI_2$ material, but rather a differnt and unique material. However the presence of small amount of AgI cannot be ruled out.

The transition temperature from the low conducting phase to a high conducting phase was also observed from DTA studies. The sample and reference material Alumina (which shows no phase change in temperature range of operation), were heated together in a furnace at a constant rate of $10°K/min$. The observed changes in the heat content of the samples were attributed to the phase transition of the sample, which was observed at $418°K$. The temperature was compared and correlated with that from the $\log \sigma T$ Vs $10^5/T$ plots from conductivity measurements.

4 TRANSPORT STUDIES

The cylindrical pellets with an area of 1.34 cm^2 and with a thickness of 2-3 mm were prepared under a pressure of $5000Kg/cm^2$ using Perkin-Elmer pelletiser. Pure silver powder mixed with electrolyte (3 : 1) was used as electrode material for better electrode-electrolyte contact and also to eliminate the interfacial resistance. The pellets were sandwitched between two silver foils to form cell of the type Ag/Ag-Electrolyte-Ag/Ag. The pellets were loaded in a glass cryostat with a Iron constantin thermocouple to measure temperature. The cryostat was kept in a glass muffle furnace with a temperature controller.

It was observed that the solid solutions were frequency independant in small range 100 Hz - 10 KHz. Hence the impedance at 1 KHz is measured and Ionic conductivity was calculated using the relation :

$$\sigma = \frac{\sigma_0}{T} \exp(-E_a/kT)$$

The experimentally determined ionic conductivity values along with their activation energies for solid solution were shown in Table 1. The best conducting ratio was obtained for x = 0.7. The log σ T Vs $10^5/T$ plot was shown in Fig.2. The curve resembles typical AgI curve with a phase transition at $426°K$. In the case of AgI a conductivity of $2.4 \times 10^{-7} \Omega^{-1}$ cm^{-1} and overall activation energy of 0.56 ev is reported. In the temperature region 300 - $420°K$ the variation of conductivity is linear and a phase transition occurs at $420°K$. In the present investigation it was observed that the conductivity increases linearly from room temperature to $426°K$, after the phase transition a steep increase in conductivity is observed. After that the conductivity decreases slightly and increases slowly. This anamolous behaviour of $Ag_{0.3} Pb_{0.7} I$ was attributed to the mobility of Pb^{2+} ions in α-phase of AgI. Hence it can be concluded that Lead doped AgI behaves as β-AgI in lower temperature region with an enhanced conduction, where as in the high temperature α-phase the conductivity is less compared to pure AgI, because of coionic conduction of Pb^{2+} and Ag^+.

5 DIELECTRIC STUDIES

The dielectric studies of the system $Ag_{1-x}Pb_xI$ were undertaken to study the dielectric response of the system and to correlate the results to those obtained from the transport studies. Hence both from the transport and dielectric studies the correlated picture of the transport mechanism can be proposed and presented. In the present paper the dielectric studies for the best composition were presented and compared with published work of pure AgI [7]. It was observed that the K' values of the present system are higher than those of pure AgI, which is consistant with the observed enhancement in ionic conductivity.

In dealing with the dielectric response of the superionic conductor, the ionic carrier (Pb^{2+}, Ag^+) lies between the usual region of Debye solid and Drude like free ion model [8]. The frequency dependence of conductivity predicts the non-existence of peak near Debye-Frequency, which scales with the inverse of ionic mass. Assuming that the ionic carriers have two basic degrees of freedom:
1) An oscillatory motion in Harmonic potential provided by the rigid lattice (2) A random walk process through which the ions can diffuse throughout the material. The frequency dependance

of conductivity gives a well correlation to the non-observation of dielectric peak.

It can be observed that the dielectric constant decreases with frequency (Fig.3) i.e. 41.9×10^3 at 100 Hz to 0.041×10^3 at 100 KHz. The loss factor (K") in the same region decreases from 16.32×10^3 to 0.032×10^3 (Fig.4). The tan δ varies from 8.347 to 0.7825 in the same region. Within this region the tan δ varies exponentially with frequency (Fig.5).

Since the dielectric response of the solid solution $Ag_{0.3}Pb_{0.7}I$ reflects the contribution of the various polarising species present, i.e permanant and induced dipoles, electrons and ions. It is well established that the dipolar responses are limited at the lower end of the frequency spectrum by a loss peak, while dielectrics with ionic carriers don't show any peaks where as the loss is dominated by a steep raise in low frequency branch of the form σ/ω, owing to the d.c. conductivity [9]. The present study reveals that for $Ag_{0.3}Pb_{0.7}I$ the log K' varies exponentially with frequency.

The presently observed form of dielectric response has been observed by Johnsher as a form of universal response in which the hopping charge carriers contribute to the dielectric and dielectric polarization [10].

6 SPECTROSCOPIC CHARACTERISATION

The FIR transmission and reflection measurements for the material $Ag_{0.3}Pb_{0.7}I$ were carried out using a Poly Tech 30 fourier spectro meter in the region 500-50 cm^{-1}, with automatic scanning controls.

Fig.6 shows the transmission spectra of $Ag_{0.3}Pb_{0.7}I$ solid solution at room temperature in the region 500-50 cm^{-1}. In the region 500 - 50 cm^{-1} two bands were observed at 456 cm^{-1} and 490 cm^{-1}. It was reported that the AgI streching modes falls in the region 80-120 cm^{-1}(11,12,13). In pure AgI (14), the silver ion is surrounded by four Iodide atoms and the stretching bands are observed in the region 80-120 cm^{-1}. In the low frequency region 45 cm^{-1}, the deformation type motions of the Iodide lattice relative to the metal ion were observed at 39 cm^{-1}, 35 cm^{-1} in the case of HgI_4^{2-} and 24 cm^{-1} and 30 cm^{-1} in the case of CdI_4^{2-}(15).

In the case AgI, Raman and IR broadening of lines due to the temperature has been attributed to the damping of phonons by the interaction with mobile ions (12,16,17). This explanation is applicable to the case of $Ag_{0.3}Pb_{0.7}I$, because the ionic conductivity experiments indicate that Ag^+ ions are mobile species, whose mobility is temperature dependant. Hence the broadening of the 50 - 120 cm^{-1} band with double peak at 75 cm^{-1} and 90 cm^{-1}, arises from a high anharmonic potential for the above explained modes. The splitting of bands is attributed to the presence of Lead metal ion. The bands at 456, 490 cm^{-1} are very weak compared to the AgI bands but are slightlty broad which are assigned to the asymmetric stretching of the PbI_4^{2-} species. These bands are in good agreement with other tetrahedral compounds of Pb, Hg, Cd and Zn in solid state (18).

Hence from far IR measurements it was observed that the solid solution $Ag_{0.3}Pb_{0.7}I$ contains AgI, a small quantity of PbI_4^{2-} anions as confirmed by the stretching mode frequencies.

7 CONCLUSION

The solid solution $Ag_{0.3}Pb_{0.7}I$ was characterised by the X-ray diffraction, DTA, transport, dielectric and spectroscopic techniques. The formed solid solution is different from the Ag_2PbI_4 stoichiometric compound as is evident from the XRD and spectroscopic characterisation. The enhanced ionic conductivity is attributeds to the high thermal disorder created in the $Ag_{0.3}Pb_{0.7}I$ because of the presence of Pb^{2+} and the lattice loosening due to the disimilar ionic species (Ag^+ and Pb^{2+}). This attribution is supplimented by the dielectric studies, where a high dielectric loss was observed, when compared to pure AgI in conformity with the enhanced ionic conduction. The phase transition observed in the conductivity studies is well correlating with the observed from the DTA.

REFERENCES

1. Funke K, Proc. Solid St. Chem.
   11, 345 (1967).
2. Yao Y F Y and Kummer J T
   J. Inorg. Nucl. Chem. 29, 2453 (1967)
3. Shahi K and Wagner J B Jr.
   Jr. Phy. Chem. Solids 43, 1713 (1982).
4. Brightwell J W, Bukley C N and Roy B
   Solid St. Commn. 42, 1715 (1982)
5. Bailer J C etal
   'Comprehensive Inorganic Chemistry'
   Vol.2, Page 135, Pergamon Press, Oxford (1973).

6. Shahi K, Wagner J B jr.
   Solid State Chemistry 42, 107 (1982).
7. Gon H B and Rao K V
   Indian Jr. Physics, 54A, 50-55 (1980)
8. Huberman B A and Sen P N
   Phy. Rev. Letters 33, 1379 (1974).
9. Johnsher A K, Phil. Magzine B
   38, 587 (1978).
10. Westphal W B and Sib A
    M.I.T. Technical Report, AFML-TR-72-39, April 1971.
11. Habbal, F, Zvirgzdr J A and Scott J F
    J Chem. Phy. 69, 111 (1978).
12. Burns G, Decol F H and Shafer M W
    Solid State Commn. 19, 287 (1976).
13. Gallagher D A and Klein M V
    Phy. Rev. B. 19, 4282 (1979).
14. Brusche P, Buhrer W and Perkins R
    Jr. Phy. C 10, 4023 (1977).
15. Sudharshanan R, Ph.D. Thesis I.I.T. Madras.
16. Bottger G L and Geddes A L
    Jr. Chem. Phy. 57, 1215 (1972).
17. Hanson R C, Fieldly T A and Hochheimer H D
    Phy. Stat. Solidi B 70, 567 (1975).
18. Iazau Nakamoto
    Infra Red Spectra of Inorganic and Coordination Compounds
    Chapter II, J. Wiley and Sons
    New York (1970).

TABLE 1

Compound	Ionic conductivity $Ohm^{-1}Cm^{-1}$	Activation Energy (ev)
AgI	5.1×10^{-7}	0.46
$Ag_{0.9}Pb_{0.1}I$	1.37×10^{-5}	0.0192
$Ag_{0.8}Pb_{0.2}I$	1.16×10^{-5}	0.0146
$Ag_{0.7}Pb_{0.3}I$	1.95×10^{-5}	0.0195
$Ag_{0.6}Pb_{0.4}I$	1.20×10^{-5}	0.0212
$Ag_{0.5}Pb_{0.5}I$	3.08×10^{-5}	0.0202
$Ag_{0.4}Pb_{0.6}I$	3.89×10^{-5}	0.0142
$Ag_{0.3}Pb_{0.7}I$	4.85×10^{-5}	0.0152
$Ag_{0.2}Pb_{0.8}I$	2.54×10^{-5}	0.0194

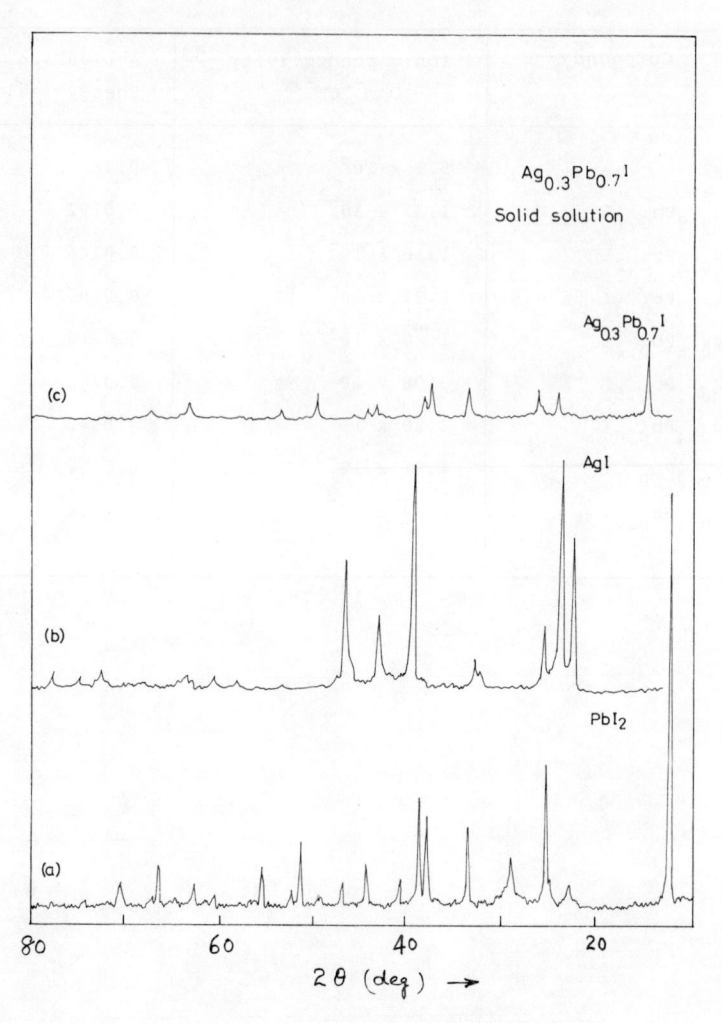

Fig.1 The x-ray diffraction patterns of pure AgI, PbI$_2$ and solid solution Ag$_{0.3}$Pb$_{0.7}$I.

Fig.2 Plot of log σT Vs 10^5/T for $Ag_{0.3}Pb_{0.7}I$.

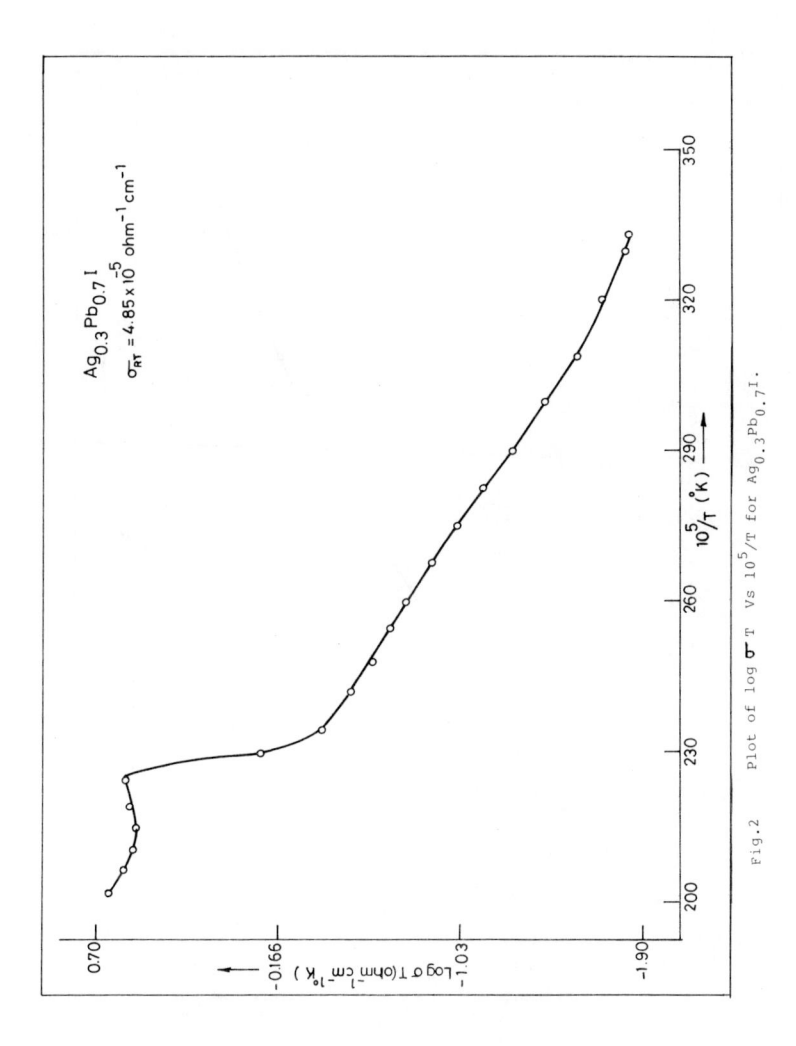

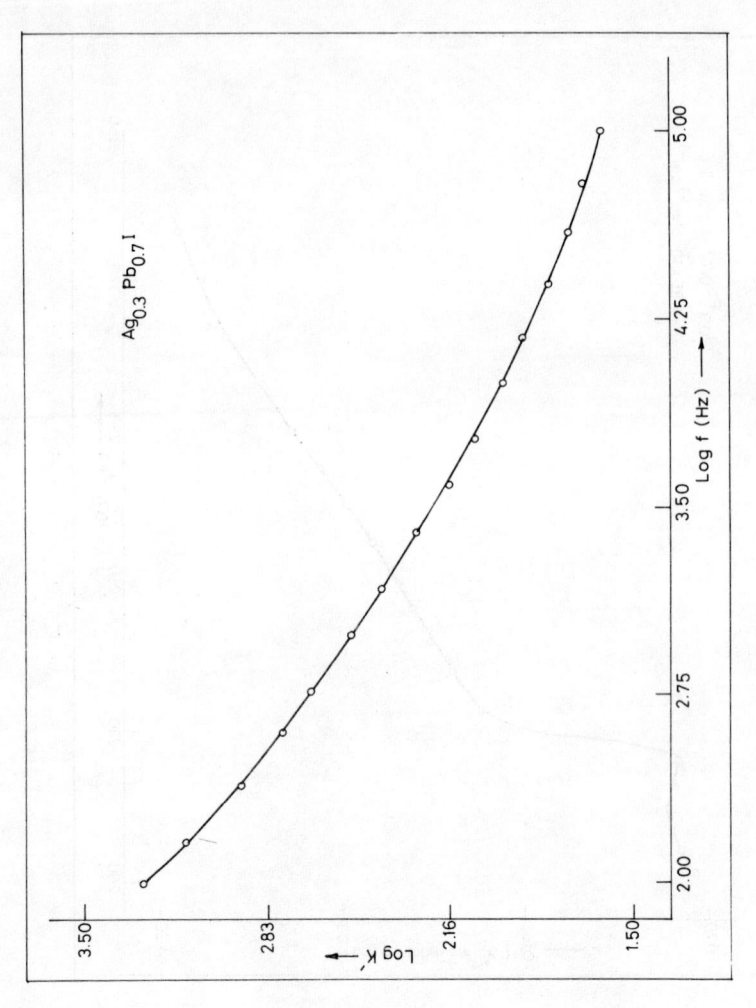

Fig.3 Plot of log f Vs log K' for $Ag_{0.3}Pb_{0.7}I$.

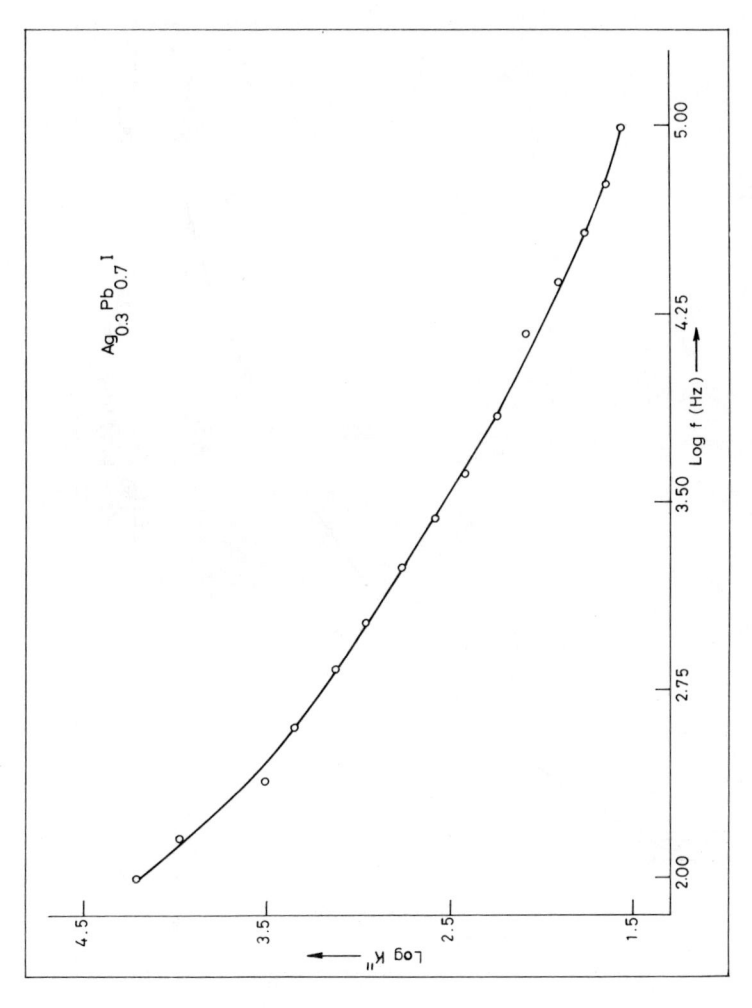

Fig.4 Plot of log f Vs log K" for $Ag_{0.3}Pb_{0.7}I$

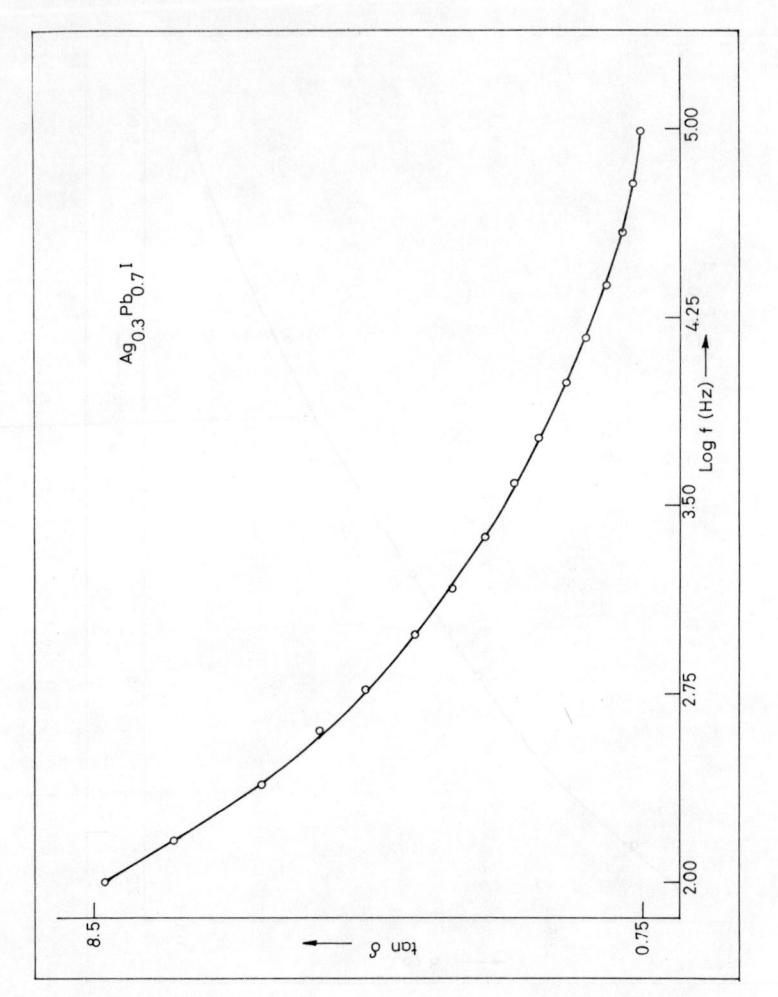

Fig.5 Plot of log f vs tan δ for $Ag_{0.3}Pb_{0.7}I$.

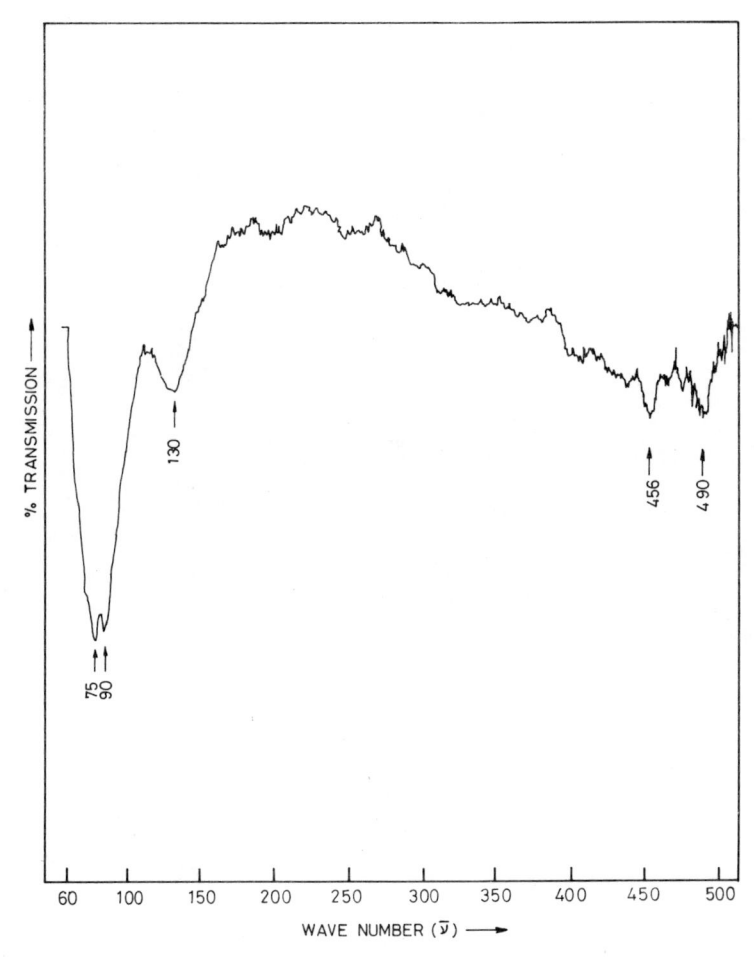

Fig.6 The FIR spectra of the sample $Ag_{0.3}Pb_{0.7}I$ in the region 500 - 50 cm^{-1}.

ADVANCES IN FOURIER TRANSFORM INFRARED SPECTROSCOPY

B. Rambabu and W. Eugene Collins, Department of Physics
Southern University, Baton Rouge, Louisiana - 70813 USA.

INTRODUCTION

Absorption and flurescence spectroscopies the UV and visible regions, Infrared and Raman spectroscopies as well as various resonance spectroscopies are widely used in the characterization of materials. Spectroscopy in the UV and visible regions probe electronic transitions of specific atoms or ions in a solid providing information about the oxidation states of ion, their local symmetry and defect centres. Infrared and Raman spectroscopy provide information on the vibrations of atoms and molecules in the solid. Infrared spectroscopy is well known for the ease of sample preparation, accuracy, cost effective-ness ness and widespread availability of reference spectra.[1] However conventional dispersive IR techniques suffer from serious limitations, for example only small fraction of the source energy reaches the detector, and recording the spectra can be slow. Fourier Transform Infrared Spectroscopy (FTIR) overcomes many of the drawbacks inherent in dispersive IR technique with increased speed and higher signal-to-noise ratios. FTIR instrumentation is currently being applied to studies of surface reactions and catalysis [2-6], biopolymer identification,[7] and interfacing to high performance liquid chromatography[8] and thin layer chromatography.[9]

Theory of FTIR

Infrared radiation can be analyzed spectroscopically either by dispersing it with a prism or grating, or by interference technique.[10] In figure 1, the dispersive IR spectrometer is shown. The source radiation passes through the sample, is geometrically dispersed by a grating and reaches the detector after exiting through a slit. The radiation must be analyzed monochromatically, and thus a long time is needed to obtain a complete spectrum. Because radiation passes through slits, only a small amount of the total source radiation is utilized. Also, the dispersive IR spectrometer is sensitive to stray light, may contain many moving parts and does not possess a method for internal calibration.

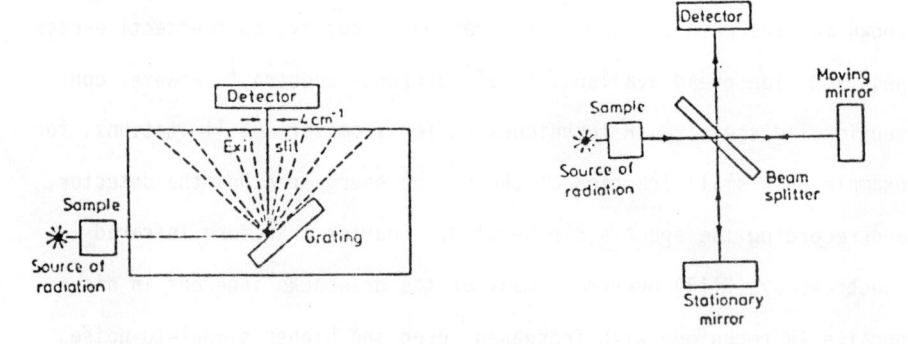

Fig. 1: Opitcal diagram of the IR dispersive spectrometer(Ref. 11, p. 89)

Fig. 2: Optical diagram of the interferometer (Ref. 11, p. 90).

Fourier Transform Infrared Spectroscopy (FTIR) is an interference technique which through recent advances in digital computer technology has demonstrated in superiority to dispersive IR techniques. In figure 2 a Michelson Interferometer is shown to contain a source, detector, stationary and moving mirror, and a beam splitter. The beam splitter the crystal of potassium bromide coated with germainium,[10] splits the incident light equally reflecting half to the stationary mirror while transmitting the other half to the moving mirror. As the moving mirror tracks an optical path length, the two beams are reflected back to the detector, where they interfere either constructively or destructively forming an interferogram. For monochromatic radiation of optical frequency , the intensity of light reaching the detector as a function of optical path difference x is found to vary cosinusoidally and can be expressed as [1, 10, 11]

$$I(x) = B(\bar{\nu}) \cos(2\pi\bar{\nu}x)$$

$B(\bar{\nu})$, the source intensity as a function of optical frequency, is equal to the product of the reflectance of the beam splitter, the transmittance of the beam splitter, and the input energy at frequency . For polychromatic radiation, the interferometer output is a sum of all the interferences as each component interferes destructively or constructively with every other component. The interferogram intensity as a function of optical path length, $I(X)$, and the source intensity as a function o optical frequency $B(\bar{\nu})$ is given by the cosine Fourier transform:[10]

$$I(x) = \int_{-\infty}^{+\infty} B(\bar{\nu}) \cos(2\pi\bar{\nu}x) \, d$$

and the inverse transform:

$$B(\bar{\nu}) = \int_{-\infty}^{+\infty} I(X) \cos(2\pi\bar{\nu}x) \, dx = 2\int_{0}^{+\infty} I(x) \cos(2\pi\bar{\nu}x) \, dx$$

which equates the optical spectrum $B(\bar{\nu})$ to the interferogram.

The Fourier transform technique, although known to Michealson in 1892,[1] was impossible to implement due to the size and complexity of the required calculations. This interference technique remained unused untill 1965, when the fast Fourier transform (FFT) algorithm was developed by Cooley and Tukey[1, 12].

The FFT method allows rapid calculations of Fourier transforms by reducing the number of computer additions and multiplications from n (n-1) operations to $n\log_2 n$ operations. For an 8192 point energy, the conventional Fourier transform method requires 8192 x 8191 = 6.7 x 10^7 operations while the FFT method reduces the time of calculation by an order of 630 to 1.6 x 10^5 operations. With modern fast array processors, these calculations can be carried out in less than a second.

Because dispersive IR systems sample one wavelength at a time, while FTIR samples all wavelengths over the interferometer pathlength, a complete IR spectrum can be measured in the time it takes a dispersive IR to measure one resolution element. For a spectrum with N resolution elements, the interferometric systems are N times faster and thus have N greater analytical sensitivity than conventional IR systems.[10] This is called the multiplex or Felgett's advantage as shown in Fig. 3a. In addition, the FTIR can scan one spectrum N times in the time it takes a dispersive IR to scan the spectrum once. This again increases the analytical sensitivity of the FTIR technique. Interferometers also have greater "throughput" than dispersive IR methods. Throughput, a measure of the optical efficiency of the system, is defined as the product of the area and solid angle of the beam passing from the source to the detector. Jacquinot's advantage, as shown in Fig. 3b, is not as important as Felgett's advantage, because optimal use of a detector's signal-to-noise ratio occurs when less than a maximum throughput is employed.[1,13] To provide internal calibration in the FTIR system, a helium-neon laser (Fig. 3c) is used. This is referred to as Conne's advantage. FTIR systems are also superior to conventional IR systems in that they are insensitive to stray light and heat effects.[10] Primary disadvantages of FTIR are high equipment cost and lack of reference spectra.

3a. FELGETT'S ADVANTAGE

3b. JACQUINOT'S ADVANTAGE

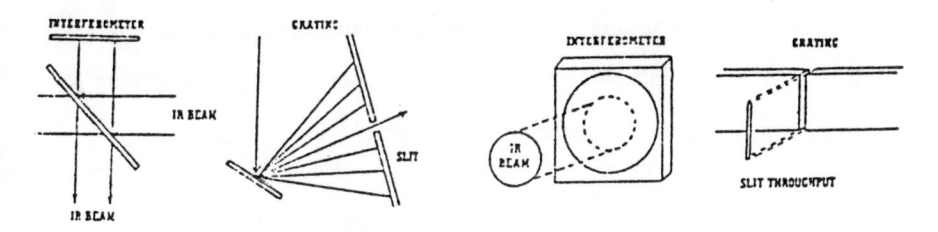

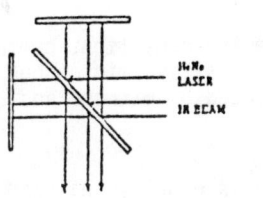

3c. CONNE'S ADVANTAGE

Fig. 3: FTIR advantages (Ref 10, p. 246).

FTIR Applications

Because of the increased sensitivity, FTIR has been interfaced with several well known spectroscopic techniques, notably photoacoustic and diffuse reflectance spectroscopies. Diffuse reflectance shown in Figure 4b, measures the distinct component of reflected light whose direction is unrelated to that of incident radiation.[14,15] With the aid of the high sensitivity of FTIR, the low intensity diffuse light isotropically scattered by the sample is analyzed. Developed in 1920's diffuse reflectance is used in the determination of exact colors for industry and analytical biochemistry.[16] In the photoacoustic experiment shown in Fig. 4 a, intensity modulated light enters a gas filled cell through a KBr window.

The incident radiation is absorbed by the sample, causing periodic pressure waves in the coupling gas which are detected by sensitive microphone.[17] The photoacoustic effect was reported by A. G. Bell in 1880.[18] Diffuse reflectance Fourier transform infrared spectroscopy (DRIFT) and Fourier transform infrared photoacoustic spectroscopy (FTIR - PAS) are complementary techniques, both capable of studying powders, dusts, surfaces and depth profiles.[19]

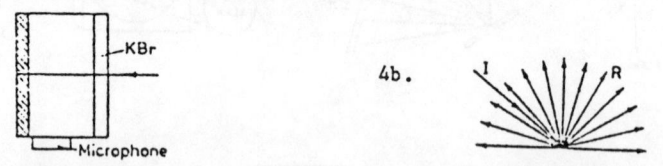

Fig. 4: a) PAS cell, the incident light produces pressure fluctuations which are detected by a sensitive microphone b) Diffuse reflectance, the scattered light is collected by mirrors and directed to the detector.
(Ref. 11, p. 109)

DRIFT Studies

Diffuse reflectance is usually analyzed by means of the Kubelka-Munk equation:[11, 14, 16]

$$f(R_\infty) = (1-R_\infty)^2/2R_\infty = k/S$$

where R_∞ is the ratio of the diffuse reflectance measured at infinite depth (the depth at which the signal remains constant) to that of a reference, K is an absorption constant and S is a scattering constant. When S remains constant and the absorption bands of the sample are of low absorptivity, the DRIFT Spectrum, as seen with poly proplene in Fig. 5a and IRGANOX (an antioxidant) in Fig. 5b. The greatest contribution to $f(R_\infty)$ appears from depths of less that 0.5mm, with in a maximum depth of 2-3 mm.[15]

In order to reduce scattering intensities, a KBr overlayer (50mg) which scatters light isotropically, is placed on top of the fibers.[20] An overlayer technique is employed in depth profiling studies, using either KBr or diamond powder.[21]

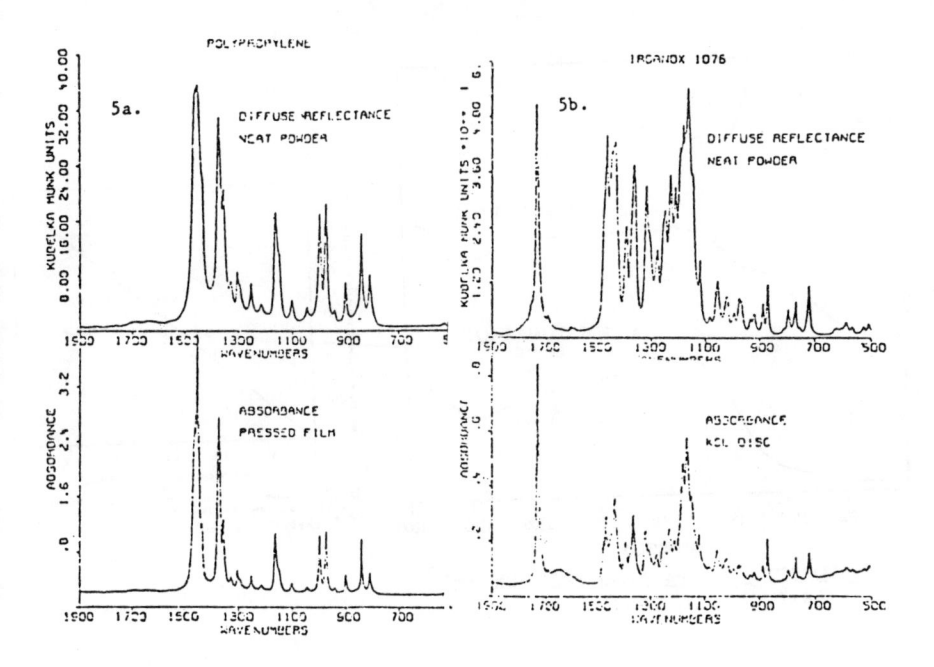

Fig. 5: Comparison of the DRIFT and absorbance spectra of a) polypropylene and b) IRGANOX (Ref. 14, p. 635.

DRIFT has been applied to quality control characterizations of an E-glass fiber/r-APS (-aminopropyltriethoxilane) coupling agent system. Silane coupling agents are applied to glass fibers to reinforce material characteristics.[20] Fig. 6 shows the DRIFT spectrum in the 900-1800 cm^{-1} region of this system with a KBr overlayer. A) is the spectrum for heat cleaned E-glass fiber B) is the spectrum for -APS treated E-glass and C) is the subtraction OF AO and B).

This difference spectrum resembles that of γ-APS. Linear combination curves are constructed as shown in Fig. 7 by equating band intensity in the C-H region to γ-APS concentration.[22] The high sensitivity of DRIFT allows characterization of small amounts of surface agent on glass fibers.[23]

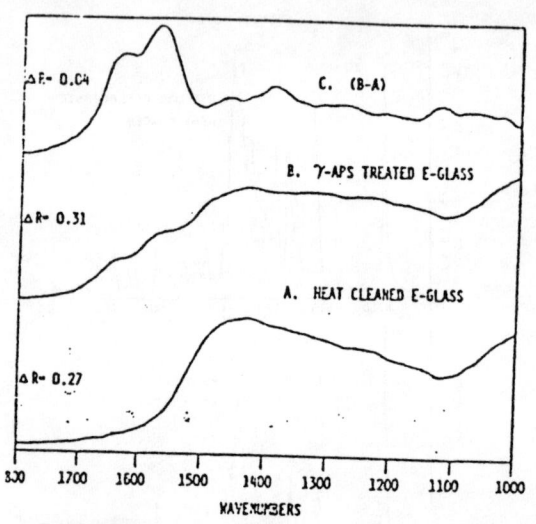

FIG. 6: DRIFT spectra from 1800-900 cm^{-1} (Ref. 19, p. 790)

FIG. 7: Calibration curve for orientation-averaged spectra. Normalized integrated intensity of C-H stretching region used to construct curve. (Ref. 20, p. 410)

FTIR-PAS Studies

Q, photoacoustic signal intensity is given by:[27]

$$Q = a\, u\, f^{-3/2}$$

where a is the optical absorptivity in cm^{-1}, U is the thermal diffusion length in cm, and f is the beam chopping frequency is given by:[27]

$$f = 2\, v\, w$$

where v is the moving mirror speed in cm sec^1 and w is the infrared frequency in cm^{-1}. By altering the interferometer mirror velocity, which is related to the light modulation frequency, an effective penetration depth can be chosen.[11]

FTIR-PAS is an excellent technique for the study of films[28] and subsurfaces[29] or any opaque or strongly scattering materials.[30] FTIR-PAS employs non-destructive techniques, instead of the conventional crushing of the sample into a KBr matrix which may introduce artifactual effects. As a result, spectra of biopolymers have been taken which yield information on secondary structure. The FTIR-PAS spectrum of poly (r-benzyl glutamate) was shown to be an r-helical structure, confirming Raman, x-ray diffraction, and conventional IR studies.[7]

FTIR-PAS is highly dependent on particle size. For particle sizes less than 20 mm, absorption is complex due to similarity insize of the thermal diffusion length and the optical length. For particles of larger size, this problem can be overcome by a suitable variation of the mirror velocity.[31] FTIR is also dependent on heating effects, as the excitation of the PAS cell gas molecules to produce pressure waves can be viewed as heating process. Randomly placing a sample of film in a PAS cell is not the best way to reproduce an absorbance spectrum. If the effect of heating from the rear surface of a film is not corrected for by proper positioning, the FTIR-PAS spectra may be structure less.[32]

Conclusion

Fourier transform infrared spectroscopy has been demonstrated to have superior analytical sensitivity when compared to conventional IR spectroscopy. As a result of this improvement FTIR spectroscopy has been successfully interfaced to a number of sensitive techniques. Surfaces and depth profiling problems are presently being elucidated with FTIR. In the future, greater emphasis will be placed on increasing the range of analytical techniques such as thin-layer chromatography and in the elucidation of catalytic mechanisms.

Bibliography

1. Griffiths, P. R. "Chemical Infrared Fourier Transform Spectroscopy", Wiley: New York, 1975.
2. Busca, G.; Zerlia, T.; Lorenzelli, V.; Girelli, A. J. Catal., 1984, 88, 125-136.
3. Kaul, D. J.; wolf, E. E. J. Catal. 1985, 91, 216-230.
4. Kaul, D. J.; Wolf, E. E. J. Catal. 1985, 93, 321-330.
5. Sayed, M. B.; Kydd, R. A.; Cooney, R. P. J. Catal. 1984, 88, 137-149.
6. Edwards, J. E.; Schrader, G. L. J. Catal. 1985, 94, 175-186.
7. Renugopalakrishnan, V.; Bhatnagar, R. S. J. Am. Chem. Soc. 1984, 106, 2217-2219.
8. Conroy, C. M.; Griffiths, P. R.; Jinno, K. Anal. Chem. 1985, 57, 822-825.
9. Zuber, G. E.; Warren, R. J.; Begosh, P. P.; O'Donnell, E. L. Anal. Chem. 1984, 56, 2935-2939.
10. Mile, C.; Guiliano, M.; Reymond, H.; Dou, H. Intern. J. Environ. Anal. Chem. 1985, 21, 239-260.
11. Koenig, J. L. Advances in Polymr Sci. 1983, 54, 89-134.
12. Cooley, J. W.; Tukey, J. W. Math. Comput. 1965, 19, 297.

13. Hirshfeld, T. Appl. Spectrosc. 1985, 39 (6), 1086-1087.

14. Frei, R. W.; MacNeil, J. D. "Diffuse Reflectance Spectroscopy in Environmental Problem Solving," CRC Press: Cleveland, Ohio, 1973.

15. Chalmers, J. M.; MacKenzie, M. W. Appl. Spectrosc. 1985, 39 (4), 634-641.

16. Wendlandt, W. W.; Hecht, H. G. "Reflectance Spectroscopy," Interscience New York, 1966.

17. Pao, Y. H. "Optoacoustic Spectroscopy and Detection," Academic Press: New York, 1977.

18. Bell, A. G. Am. J. Sci. 1880, 20, 305.

19. Rosencwaig. A. "Photoacoustics and Photoacoustic Spectroscopy," Wiley-Interscience: New York, 1980.

20. McKenzie, M. T.; Culler, S. R.; Koenig, J. L. Appl. Spectrosc. 1984, 38 (6), 786-790.

21. Brackett, J. M.; Azarraga, L. V.; Castles, M. A.; Rogers, L. B. Anal. Chem. 1984, 56, 2007-2010.

22. Mckenzie, M. T.; Koenig, J. L. Appl. Spectrosc. 1985, 39 (3), 408-412.

23. Culler, S. R.; McKenzie, M. T.; Fina, L. J.; Ishida, H.; Koenig, J. L. Appl. Spectrosc. 1984, 38 (6), 791-795.

24. Fuller, M. P.; Griffiths, P. R.. Appl. Spectrosc. 1980, 34 (5), 533-539.

25. Schultz, T. P.; Templeton, M. C.; McGinnis, G. D. Anal. Chem. 1985, 57, 2867-2869.

26. Meldrum, B. J.; Orr, J. C.; Rochester, C. H. J. Chem. Soc., Chem. Comm. 1985, 17, 1176-1177.

27. Koenig, J. L. Pure and Appl. Chem. 1985, 57, 971-976.

28. Rockley, M. G. Chem. Phys. Lett. 1979, 68 (2,3), 455-456.

29. Teramae, N.; Tanaka, S. Appl. Spectrosc. 1985, 39 (5), 797-799.

30. Highfield, J. G.; Moffat, J. B. Appl. Spectrosc. 1985, 550-552.

31. Rockley, N. L.; Woodward, M. K.; Rockley, M. G. Appl. Spectrosc. 1984, 38 (3), 329-334.

32. Teremae, N.; Tanaka, S. Anal. Chem. 1985, 57, 95-99.

CHARACTERISATION OF CRYSTALLINE PHASES FORMED DURING
ANNEALING OF SOME METALLIC GLASSES USING X-RAY DIFFRACTION.

S.B. RAJU AND G. SURYA PRAKASA RAO*
DEPARTMENT OF PHYSICS,
ANDHRA UNIVERSITY,
WALTAIR::530 003, INDIA.

X-ray diffraction (XRD) technique is used to characterise the crystalline phases formed during annealing at temperatures 323, 373 473, 573, 673, 773, 873 and 973 K for 2 hours duration in matallic glasses like $Fe_{40}Ni_{40}B_{20}$, $Fe_{82}B_{12}Si_6$, $Fe_{78}B_{13}Si_9$, $Fe_{81}B_{13.5}Si_{3.5}C_2$ and $(Fe\ Co)_{70}(MoSiB)_{30}$. The formation of phases like α-Fe, Fe_2B, Fe_3B, Fesi, Fe_3C, (Fe, Ni) and (Fe, Co) are identified and crystallisation kinetics of amorphous to crystalline transformations are discussed.

INTRODUCTION:

Metallic glasses have attracted a lot of attention in recent years, since they have a combination of remarkable properties, which are directly derived from their glassy state(1), namely soft ferromagnetism, relatively high electrical resistivity, very high tensile strength and excellent corrosion behaviour. However, crystallisation studies of metallic glasses (2,3) revealed that some partially or even fully crystallised metallic glasses possess even improved properties, e.g. higher ductility, less magnetic core losses for high frequency applications or stronger flux pinning interactions in super conducting metallic glasses. The crystallisation studies are important in understanding glass formation and assessing the thermal stability of metallic glasses. Crystallisation of metallic glasses well below the glass transition temperature is found to proceed by nucleation and by growth processes. The driving force is the difference in free energy between the glass and the appropriate crystalline phase(s) and the growth may be primary, eutectic or polymorphic and each with a single activation energy similar to that of diffusion.

2. EXPERIMENTAL:

Metallic glasses of the present study have been obtained from Allied Corporation and Vitrovac Companies. After annealing the samples, the X.ray diffraction plots were taken with a Philips XRD system PW 1730/PW 1390.

3. RESULTS AND DISCUSSION:

Some typical XRD plots obtained for annealed and as cast metallic glasses are shown in Fig.1 and 2. The formation of crystalline phases on annealing for metallic glass $Fe_{82}B_{12}Si_6$ is shown in Fig.1 and the as cast samples of metallic glasses exhibit no peaks or a broad maximum, characteristic of amorphous nature (Fig. 2). The formation of phases is discussed separately for each glass below.

---

*Lecturer in Physics, Department of Physics, Dr.V.S.Krishna Govt. Degree College, Visakhapatnam - 530 003 (A.P.) INDIA.

a) $Fe_{40} Ni_{40} B_{20}$

Annealing at 323 K, 373 K and 473 K for 2 hours also give a broad diffraction maximum in each plot without any indication of formation of crystalline phases. But XRD plots for the sample annealed at 573 K, 673 K, 773 K and 973 K exhibit different peaks, and the possible crystalline phases which can be formed from amorphous $Fe_{40} Ni_{40} B_{20}$ are indicated in Table 1. In a recent study (4) of crystalization of $Fe_{40} Ni_{40} B_{20}$, four types of crystals are identified namely (i) γ (Fe, Ni) + orthorhombic (Fe Ni)$_3$ B (eutectic) (ii) γ (Fe, Ni)+ FCC (Fe, Ni)$_{23}$ B_6 (eutectic) (iii) γ (Fe, Ni) primary and (iv) unknown Fe-rich phases. However in an early study, using Mossbauer Specroscopy for $(Fe_{1-x} T_x)_{80} B_{20}$ glasses (5), where (T=Co, Ni), the crystallisation has been found to occur in two steps.

$$(Fe_{1-x} T_x)_{80} B_{20} \rightarrow (Fe_{1-x} T_x) + (Fe_{1-x} T_x)_3 B.$$
$$\rightarrow (Fe_{1-x} T_x) + (Fe_{1-x} T_x)_2 B.$$

From the Table 1, it is evident that γ (Fe, Ni) is crystallising along with (Fe, B), Fe_3B and (Fe, Ni)$_{23}$ B_6 phases in confirmation with the studies described above.

b) $Fe_{82} B_{12} Si_6$

The annealing of the sample was done at 373 K, 473 K, 573 K, 673 K, 773 K, 973 K. The crystallisation of the sample was found to occur at a lower temperature compared to the earlier Fe-Ni-B metallic glass. A clear formation of Fe phase is observed at 373 K and whose concentration is found to increase gradually with annealing temparature. The formation of a second phase is observed only after annealing at 773 K or above. The formation of phases namely Fe B, Fe_2B, Fe_3B, Fe Si, Fe_3 Si is characterised (Table 1) by comparing the 'd' values obtained with the standard 'd' values. An earlier study of $Fe_{82} B_{12} Si_6$, annealed at temperature of 710 K (6), revealed the formation of Fe-9 at % Si, Fe_2B and $Fe_{77} B_{18} Si_5$. It is also observed that the intesity of the lines corresponding to 20 33° and 36° increase when the sample is annealed above 673 K, indicating the formation of Fe_3B and Fe-Si phases at higher temperature annealing.

(c) $Fe_{78} B_{13} Si_9$

After annealing at 323 K, the XRD plot exhibits only a broad maximum around 20 46°, indicating atomic rearrangement towards formation of Fe_3B in contrast to the earlier sample where the formation of this phase was observed only after α-Fe phase was formed. XRD plots of the sample annealed at 773 K, exhibited a single line whose 'd' value of 2.04Å, can be attributed to Fe_3B phase. By raising the annealing temperature to 973 K, three phases namely α-Fe, Fe_3-Si and Fe_3 B are observed. The change in the crystallisation process in the two Fe-B-Si metallic glasses studied can be attributed to their metalloid concentration being on either side of eutectic metalloid concentration of 20%. Different crystallisation process were reported in our earlier differential thermal analysis studies of the same two glasses (7), which also supported the above observation.

(d) $Fe_{81} B_{13.5} Si_{3.5} C_2$

As in metallic glass $Fe_{82} B_{12} Si_6$, the crystallisation is started for low temperature annealing ie., at 323 K and the phase is identified as $Fe_3 C$ similar to the metallic glass, $Fe_{78} B_{13} Si_9$, where the crystallisation of $Fe_3 B$ is found to happen initially. The formation of $Fe_3 B$ phase is observed only after annealing at 773 K or above. However the X-ray investigations (8) of annealed $Fe_{81} B_{13.5} Si_{3.5} C_2$ at 1073 K for 2 hrs., showed the formation of α-Fe, Fe_2B, Fe_3B, Fe_3C and $Fe_{23}(C, B)_6$. Swartz et al (9) also observed a pattern leargely crystalline in nature and includes the spectrum of Fe_3B and addition of main α-Fe lines to the amorphous spectra. In an another study of the same sample by Roy and Muzumdar (10) the presence of α-Fe and Fe_2B are only reported.

(e) $Fe_4 Co_{66} Mo_5 Si_{15} B_{10}$:

XRD plots of the samples annealed at 323 K, 373 K, 473 K, 573 K indicate no presence of lines indicating the formation of crystalline phases. For the sample annealed at 673 K, a simultaneous formation of $Fe B$, $Fe_2 B$, $Fe_3 B$ is observed for 2θ values around $45°$. When the annealing temperature is raised to 773 K, the intensities of these phases only increase without the formation of new phases. But when the temperature is increased to 973 K, α-Fe phase is observed. This is in confirmation with the general tendencity of the formation of α-Fe phase as a second step when metalloid fraction is above 20%.

In conclusion the peristence of $Fe_3 B$ phase in all the samples and after different annealing treatments indicates an extra stability of this phase around 20 at % of B, in accordance to thermomagnatic measurements by Tarnozi et al (11).

T A B L E - I

S.No.	Sample	Crystalline phases Identified
1.	$Fe_{40} Ni_{40} B_{20}$	$\gamma(Fe, Ni)$; $Fe_3 B$; Fe_2B; $(Fe Ni)_{23}B_6$
2.	$Fe_{82} B_{12} Si_6$	α-Fe ; Fe_2B; $Fe B$; Fe_3B; $Fe_3 Si$; $Fe Si$
3.	$Fe_{78} B_{12} Si_9$	α-Fe ; Fe_3Si; Fe_3B; (Fe, Si)
4.	$Fe_{81} B_{13.5} Si_{3.5}C_2$	Fe_3B; Fe_3C
5.	$Fe_4 Co_{66} Mo_5 Si_{15} B_{10}$	α-Fe; FeB, Fe_2B; Fe_3B

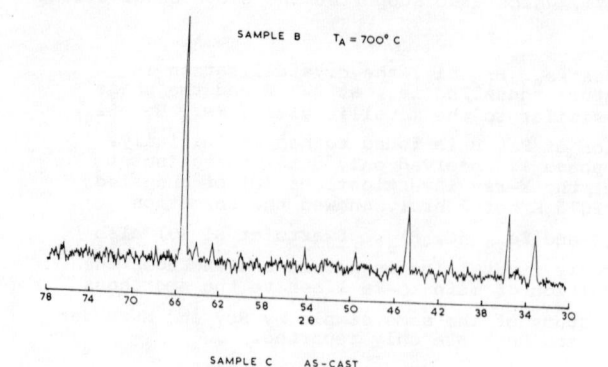

Fig. 1: XRD of annealed $Fe_{82} B_{12} Si_6$

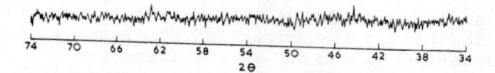

Fig. 2 XRD of as cast metallic glass $Fe_{78} B_{13} Si_9$

REFERENCES:

(1) R.W. Cah. contemp. phys. 21, 43 (1980).

(2) A. Datta, N.J. Decristofero, L.A. Davis, Proc. Rapidly Quenched Metals (Sendai) 4, 1007, 1981.

(3) W.L. Johnson, in 'Glass Metals-1, ed. H,-J. Guntherodt, H. Beck, Topics in Applied physics vol 46, 191, (1981).

(4) V.S. Raja Kishore and S. Ranganathan Proc. Int. Conf. on Metallic and Semiconducting glass, Hyderabad, India (1986).

(5) H.P. Klein, M. Ghafari, A. Ackermann, U. Gonser and H.G. Wagner, Nucl. Instr. Meth. 199 (1982) 159.

(6) Hang Nam OK., A.H. Morrish, Phys. Rev. B 22, 3471 (1980).

(7) G. Suryaprakasa Rao, and S.B. Raju. To appear in Hyperfine Interaction, May, 1987. Proc. of Int. Conf. on Metallic and Semiconducting glasses. Universities of Hyderabad (Hyderabad) India (1986).

(8) B. Bhanu Prasad, Anil K. Bhatnagar, and R. Jagannathan, J. Appl. Phys. 54, 2019 (1983).

(9) J.C. Swartz, R. Kossowsky, J.J. Hangh and R.F. Krause, J. Appl. Phys. vol. 52 (5) 3325 (1981).

(10) Ratnamala Roy, and A.K. Majumdar, J. Magn. Magn. Mater. 25, 83 (1981).

(11) T. Tarnoczi, I. Nagy, C. Hargitai, M. Hosso, IEEE. Trans. Magn. MAG-14. (1978) 1025.

Thermoluminescence Induced by TSFE - a new analytical tool.

S. Murali Dhara Rao[*], K.S.V. Nambi[+] and M.P. Chougaonkar[+]

[*] Technical Physics & Prototype Engineering Division
[+] Health Physics Division
Bhabha Atomic Research Centre, Trombay, Bombay-400 085.

ABSTRACT :

Thermoluminescence was observed upon exposing thin single crystalline CaF_2 to the field emission electrons from a pyroelectric material subjected to heating and cooling cycles. The glow curves of CaF_2 are observed to be similar to those obtained on γ-irradiation. Since the field electrons from the materials are available at elevated temperatures depending on the pyroelectric material, it is possible to irradiate crystals at various temperatures and analyse the resulting glow curves. Typical results obtained with CaF_2 using Lithium Niobate as the TSFE source are presented. The activation energies of the peaks obtained with this method are compared with the published data on γ-irradiated crystals.

TSFE - Thermally Stimulated Field Emission.

INTRODUCTION

Light emission from pyroelectric materials (PEM) during heating and cooling cycles was observed by earlier workers [1,2]. Nambi [3] investigated this phenomenon in detailed; and concluded that the bursts of light are caused by the ionisation of the air or gas surrounding the PEM. Reosenblum et al [4] observed that free electrons are emitted when $LiNbO_3$ is heated and called this 'Thermally stimulated Field Emission' (TSFE). Later it was shown [5,6] that CaF_2 crystals exposed to the field emission from green tourmaline exhibited thermoluminescence (TL) on subsequent heating. The present paper briefly reviews the earlier work and examins the possibility of using this as an analytical tool in the study of crystalline defects.

EXPERIMENTAL

All the materials used in the present investigations namely CaF_2, CaF_2 Dy, LiF;Mg used as TL samples and the PEM crystal, $LiNbO_3$ were grown in the laboratory using highly pure (99.99%) starting materials.

The TL glow curve recorder used for the experiments is described elsewhere [7] and employs a flat heater to accommodate crystalline samples measuring 8mm X 8mm X 0.5mm. Thin samples are used in order to transfer heat to the PEM without over heating the sample. Since $LiNbO_3$ has a high curie temperature, it can be used for high temperature TSFE exposure of the samples. It would then permit the study of materials exhibiting complex peak structures. The samples in this case are heated with the PEM on top, to the temperature when the desired low temperature peak is absebt. The PEM is removed from the sample surface while holding the heater at this temperature. After this the sample is allowed to cool by switching off power to the heater. It is thus ensured that the sample is not exposed to the TSFE during the cooling cycle of the PEM. If the lower temperature peaks are to be observed, the sample is cooled to room temperature with the PEM and thereby filling the traps causing low temperature peaks.

RESULTS AND DISCUSSION

Fig. 1(a) and (b) show the TL glow curves obtained on a CaF_2 sample subjected to two and three cycles of TSFE exposure. Fig.1(c) shows the heights of different peaks as a function of temperature. This shows that the TSFE dose is linear i.e. the amount of dose is proportional to the number of cycles.

Thus for a material exhibiting poor TL response or for enhancing the intensity of a particular TL peak one has to give more TSFE cycles.

Fig. 2 gives the TL glow curve of a CaF_2 sample exposed to one cycle of TSFE from a $LiNbO_3$ crystal. It is observed that $LiNbO_3$ with higher pyroelectric coefficient produces higher TSFE thereby producing in one cycle the doese given by six exposures with green tourmaline. Six distinct peaks are obtained between room temperature and 500°C. Fig.3b gives the TL glow curve obtained on the same sample subjected to a 100 R γ- dose. The number of peaks and their position is identical. The peak intentities are different, the 80°C peak in particular. This is expected as the crystal in the case of TSFE is at a higher temperature for longer duration of the irradiation and the lower temperature peak is filled only during the cooling cycle while the high temperature peaks are filled during both heating and cooling cycles.

The ability of irradiation at high temperatures is made use of in the present case to evaluate the complex peak structure exhibited by CaF_2 samples. The result of these experiments are shown in fig. 3. Fig. 3 a shows the glow curve recorded on heating the PEM and sample upto 100°C. It is observed that the 80°C peak is absent. Heating upto 190°C removes the 200°C peak and the 260°C peak is removed on heating upto 380°C. The 316°C peak is free from the interference of the other peaks and lends itself to analysis. Assuming first order kintetics the following equation was used to evaluate the activation energies E, of the glow curves. $E = \dfrac{1.51 \; R. \; Tm \; . \; T1}{(Tm - T1)}$

Where k is the Boltzman constant,
   Tm is the peak temperature
and Tl is the temperature at half intensity of the
   ascending portion of the glow curve.

The activation energies so obtained are tabulated in Table I. The values obtained by Kolaly[8] are also shown alongsid for comparison. The present values are higher but the peak temperatures are also higher. More accurate values may be obtained by using curve fitting methods which are being carried out.

CONCLUSION

It is demonstrated that TSFE can be effectively used for the analysis of defects in crystals in a similar manner as X- and γ- radiations. The added advantage with TSFE is that the sample need not be removed from the heater strip during the entire experiment. If the same, has to be done with X and γ-radiations, it would require complex heating cells which can be put into the irradiating cells. If the TL material is not sufficiently sensitive, it is necessary to increase the number of TSFE cycles to obtain sufficient dose.

Acknowledgements : The authors are greatful to Shri S.D. Soman, Associate Director, Radiological Group and Shri M.K. Gupta, Head, Technical Physics & Prototype Engineering

Division for their constant encouragement and interest in the present work.

References :

1. G.D. Robertson and N.A. Baily. J. Appl. Physics 39 (1968) 2905.

2. W.A. Syslo. Proc. V. int. symp. on Exoelectron emission and desimetry. Z vikor (eds. A. Bohum and A. Scharman) (1976) 207.

3. K.S.V. Nambi Phys. stat. Solidi(a) 82 (1984) K-71.

4. B. Rosenblum, P. Braunlich and J.P. Carrico. Appl Physics. Letter 25 (1974) 17.

5. S. Murali Dhara Rao, K.S.V. Nambi and M.P. Chougaonkar. National Conference on Thermoluminescence and its applications Ahemadabad 1984.

6. K.S.V. Nambi, S.M.D. Rao and M.P. Chongaonkar Nuclear Tracks, 10 (1985) 243

7. C.M. Sunta. Ph D Agra University (1971)

8. M.A. El-Kolaly Ph D Thesis Bombay University (1977).

TABLE I

Activation Energies of the TL peaks

Peak No.	Tm°C	T1°C	E(ev)	E(ev)Ref.8
2	202	182	1.09	1.2
3	262	240	1.62	1.7
4	316	292	1.80	1.9

358

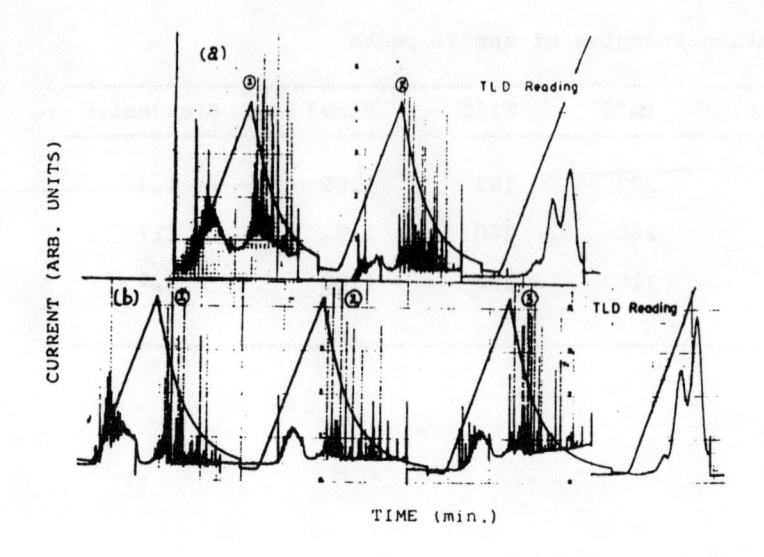

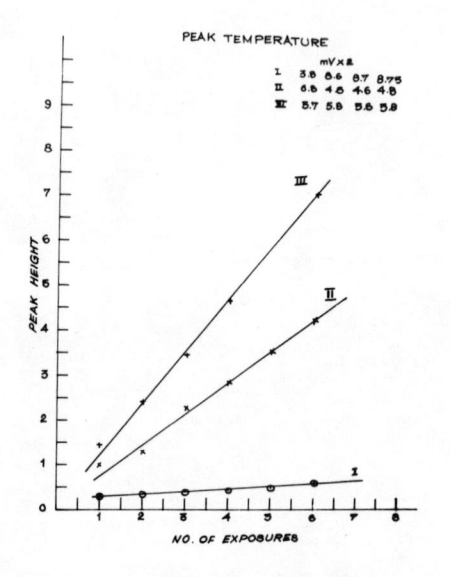

Fig.1 (a) and (b) TL Glow curves of CaF_2 exposed to two and three TSFE cycles respectively from green tourmaline

(c) TL glow peak height Vs. number of TSFE cycles from green tourmaline.

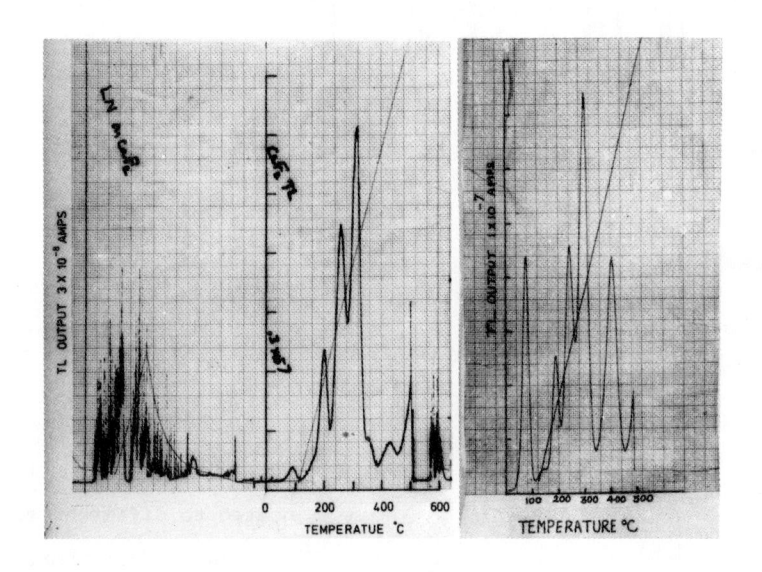

Fig 2. a) TL glor curve of CaF$_2$ subjected to TSFE - dose from LiNbO3.

b) TL glow curve of CaF$_2$ subjected to 100R r-dose from a 60 Co source.

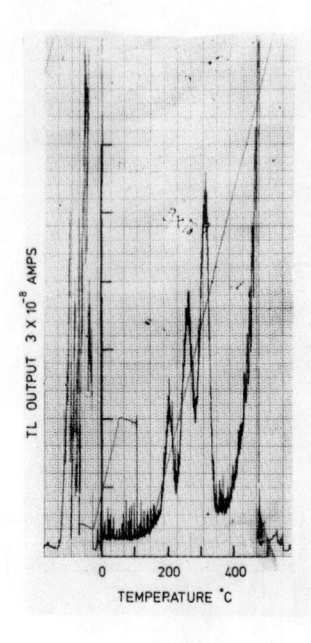

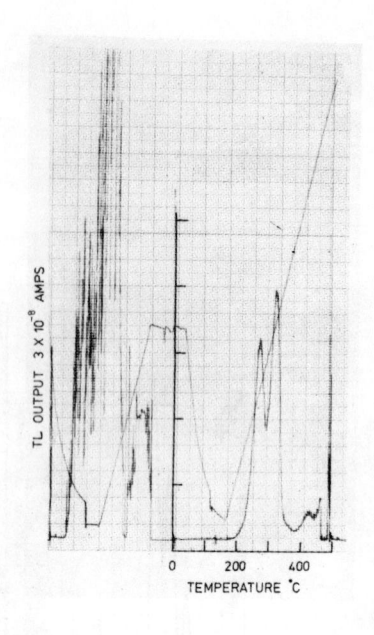

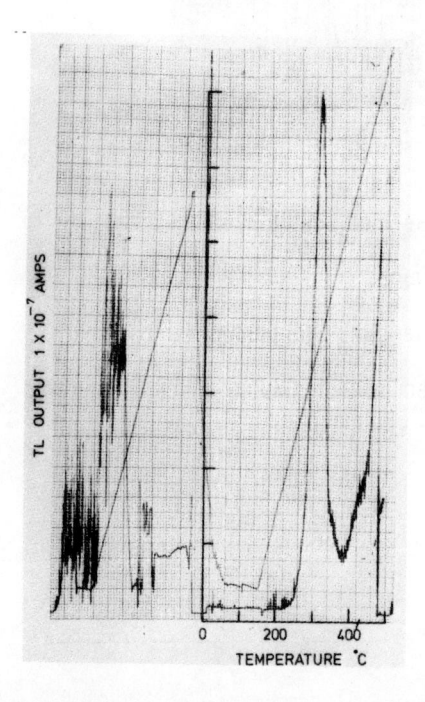

Fig. 3 CaF$_2$ with LiNbO3 on top heated to different temperatures before LiNbO3 is removed and CaF$_2$ cooled to room temperature a) 120°C b) 180°C and c) 280°C.

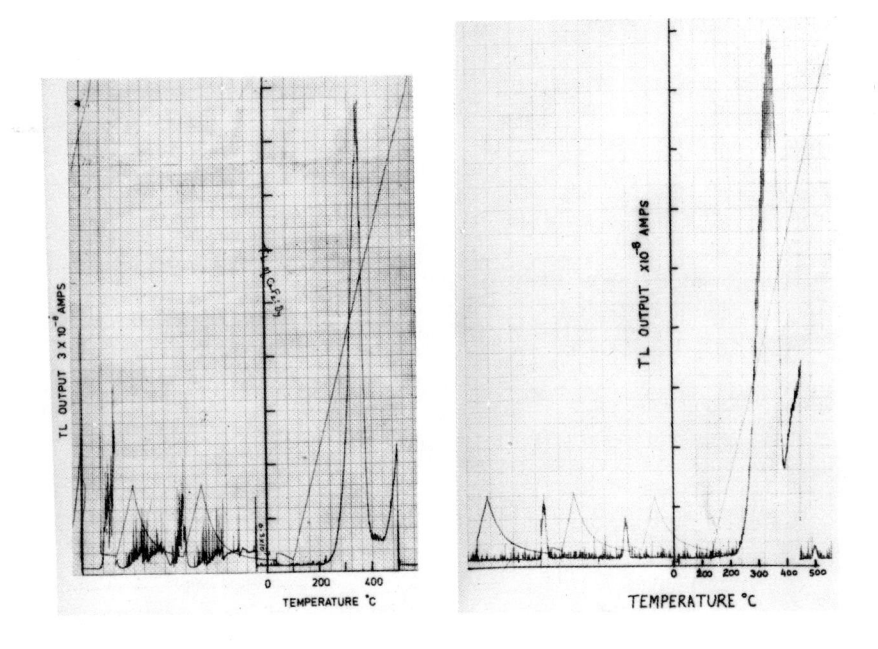

Fig. 4 TL glow curves recorded after TSFE dose from
LiNbO3 a) CaF$_2$: Dy (0.2 wt%) b) LiF:Mg (0.04 wt%)

APPLICATION OF FOURIER TRANSFORM INFRARED (FT-IR)
SPECTROSCOPY BY PYROLYSIS TECHNIQUE TO POLYMERIC SYSTEMS

Dr Seow P.K., Dr Yong, W.M. & Dr Mohinder Singh, M.
Rubber Research Institute of Malaysia
Kuala Lumpur

The recent acquisition of an important research-grade FT-IR instrument by the Rubber Research Institute of Malaysia has enabled the Institute to harness this powerful technique to better serve the Rubber Industry in Malaysia. Using this technique assisted by powerful Library Search Programme and advanced computation techniques, preliminary investigations on chemical and compositional characterization of polymeric materials found in the local industry were made.

This paper describes the application of FT-IR Spectroscopy using the currently familiar sampling technique of pyrolysis, (viz Pyrolysis FT-IR) to commonly encountered commercial polyblends of NR-SBR vulcanizates. The computer-assisted chemical and compositional characterization of polyblends is shown to provide faster identification and quantitative analysis of polymeric materials for the Rubber Industry.

INTRODUCTION

The modern FOURIER TRANSFORM INFRA-RED (FT-IR) Spectrometer is a more sophisticated instrument using the powerful Michelson Interferometer (Fig.2A) to cause an interference pattern of constructive and destructive bands. Thus, in place of the diffraction type of monochromator (Fig.1), the FT-IR Spectrometer requires very sophisticated Michelson Interferometry where one mirror is fixed while the other mirros is mobile: thus generating an optical path difference leading to the resultant interferogram in figure 2B. Each spectrum has its own interferogram. Each interferogram is a unique waveform. Each data point on the waveform represents all the wavelengths: i.e. all the spectral information we require are trapped in the interferogram. By the process of a complicated mathematical manipulation aided by computer, these data points can be decoded to give the spectral waveforms. Thus Fourier Transform is the process by which a normal waveform spectrum could be generated from the corresponding interferograms produced by the Michelson Interferometer.

Through the use of Michelson Interferometer, the FT-IR spectrometer modulates polychromatic radiation in time, rather than in space. By virtue of its Interferometric principle, the FT-IR instrument attains high resolution, enchanced signal-to-noise ratio and high optical energy throughout. On-line data manipulation via sophisticated computer system enables many tedious data-handling jobs to be accomplished smoothly. With the availability of Difference Spectroscopy made possible by appropriate softwares, the FT-IR spectrometer makes it possible even to quantify components present in minute quantities.

FT-IR INSTRUMENTATION IN RRIM

There has been significant advancement in the state-of art in FT-IR instrumentation[1&2] and in the speed of IR data acquisition and processing. Depending on the resolution desired, most modern FT-IR instruments can produce in a matter of seconds a complete IR spectrum as displayed on a VDU. Besides the remarkably lower signal -to-noise ratio, the Jaquinots' advantage[3], the Multiplex advantage[4] and the Cones advantage[5] and the fast Fourier Transform process (fft)[6] are features common to all well-designed FTIR systems.

However, the precision of the spectroscopic measurements vis-a-vis the optical efficiency of the system is very much a function of the stability of the Michelson Interferometer system. Thus it is critical that the moving-mirror mounting in the Michelson Interferometer should remain strictly in optical alignment for good resolution. The FTIR instrumentation SIRIUS 100, acquired by the RRIM is based on Mattsons cube-corner retroreflector design which automatically compensates for any mirror tilt[7]. Besides being equipped with a Dual Detector System (MCT type A and B contained in one) the SIRIUS 100 FTIR is supported by a STARLAB Computer Network, which subscribes fully to the UNIX operating system developed by Bell Laboratories. The advantages of UNIX based laboratory data processing have been expertly dealt with by White, R.L.[8]. In essence a variety of data manipulation and transfer operations can be simplified by implementing the UNIX operating system.

PROBLEMS OF POLYMER ANALYSIS AND CHARACTERIZATION

Infrared Spectroscopy has long been applied to the study of polymers. It provides valuable informations on three important aspects of polymers and their blends: chemical composition, molecular structure (including taticity and crystallinity) and sequence distributions[9]. With the advent of Fourier Transform Infrared (FT-IR) Spectroscopy, numerous problems in the field of polymer characterization became easily achieved on account of all the advantages of FT-IR cited in previous paragraphs.

Like many laboratories dealing with a wide range of problems the general emphasis has to be on obtaining adequate rather than exhaustive solutions and often the equipment is kept for general use rather than for specific configurations requiring special techniques. Polyblends and composite materials are the most commonly encountered commercial samples in our laboratory. The raison d'etre behind present-day market requirement for minimal developmental cost has enhanced the demand on polymer identification and composition of polyblends of interest to the industry: so much so that there is relatively little work devoted to structural and other polymeric studies. As we often encounter vulcanizate samples the sampling method of choice for the moment is centered on Transmission FT-IR of pyrolyzates obtained from the polymers.

EXPERIMENTAL

Typical to the vulcanized nature of most products received from the Industry the "non-integrated" Py + (FT-IR)[10] is the method of choice for producing the desired "finger print" Spectrum of the polymer. The pyrolyzates of the polymers and their blends or composite materials were obtained from their acetone-extracted materials by gas-flame pyrolysis according to the method 7.2 of the International Standards ISO/Dis 4650.2. In most cases it is sufficient to scan the liquid pyrolyzates mounted onto NaCl sample-cell by means of absorption mode of the SIRIUS 100 FT-IR. All the runs were generally made at 4cm-1 wavenumber resolutions with iris opening at 30%. The wavenumber reproducibility between 2000 and 2200 cm-1 is circa. 0.05cm-1.

COMPUTER ASSISTED POLYMER IDENTIFICATION

The impact of computer data processing on polymer identification and polymer characterization is no doubt remarkable. As shown in figure 4 representing a flow-chart of Computer-aided Analysis of Polymers, polymer identification encompases essentially the three steps on the right-hand side of the chart. For those having the Library Search Programme the left-hand side of the chart applies.

By means of the powerful library-search programme, the first step of the polymer identification was made. Thus the print-out of the library-search for a commercial, solid black-filled sample pyTT20286 for the best five matches are as follow:

SEARCH METRIC USING DERIVATIVE ABSOLUTE DIFFERENCE
Search Report for pyTT20286

Rank	Match-factor	Compound Name
1	181	acetone-extracted SBR1712 rubber
2	236	BR polybutadiene resin
3	282	NBR Krynac 800 rubber
4	286	SBR 1712 rubber in-situ
5	295	NR black vulcanizate

The multiple over-lay of spectra between the SBR 1712 "standard" and the unknown TT20286 (figure 3) confirms the status of the latter. Thus SBR major is immediately recognised but careful observations show the lack of a complete match. Therefore we proceed to the next step: comparing the pick-peak file of the unknown with the standards. Before proceeding further we checked on the reproducibility of the diagnostic peaks derived from spectra of pyrolyzates done on different days. These are presented in Table 1 on Pick-peak Identifications.

TABLE 1. PICK-PEAK IDENTIFICATIONS
Diagnostic peaks in cm-1

pySBR712	698.26	752.26	908.04	966.45	990.79
NRSBRO100	697.29	751.88	907.55	966.44	991.36
pyTT20286	696.98	[741.10]	905.74	966.69	990.07
					1375.66
pyNRwhlx	798.35	887.33	911.07	964.08	1375.95
NRSBR1000	[810.50]	887.21	[ ]	964.22	1375.71
nsNRwhlx	798.87	887.38	910.25	963.95	1375.64

As can be seen from Table 1 the reproducibility of the wave-number position of the dignostic peaks for raw resin SBR1712 as compared to compounded SBR labelled as NRSBRO100 is good and their differences are within limits of 1 wavenumber. For the case of NR, while the correlation between the two samples of raw rubbers were very good, the compounded NR designated as NRSBR1000 showed good reproducibility for the major diagnostic peaks like circa. 887cm-1, but characterized by the presence of adventitious peak or absence of certain dignostic peak. In fact it has been recognised that for compounded and vulcanized rubber there is always this characteristic absence of dignostic peak at circa. 910cm-1. The absorptions at 905.74cm-1 and the adventitious peak at 741.10cm-1 are probably the result of interaction with the presence of compounding agents, etc. It is the presence of a peak of 1375.64cm-1 that cast suspicion on the presence of NR (or IR).

Using Difference Spectroscopy, pySBR712 spectrum was subtracted from TT20286 spectrum giving a resultant difference spectrum. When this difference spectrum was searched against the Library Search Programme, using the Derivative Absolute Difference Mode, NR whole latex was accorded as the best match with a factor of 241, thus confirming the presence of NR in the sample of TT20286.

TABLE 2. FT-IR: PRINCIPAL ABSORPTIONS IN ORDER OF
DIAGNOSTIC VALUE FOR PYROLYZATES OF POLYBLENDS
NR/SBR

sbr cm-1	698	769	908	991	1495	966
	vs	s-fs	s	m	m	m
NRSBR8020	694m	760m		991wm		965m
	887vs	1375s	887vs		1646m	
NRSBR7030	695vs	767m		991m	1498m	965m
	887vs	1375s	887vs		1645m	
NRSBR5050	697vs		907s	991m	1495s	966s
	888vs	1375s	888vs		1645s	
NRSBR3070	697vs	752m	908s	992m	1496∿s	966m
	889m	1375m	889m			
NRSBR2080	697vs	752s	908s	992m	1495s	966∿s
		1375m				
	vs	s-fs	m	m	m	m
NR cm-1	887	1376	798	911	1645	964

In table 2 is presented the absorption peak values of pyrolyzates for various black-filled vulcanizates of NR/SBR compounded rubber. For convenience the vulcanizates for NR and SBR were also given. As can be recognized from the table, there is normally no problem in deciding the presence of SBR and NR in the blends when NR composition is high. This is especially true when the most pronounced absorptions at circa 698cm^{-1}(vs) and 887cm^{-1}(vs) respectively for SBR and NR were taken as diagnostic peaks. But problems only arise when NR constituted only a small proportion in the blends. As for instance, in the polyblend designated as NRSBR2080 887cm^{-1} appears to be absent, (however, sometimes it does appear as a shoulder). A priori, this could lead to the conclusion that the polymer concerned is SBR only. However, the unmistakable presence of absorption peak at 1375cm^{-1} (instead of 1374cm^{-1} for SBR pure), cast doubt as to the validity of the above interpretation. The Absorption subtract spectroscopy as elaborated under the section on quantitative analysis showed that the Difference Spectrum resultant from the subtraction is that of NR.

QUANTITATIVE ANALYSIS BY DIFFERENCE SPECTROSCOPY

There are three methods of approach towards quantifying the polymeric components present in a polyblend (figure 4):

> The Ratio Method
> The Least Square Method
> The Subtraction Method.

The latter method, more commonly known as Difference Spectroscopy has been the motivation of many quantitative analysis in FT-IR. This method is founded on the additive property of absorbances of the components in a blend[11 & 12]. For simplicity consider a two-component system, the absorbance at any wavenumber W can be written as follow:

$$A_{xy} = A^W_x + A^W_y$$

And for a pure component:

$$A_y = A^W_p$$

Then subtraction is obtained as

$$A_{xy} - A_y = A^W_x + (A^W_y - kA^W_p)$$

where k is the subtraction factor.

It is noted that Difference Spectroscopy can identify subtle differences between spectra. In the case of pure blends or copolymers exempt of any processing agents, the subtraction gives a straightforward correspondence between scaling factor and percentage composition of the particular component. Thus for a graft copolymer NR/PMMA consisting of 40% PMMA the following is obtained from the spectra of their pyrolyzates:

	auto scaling factor	
Subtract cxpmma 64 pmma 76	0.3986	
Subtract cxpmma 64 CX 305	0.6010	(ref.fig.5)

In this case no base correction was required to give the desired result.

However in the case of black-filled vulcanizates of NR/SBR polyblends base-correction is required to obtain the best results -

	auto scaling factor
Subtract bcNRSBR8020 bcNR	0.6199
Subtract bcNRSBR8020 bcSBR	0.1588

giving 79.6% NR.

By similar operation using Difference Spectroscopy on the black-filled compounded rubber TT20286 with base-line correction, the following is obtained.

	auto scaling factor
Subtract bcTT20286 bcSBR	0.68
Subtract bcTT20286 bcNR	0.08

Thus giving 89.5% SBR and 10.5% NR.

Thus by Difference Spectroscopy, the subtraction method has been successfully applied to evaluate the polymeric composition of polyblends from spectra of their pyrolyzates. This evaluation of a relatively small amount of NR in a blend was previously not possible with the dispersive IR. The validity of this result is supported by the Difference Spectroscopic data presented above for NRSBR8020 blend.

CONCLUSION

Thus in this introductory contribution of our laboratory, in the field of Fourier Transform Infrared Spectroscopy, we have attempted to give a clear insight to the potentialities of FT-IR applications in the analysis and characterization of polymers. The significant advantage of FT-IR lies probably in their high wave-number reproducibility and the much improved signal-to-noise ratio.

We have yet to tap the potentialities of ATR sampling techniques in FT-IR for better and faster services to the Rubber Industry.

ACKNOWLEDGEMENTS

The authors wish to thank the RRIM for the opportunity to present this paper. The valuable assistance of the followings are recognised: the able assistance of Mr. Lim Tiang Chu, the photounit, Mr. Lai Peng Thoo and Mrs. Mong. Last but not least, thanks to Dr Mohinder Singh, Head of Analytical Chemistry Division, for his enthusiastic support.

REFERENCES

1. McDONALD, ROBERTS INFRARED SPECTROMETRY Anal.Chem. 565 340-363 (April 84).

2. SEOW, P.K. Application of FT-IR to the Aanalysis and Characterization of Polymers and Rubbers. Proc. Malaysian Chemical Congress Kuala Lumpur Nov. (1986).

3. BELL, R.J. Introductory Fourier Transform Spectroscopy Academic Press, New York 1982.

4. GREEN, DAVID & REEDY, GERALD, T. Fourier Transform Infrared Spectroscopy Vol.I ed. FERRARO, JOHN, R. & BASILE, LOUIS, J. Acad. Press 1978.

5. As referred in 3 and 4.

6. JASSE, B. Fourier Transform Infrared Spectroscopy of Synthetic Polymers. Ed. DAWKINS, J.V. in Developments in Polymer Characterization. Appl. Sci. Pub. 1983.

7. WHITE, R.L., COFFEY, D.J., COVEY, J.P. & MATTSON, D.R. High Performance FT-IR using a Corner-Cube Interferometer. Appl. Spectroscopy 39(2) 1985, 320-326.

8. WHITE, R.L. Advantages of UNIZ-based Laboratory Data Processing - Computer Enhanced Spectroscopy 1, (4) 1983 185-189.

9. HAMPTON, ROBERT, R. Applied Infrared Spectroscopy in the Rubber Industry. Rubb.Chem. & Technol. 45(3) 548-612 (1972).

10. SEOW, P.K., MOHD.ROUYAN HJ.BAKAR, TAN, A.S., YONG W.M. & MOHINDER SINGH, M. Advances in Instrumentation in the Analysis of Polymers. Proc. Analab Asia 87 Conference Singapore May 87.

11. KOENIG, J.L. Application of Fourier Transform Infrared Spectroscopy to Chemical Systems. Appl. Spectrosc. 29 293-308 (4) 1975.

12. HIRSCHFELD, T. Instrument requirement for Subtraction Method. Appl. Spectrosc. 30 550 (1976).

370

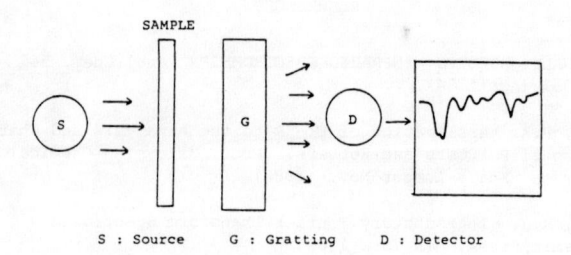

S : Source   G : Gratting   D : Detector

Fig.1. DISPERSIVE INFRARED INFRARED SPECTROMETTER

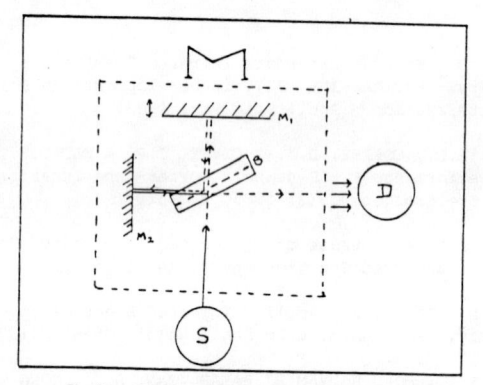

Fig.2A. FOURIER TRANSFORM INFRARED SPECTROMETER

S = source   M = interferometer
D = detector M_1 = mobile mirror
M_2 = fixed   B = beam splitter
      mirror

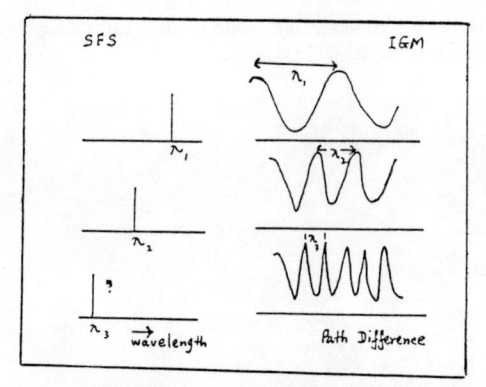

Fig.2B. SFS : SINGLE FREQUENCY SOURCE
        IGM : INTERFEROGRAM

371

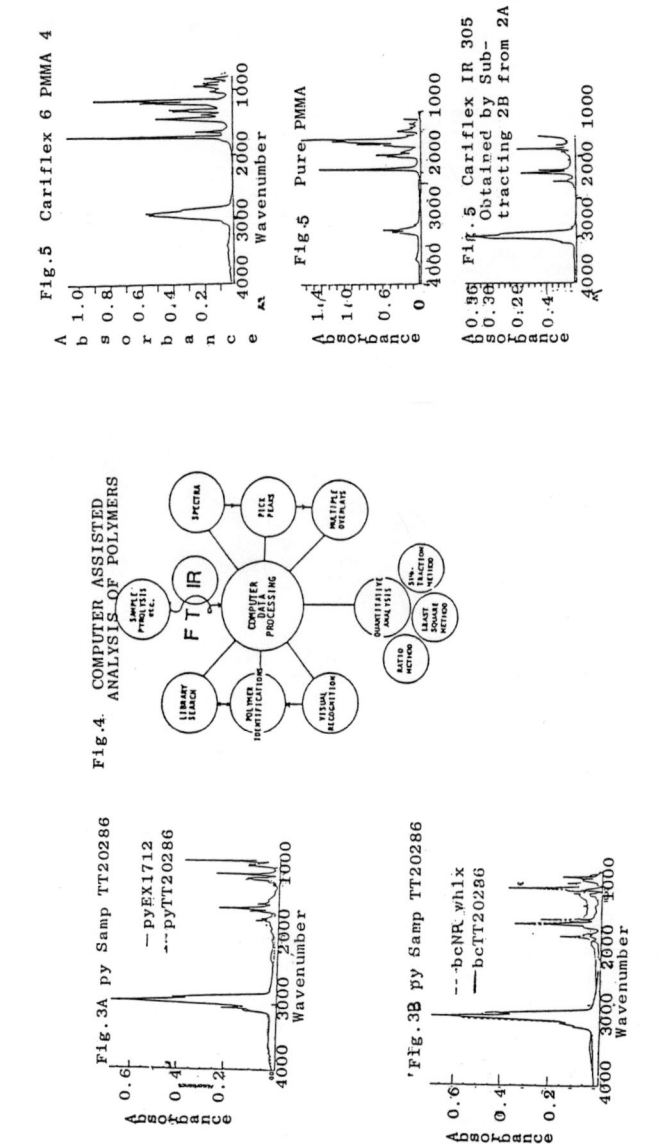

Fig.5 Cariflex 6 PMMA 4

Fig.5 Pure PMMA

Fig.5 Cariflex IR 305 Obtained by Subtracting 2B from 2A

Fig.4. COMPUTER ASSISTED ANALYSIS OF POLYMERS

Fig.3A py Samp TT20286
— pyEX1712
--- pyTT20286

Fig.3B py Samp TT20286
--bcNR.wh1x
bcTT20286

ELECTRICAL CHARACTERIZATION OF $Ag_4I_2WO_4$ SOLID ELECTROLYTE

S. AUSTIN SUTHANTHIRARAJ
Department of Energy
University of Madras
Guindy Campus
Madras-600 025 - INDIA.

Abstract:

The silver ion conducting solid electrolyte $Ag_4I_2WO_4$ in the mixed system $AgI-Ag_2WO_4$ has been taken for the study. Detailed investigations of the temperature dependent electrical conductivity (σ) and electronic conductivity (σ_e) on pellet samples of the solid have been carried out over the temperature range 300-500K. Powder X-Ray Diffraction analysis has been made as to identify the stoichiometry of the samples prepared. The solid has been found to possess a silver ion conductivity of 4.08×10^{-2} $\Omega^{-1}cm^{-1}$ with a negligible electronic conductivity of $4 \times 10^{-11} \Omega^{-1}cm^{-1}$ at room temperature. The present investigation has also indicated that $Ag_4I_2WO_4$ undergoes a phase transition around 432K.

Introduction: Solid electrolytes are most often characterized by their very large possibilities of high ionic conductivity ($\sim 10^{-2} \Omega^{-1}cm^{-1}$) at room temperature), low electronic conductivity ($\sim 10^{-8} \Omega^{-1}cm^{-1}$), stability against decomposition at ambient thermodynamic conditions and mechanical stability. Therefore, there has been an increased interest in the study of both fundamental and technological aspects of solid electrolytes in recent years. Although the science of fast ionic conduction in most of these solids has not yet been clearly understood, many workers have been contemplating new ideas and attempting to go into the structural details of the lattices which are responsible for many a property of superionic conductors.

For instance, the phase diagram of the system $AgI-Ag_2WO_4$ was determined by Takahashi et al (1) and they identified three intermediate compounds namely $Ag_6I_4WO_4$, $Ag_4I_2WO_4$ and $Ag_5IW_2O_8$ having the components of 20, 33.3 and 67m/0 of Ag_2WO_4 respectively. From the practical point of view two of these systems $Ag_4I_2WO_4$ and $Ag_6I_4WO_4$ appear as promising candidates. Our earlier studies on an all-solid-state battery based on $Ag_6I_4WO_4$ have indicated the possible applications of similar compounds in solid state batteries (2). The present investigation concerns with the electrical characteristics of $Ag_4I_2WO_4$ which is known to exhibit an electrical conductivity of the order of $10^{-2} \, \Omega^{-1} cm^{-1}$ at room temperature.

2. **Experimental Considerations:**

2.1 *Synthesis and Analysis:* Initially, Ag_2WO_4 is obtained from precipitation of aqueous solutions of $AgNO_3$ and $Na_2WO_4 \cdot 2H_2O$. The precipitate is first washed with distilled water a number of times and then dried under nitrogen flow at 120°C for 24 hours. Analar grade AgI is then mixed with Ag_2WO_4 in the molar ratio 66.7 : 33.3 and the mixture is sealed in a pyrex tube under vacuum. It is then heated at 400°C and annealed for about 18 hours. The material obtained is allowed to cool down to room temperature. The powder X-Ray Diffraction (XRD) analysis has been made using a Philips X-Ray Generator Unit (model PW 1130) with CuK radiation of wavelength λ = 1.5418 Å.

2.2 *Electrical Conductivity Measurements:* Details of the apparatus fabricated for the electrical conductivity as well as thermoelectric power measurements on silver solid electrolytes have been reported in our earlier work (3). The sample pellet is stacked between two silver electrodes. A copper constantan thermocouple is attached to the sample to measure the temperature. The whole assembly is kept in a metallic chamber whose temperature could be raised to a desired value with the help of an external heater. The internal pressure within the chamber is reduced to 10^{-2} torr. Circular pellets having a thickness of 3mm and a diameter of 10mm have been used during the study and the measurements made over a temperature range 300-500K using a 1 KHz Impedance Bridge (model GR 1650B).

From the measured values of the resistance (R) the total conductivity has been evaluated at various temperatures using the relation:

$$\sigma_{total} = \frac{t}{RA} \qquad (1)$$

where t is the thickness of the specimen pellet and A is the cross-sectional area of the pellet.

2.3 **Electronic Conductivity Measurements**: In the present case, a Wagner's d.c. polarisation cell (4) has been constructed with $Ag_4I_2WO_4$ as the specimen with silver powder and graphite powder as reversible and blocking electrodes respectively. A Keithley (model 160C) Electrometer has been employed in the measurement of electronic current, I.

Generally the factor I is given by

$$I = \frac{RTA}{LF} \sigma_e [1 - \exp(-\frac{EF}{RT})] \qquad (2)$$

where R is gas constant; T, the absolute temperature; A, the cross-sectional area; L, the length of the specimen; F, Faraday constant and E is the electric potential applied across the specimen.

When E is very high as in the present case (E = 0.4V) EF ≫ RT and thus

$$\sigma_e = \frac{LF \; I}{RTA} \qquad\qquad (3)$$

From the values of L, F, I, R, T and A, σ_e values have been estimated for various temperatures.

3. Results and Discussion:

3.1 Characterization of the Compound:

Table I shows the comparison between the present XRD results and reported data.

Table I: X-Ray Results of $Ag_4I_2WO_4$

Present Study		Reported Data	
d(Å)	I(%)	d(Å)	I
3.77	35	4.31	W
3.68	58	3.79	S
3.43	38	3.48	W
3.08	42	3.02	M
2.94	88	2.94	M
2.88	85	2.87	MW
2.81	81	2.71	M
2.69	100	2.31	M
2.65	88	2.25	VW
2.52	30	1.97	MW
2.47	30	1.92	W
2.37	30		
2.26	42		
2.22	38		
1.95	19		
1.88	19		

W — Weak
S — Strong
M — Medium

As can be seen from the table most of the values of d-spacing and diffracted intensities are in good agreement with those reported earlier (1). Thus the material prepared in the present investigation has been identified as the same stoichiometric compound $Ag_4I_2WO_4$.

3.2 *Electrical Conductivity data and Analysis*: Electrical conductivity (σ) of solid electrolytes is expressed in the Arrhenius form as

$$\sigma = \frac{\sigma_o}{T} \exp(-\Delta E/kT) \quad\quad\quad (4)$$

where ΔE is the activation energy of ion motion, k the Boltzmann constant and σ_o, the proportionality constant.

The temperature dependence of σ of various pellets of pelletizing pressure (P_p) values 3,4,5,6 and 7 ton/cm^2 respectively is presented, in figure 1.

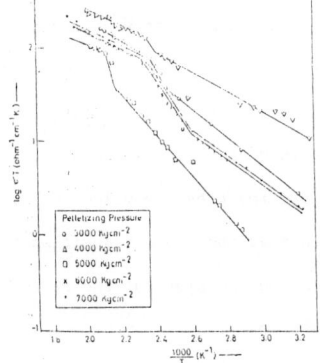

Fig 1: Temperature dependence of conductivity for various pelletizing pressures.

σ has been found to show an increase with temperature in accordance with the relation (4) and such an Arrhenius behaviour suggests that $Ag_4I_2WO_4$ is also an ionic solid. From figure 1 it is seen that the maximum room temperature conductivity of $4.08 \times 10^{-2} \Omega^{-1} cm^{-1}$, comparable with that reported (1) is exhibited by the pellet prepared at a pressure of 4 ton/cm^2, while the variation of σ with T is similar in all the cases. The reason for the observed maximum conductivity for the pellet prepared at a P_p of 4 ton/cm^2 can be given as follows:

In the case of pellets, the package of fine powder particles is highly dependent on the pressure at which they are pressed together. Upto a certain optimum value of Pp the package density and hence the pellet density also would show an increasing behaviour. Beyond this optimum value, the denisty of the pellet will remain unaltered irrespective of the applied pressure, excepting cracks at high pressures, and hence at the optimum pressure value the pellet density becomes equal to the bulk density. In the case of $Ag_4I_2WO_4$ solid the optimum pressure has been found to be $4ton/cm^2$. As a result the measured σ value for the above pellet sample is comparable with reported data whereas in the other cases there have been considerable differences observed in σ values compared to the reported one.

Another interesting observation in the case of σ plots is that two well-defined temperature regions are noticed in each curve, corresponding to the low and high temperature phases of the solid $Ag_4I_2WO_4$. Though the phase transition temperatures values are found to be 426, 432, 467, 435 and 431 K respectively for pelletizing pressure values of 3,4,5,6 and 7 ton/cm^2 it can be safely concluded that the phase transition temperature of $Ag_4I_2WO_4$ is around 432K as revealed by the plot for the Pp value of $4ton/cm^2$.

From these σ plots values of activation energy (ΔE) for Ag^+ ion migration in the solid corresponding to both low and high temperature phases are estimated and are presented in Table II.

Table II: Values of Activation Energy in $Ag_4I_2WO_4$ Solid:

Pelletizing Pressure 'Pp' (Ton/cm^2)	Activation Energy ' ΔE' (eV)			
	Present Study		Reported Data	
	Low temp. region	High temp. region	Low temp. region	High temp. region
3	0.14	0.1	Not available	Not avail.
4	0.08	0.08	0.06	0.07
5	0.19	0.07	Not avail.	Not avail.
6	0.12	0.09	"	"
7	0.12	0.09	"	"

Electronic Conductivity Data and Analysis: Using equation (3) the room temperature σ_e has been estimated as $3.97 \times 10^{-11}\ \Omega^{-1} cm^{-1}$ and it is obviously very negligible compared to the measured σ_{tot} of $4.08 \times 10^{-2}\ \Omega^{-1} cm^{-1}$ for the pellet of $4 ton/cm^2$. Such a small value of σ_e predicts the material $Ag_4I_2WO_4$ as a perfect solid electrolyte suitable for applications in solid state energy devices.

3.4 <u>Summary</u>: A fast silver ion solid electrolyte $Ag_4I_2WO_4$ has been investigated. This material has been found to exhibit a conductivity of $4.08 \times 10^{-2}\ \Omega^{-1} cm^{-1}$ and a negligible electronic conductivity of $4 \times 10^{-11}\ \Omega^{-1} cm^{-1}$ at RT. A phase transition has also been observed around 432K.

<u>References:</u>

1. T. Takahashi, S. Ikeda and O. Yamamoto,
   J. Electrochem. Soc. <u>120</u>, 647 (1973).

2. S.A. Suthanthiraraj,
   Proc. Region. Workshop. Mater. Solid State Batteries, Singapore; Edited by BVR Chowdari and S. Radhakrishna, World Sci. Publ. Co., Singapore p.423 (1986).

3. S.A. Suthanthiraraj and S. Radhakrishna,
   J. Instrum. Soc. India <u>15</u>, 227 (1985).

4. J.B. Wagner and C. Wagner,
   J. Chem. Phys. <u>26</u>, 1597 (1957).

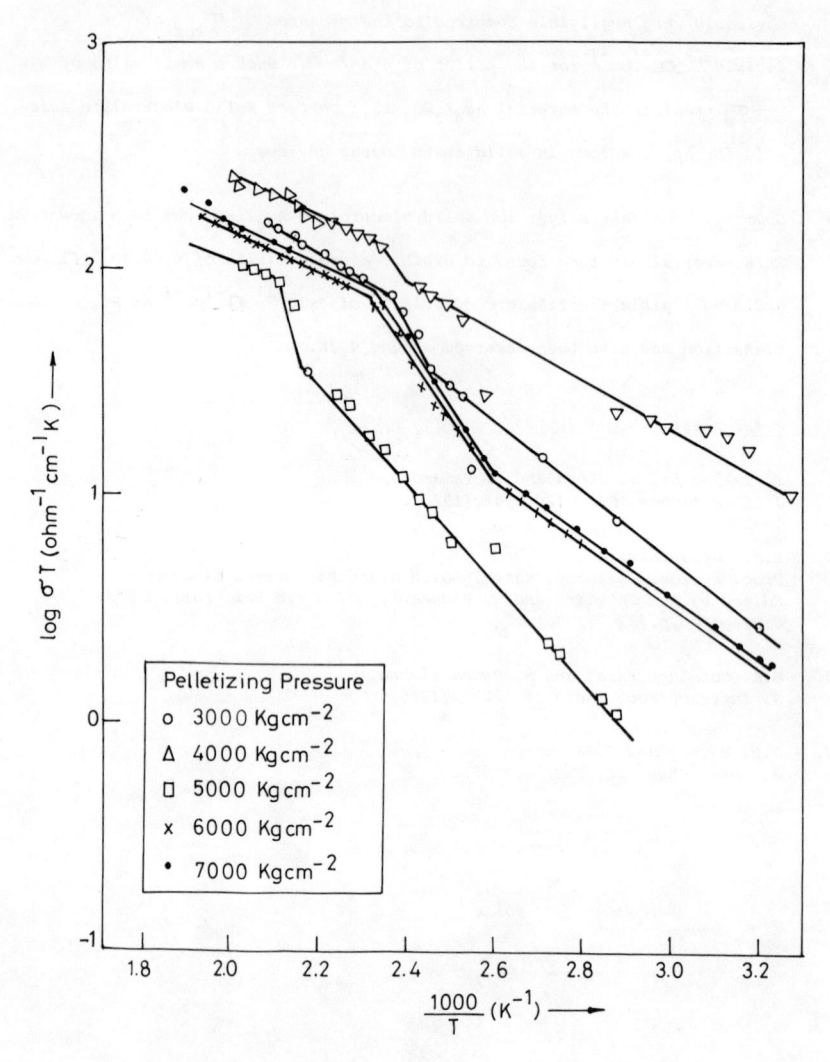

Fig.1 Temperature dependence of conductivity for various pelletizing pressure

SOLID STATE ELECTROLYTES AND INTERCALATION COMPOUNDS OF HIGH-VALENCE IONS[±]

YU Wen-hai, WANG Da-zhi and ZHU Bin[*]

University of Science and Technology of China, Hefei, Anhui, China

ABSTRACT:

It is a short review of the study on solid state electrolytes and intercalation compounds of high-valence ions. It has been found that some high-valence cations such as Mg^{2+}, Zn^{2+}, Fe^{3+} etc can be exchanged into some minerals such as montmorillonite and mordenite and changed into conducting cations, so that these denatural minerals are changed into electrolytes. On the other hand, these high-valence ions can be also inserted into some interlayer compounds as the cathodes of batteries and form several intercalation compounds.

---

[*] University of Science and Technology of Staff and Workers of Hefei, Hefei, Anhui, China.
[±] Projects Supported by the Science Fund of the Chinese Academy of Sciences.

With development of the science of energy sources, the studies of all-solid state batteries are attracting one's attention. To prepare fine all-solid state batteries, it is necessary to find fine solid state electrolytes and cathode materials. It is known that the fast ionic conductors can be used as the electrolytes, but the reports published about them were practically concentrated on that of single-valence cations such as Li^+, Na^+, Ag^+, Cu^+ etc.

In the study on the thin film batteries, we found that[1] some high-valence cations are possible to change into the conducting cations of the electrolytes, for example, in Mg/Cu cell the mixed phase MgI_2-$MgCl_2$ is just the electrolyte of bi-valence cation Mg^{2+}. This battery was possessed with an open circuit voltage of 1.30 V, a capacity of more than 1.6 mWh and an energy density of more than 0.12 Wh/cm^3, then its property was better than that of Ag/Te cell reported[2].

For finding out other solid state electrolytes with high-valence cations, we carried on chemical denaturation of the natural montmorillonite and mordenite[3,4]. The ionic conductivities of some denatural materials are listed in Table 1.

Table 1 Ionic conductivities of some denatural minerals

material	montmorillonite			mordenite
	Mg-	Zn-	Fe-	Mg-
ionic conductivity $10^{-4} (\Omega cm)^{-1}$	1.04	2.0	0.16	0.75

Various primary and rechargeable cells have been assembled by the denatural montmorillonite and mordenite, and the results are shown in Table 2. We found that the primary cell Mg/Mg-montmorillonite/I_2 was possessed with an open circuit voltage of 1.86~1.92 V and an average discharge voltage of 1.2 V corresponding to a discharge current of 170 $\mu A/cm^2$, as well as an energy density of 130 Wh/kg and a capacity of 30 mAh for a cell weighted 0.25 g. We found also that the rechargeable cell Mg(or Zn)/Mg-(or Zn-)montmorillonite/V_2O_5 displayed a good electrical performance whose charge-discharge cycles achieved 15 times or more.

Table 2 Properties of some all-solid state batteries

anode	cell electrolyte	cathode	open circuit voltage (V)	energy density (Wh/kg)	type
Mg	Mg-mont.	CuCl	1.4~1.7	10.2	p.
Mg	Mg-mor.	CuCl	2.0~2.15	61	p.
Mg	Mg-mor.	MnO_2	1.8~2.2	96	p.
Mg	Mg-mont.	I_2	1.86~1.92	130	p.
Mg	Mg-mor.	V_2O_5	2.0	20	r.
Mg	Mg-mont.	V_2O_5	2.0	7	r.
Zn	Zn-mont.	V_2O_5	1.5~1.37	11.4	r.

note: p.-primary; r.-rechargeable.

In the study of rechargeable solid state batteries, it has been found that some high-valence ions may be inserted into the layered compound e. g. V_2O_5 etc and form intercalation compounds. This cathode reaction i.e. insertion reaction or topochemical reaction resulted in an excellent charge-discharge performance of batteries.

To investigate the insertion reaction of high-valence cation into cathode material, the cell was constructed as following:

M / M-montmorillonite / cathode

where M was an anode metal such as Mg, Zn, Fe etc, cathode was a layered or spinel structure material such as V_2O_5, V_6O_{13}, TiS_2, MoS_2, Fe_3O_4 etc. In this cell the high-valence cations emerged from the anode were be migrated through the M-montmorillonite as the solid state electrolyte and inserted into the cathode material under the action of the cell voltage itself or the external electrical field applied.

The intercalation of Mg^{2+} and Zn^{2+} into V_2O_5 cathode has been firstly confirmed by SER[5], and it was clearly shown that the high-valence ions had been inserted into V_2O_5. Then the intercalation of Mg^{2+} into V_2O_5 has been investigated by XRD furtherly[6]. Three inserted phases called as phase I, II and III were observed and the XRD data of the phases were very similar to that of the intercalation of Li^+ into V_2O_5. It was meant that the insertion positions both of Mg and Li atoms in V_2O_5 were the same. Moreover, the intensities of XRD peaks about the inserted phases and an amorphous background of Mg in V_2O_5 were decreased as time except the phase II. It was shown that phase II was the most stable.

The products of Mg in V_2O_5 have been investigated furtherly by ESR. It was found that there were not hyperfine structure arised from the lower valence vanadium such as V^{4+}, V^{3+} etc in the spectrum even at 110 K temperature. Then it was meant that the mechanism of the intercalation of Mg^{2+} into V_2O_5 could be not similar to that of Li^+ into V_2O_5 accompanied by the reduce of V^{5+} to V^{4+}. Therefore, it was possible that some positive centers which could catch the electrons might be formed to compensate the local charge inbalance in the process of Mg^{2+} into V_2O_5. Namely, it was perhaps similar to a physical doping process.

The intercalations of Mg^{2+}, Zn^{2+} and Fe^{3+} in V_6O_{13}, TiS_2, MoS_2, Fe_3O_4 etc have been investigated initially by XRD. It has been observed that there wasn't any new phases but some dispersion and broaden of XRD peaks caused by the lattice distortion.

From above results it is shown that the high-valence cations can be inserted into many cathode materials, which is very interesting evidently in studing solid state batteries and insertion reaction.

REFERENCES:
1. YU Wen-hai et al, in "Materials for Solid State Batteries" ed. by B. V. R. Chowdari and S. Radhakrishna, World Scientific Publishing, Singapore, 1986, p. 481.
2. Arora M. R. et al, Thin Solid Film, 71 (1980), 103.
3. WANG Da-zhi et al, idem ref. 1, p. 461.
4. Yuan Wang-zhi et al, "Phase Transformation of Mordenite", to be reported in this conference.
5. YU Wen-hai et al, Solid State Commu. 61 (1986), 271.
6. YU Wen-hai et al, "Intercalation of Mg in V_2O_5", to be reported in this conference.

STUDY OF PHASE TRANSITION OF MORDENITE BY IR AND XRD

YUAN Wang-zhi[*], WANG Da-zhi, ZHU Bin[**] and YU Wen-hai
University of Science and Technology of China, Hefei, Anhui, People's Republic of China

ABSTRACT:

The phase transition of thermal treated mordenite has been studied by XRD, IR, DTA and water absorptivity. The characteristics of the phase transition of crystalline to non-crystalline has been discussed.

---

[*]Guizhou Engineering Institute, Guiyang, Guizhou, China.
[**]University of Science and Technology of Staff and Workers of Hefei, Hefei, Anhui, China

1. INTRODUCTION

The natural mordenite with many good properties, such as ion exchange, adsorption, catalysis and etc, has been applied in many different industries[1]. These properties can be changed greatly during thermal treatment. It is due to the changes of structure, So many works have been done on the changes of structure and properties of zeolite.[2-4] This paper will report our result about the phase transition of crystalline mordenite to non-crystalline.

2. EXPERIMENT

The mordenite used in this work was from Anhui, China. Fig.1 shows the X-ray powder crystal diffraction(XRD) pattern of the raw material. The principal phase is mordenite and the accompanying minerals are quartz and labradorite. The chemical composition of the raw material is given in Table1.

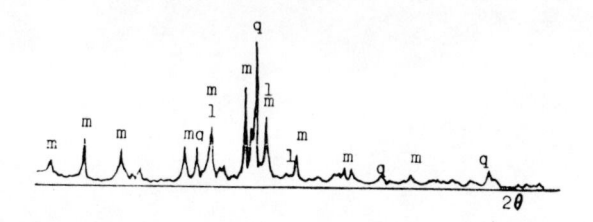

Fig. 1. Pattern of X-ray powder crystal diffraction of raw material
m— mordenite, q - quartz, l -labratorite

composition	SiO_2	Al_2O_3	CaO	Na_2O	K_2O	Fe_2O_3	MgO	TiO_2	others
content w%	77.8	11.3	1.92	2.60	2.31	1.73	0.30	0.13	1.91

Table 1 Chemical composition of raw materials

The specimens were treated in a box-furnace at various indicated temperature for two hours. X-ray diffraction traces of these specimens obtained after cooling to room temperature in air, using a Rigaku D/Max-rA diffractometer. IR spectra were recorded in a Nicolet 170 SX spectrophotometer on samples dispersed in KBr pellets.

Simultaneous DTA and TG curves were obtained in air using thermoflex apparatus.

The water loss was measured by comparison between the weight of the specimens before and after thermal treatment. The water absorptivity of specimens treated at different temperature was measured by comparison between the weights of specimens before and after absorbing water in 100 percent relative humidity at $25°c$.

3. RESULTS AND DISCUSS

X-ray diffraction traces of the specimens obtained after heating at various temperatures are shown in Fig.2. The characteristic reflections of quartz can be an internal standard for intensity of reflections of mordenite. The intensities of the reflections of mordenite were no changed when the treatment temperature was lower than $700°c$, weaked at $800°c$ and $850°c$, disappeared at $900°c$. It is shown that a phase transition of crystalline to non-crystalline occurred above $800°c$.

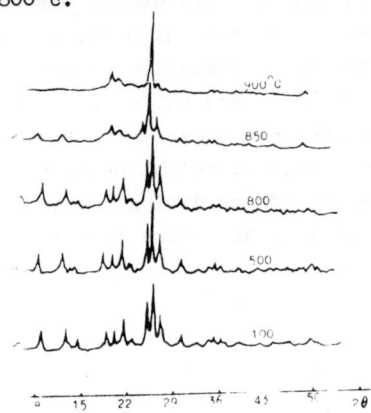

Fig.2 X-ray diffraction traces of mordenite after heating at indicated temperatures

Fig.3 shows the water loss and water absorption. It is shown that the water loss of mordenite and water absorption is reversible when the temperature of treatment was below $400°c$. But the water loss was larger than the water absorption during the $500°c - 800°c$ thermal treatment. This means that some changes of the structure of mordenite happened. The water absorption tended to zero when the treatment temperature was above 800°c. It is corresponding to the phase of non-crystalline.

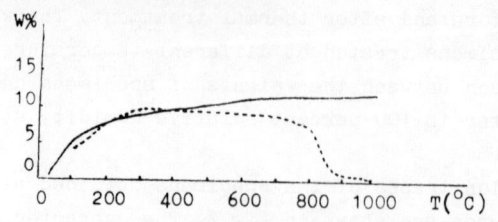

Fig. 3 The weight loss and water absorption of mordenite specimens treated in different temperature
———————— weight loss curve
- - - - - - - - water absorption curve

DTA and TG curves of the mordenite are shown in Fig.4. The DTA curve shows three endothermic effects, the first of which (20-400°c) corresponds to the loss of zeolite water and the other two, with peak teperature of 583°c and 690°c, correspond to the quartz and labradorite. The TG curve shows a large water loss between 20 and 400°c and constance of weight above 600°c. It is clear that there is not thermal effect during the crystalline to non-crystalline phase transition, and there is not fixed temperature for this phase transition.

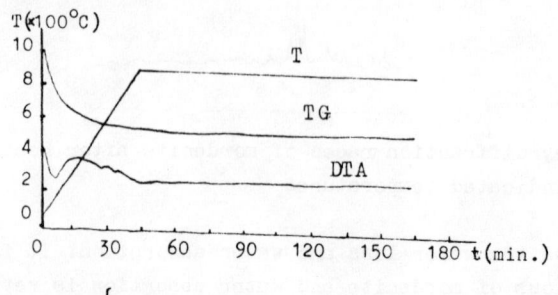

Fig. 4 DTA and TG curves of mordenite. Heating rate 10°c/min

The Fig.5 shows the IR spectra of mordenite treated at different temperature. The observed frequencies are summarized in Tab.2

Tem.	wave number cm^{-1}											
25°C	3594$_s$	3453$_s$	1641$_m$	1219$_w$	1039$_s$	797$_w$	778$_{sh}$		695$_w$	620$_w$	525$_{mw}$	471$_{ms}$
100	3597$_s$	3438$_s$	1641$_m$	1219$_w$	1039s	796$_w$	778$_{sh}$		693$_w$	619$_w$	520$_{mw}$	466$_{ms}$
200	3600$_s$	3453$_s$	1642$_m$	1220$_w$	1039$_s$	797$_w$	779$_{sh}$		694$_w$	620$_w$	527$_{mw}$	468$_{ms}$
300	3600$_s$	3438$_s$	1641$_m$	1220$_w$	1039$_s$	797$_w$	779$_{sh}$		695$_w$	620$_w$	524$_{mw}$	468$_{ms}$
400	3603$_s$	3438$_s$	1641$_m$	1219$_w$	1042$_s$	797$_w$	779$_{sh}$		694$_w$	620$_w$	521$_{mw}$	409$_{ms}$
500	3606$_s$	3453$_s$	1642$_m$	1220$_w$	1040$_s$	796$_w$	778$_{sh}$	721$_w$	694$_w$	620$_w$	521$_{sh}$	466$_{ms}$
600	3606$_s$	3453$_s$	1642$_m$	1220$_w$	1046$_s$	796$_w$	779$_{sh}$	721$_w$	695$_w$	620$_w$	543$_{ws}$	469$_{ms}$
700	3606$_s$	3453$_s$	1642$_m$	1220$_w$	1046$_s$	795$_w$	779$_{sh}$	721$_w$	694$_w$	621$_w$	543ws	469$_{ms}$
800	3616$_{ms}$	3453$_{ms}$	1641$_m$	1220$_w$	1050$_s$	796$_w$	779$_{sh}$	721$_w$	694$_w$	623$_w$	555$_{mw}$	468$_{ms}$
850	3624$_w$	3453$_{vw}$	1641$_w$		1088$_s$	796$_w$	780$_{sh}$	718$_{sh}$	694$_w$	623$_w$	560$_w$	469$_{ms}$
900		3530$_w$	1641$_w$		1088$_s$	796$_w$	778$_{sh}$		696$_w$	640$_w$	579$_w$	466$_{ms}$

Tab. 2 Infrared absorption bands and intensities of
mordenite treated in different temperature
intensity: s - strong, ms -medium strong, m -medium,
mw - medium weak, w- weak vw - very week,
sh - shoulder.

In Fig.5 above 800°c the strongest band at 1040 cm^{-1} shows an obvious shift to high frequency and the band at 1216 cm^{-1} assigned to assymmetic stretching made of external tetrahedron vibration disappeared. It is corresponding to the phase transition of crystalline to non-crystalline

The results of XRD, IR gave evidences of phase transition of crystalline mordenite to non-crystalline. The absorptivity of crystalline mordenite affected by this phase transition greatly. But there is no effect in DTA and TG for this phase transition. It shows that this phase transition has not thermal effect and chemical composition change.

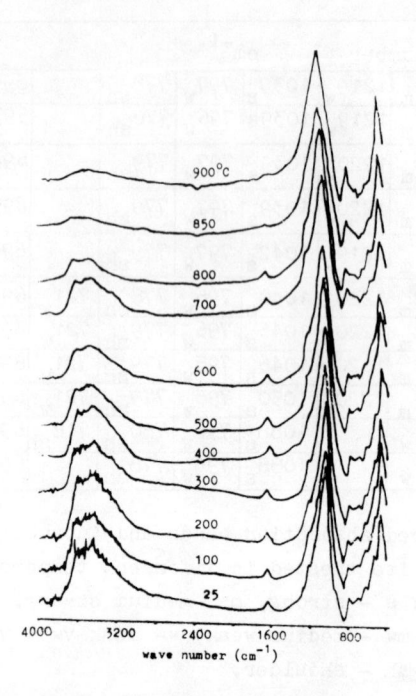

Fig.5 Infrared absorption spectra of mordenite treated at different temperature

References:
1. Xu Bangliang, ZEOLITE, Geology Publishing house, Beijing (1979), 2.
2. Han Cheng, and Zhang Quanchang, Science Bulletin(in China), 5 (1983) 288
3. J.W. Ward, J. Phys. Chem., 72, (1968) 4211
4. J. W. Ward, Molecular Sieve Zeolites I , edit. by R. F. Gould, American Chemical Society, Washington D.C., (1971) 381.

Composite Pulse Magic Echo Sequence for the Excitation of Three-level systems in Solid-State NMR

Wang Dongsheng, Li Gengying and Wu Xuewen
Department of Physics, East China Normal University
Shanghai, P.R.China

Recently, NMR spectroscopy has become a powerful technique for probing charaterization of polymeric materials and blends. Unfortunately, the traditional excitation methods in three-level systems become uneffective. The composite pulse quadrupole echo sequence (CPQES) (1) designed for the excitation of spin I=1 systems greatly reduce finite pulse width effects (2), but they also introduce new distortions namely the phase distortions into the powder lineshape (3). To overcome this difficulties, one approach is to find the higher-order composite pulse, which is obviously at the expense of increasing the excitation period, and intermediate exchange processes occurring during this period will introduce an additional complication in interpretation of lineshape (4). Another approach presented in this paper, is to use the composite pulse to construct the magic echo sequence (5).

In this paper we present a composite pulse $45_x 90_{-x} 135_x$ (indicated by $R_x(90)$) to demonstrate that the effectiveness of uniform excitation of the composite pulse magic echo sequence (CPMES) is much more superior to CPQES. By using a simplified model, we show that the phase error caused by $R_x(90)$ may be corrected by applying it to the magic echo sequence. The distortions due to the finite pulse width effects can be alleviated to any desired degree by using a CPMES.

1. Simplified Model

Siminovith et al (4) have demonstrated that the origin of the phase distortions lies in the asynchronous refocusing of magnetization due to individual spin packets. Hence, we think that the composite pulse $R_x(90)$ may be replaced by a $\delta_x(90)$ pulse appeared at time t_d, where $t_d=t_d(\omega_Q)$ (here $2\omega_Q$ is the quadrupole splitting), as shown in Fig.1.

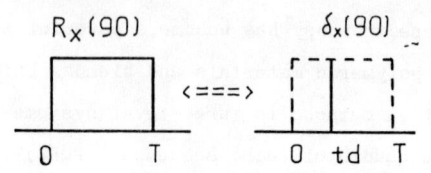

Fig.1 $R_x(90)$ composite pulse may be replaced by a 90_x pulse appeared at time t_d, where $t_d = t_d\,(\omega_Q)$, and T is the total duration of $R_x(90)$.

This remains true because the amplitude error of a composite pulse is quiet small (1) and is negectable in the range of $|\omega_Q/(2\omega_1)| \leqslant 1.0$ (here ω_1 is the rf field strength).

Fig.2 shows the CPMES employed in this paper, θ_x is a single pulse And Fig.3 shows how the Hamiltonian of spin system in toggling frame evolve when replacing $R_x(90)$ by the simplified model.

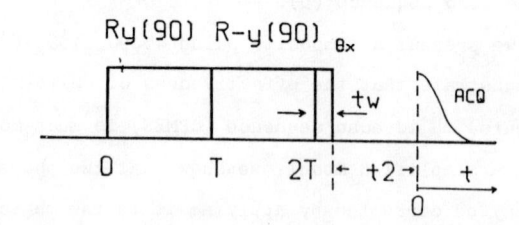

Fig.2 The CPMES discussed in the present paper, t_w is the duration of θ_x pulse.

Hence, the propagator of spin system at time $0 < t \leq 2T$ may be written as
$$U(2T) = D\exp(-i\int_0^{2T} H(t)\,dt)$$
$$= \exp(-i(-\tfrac{1}{2}Y)2T) \qquad [1]$$
where $P = \omega_q(I_p^2 - I(I+1)/3)$, and D is the dyson time-ordering operator

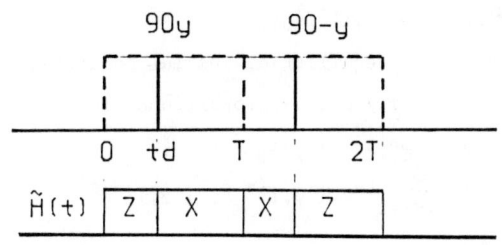

Fig.3 Schematic representation of Hamiltonian of system in toggling frame after replaced $R_y(90)$ composite pulse by a $\delta_y(90)$ pulse.

It is clear that the effect of $t_d(\omega_Q)$ so the phase error has been eliminated.

II. Theory

The analysis of the present paper makes use of the set of fictitious spin-½ operators first given by Vega and Pines (7). The density matrix $\rho(t)$ for an $I=1$ system may be expended in terms of the unit operator 1_{op} and other symmetric operators $I_{p,i}$, where $p = x,y,z$, and $i = 1,2,3,4$, given by:

$$1_{p,1} = I_p/2 \qquad [2]$$
$$1_{p,2} = (I_q I_r + I_r I_q)/2 \qquad [3]$$
$$1_{p,3} = (I_r^2 - I_q^2)/2 \qquad [4]$$
$$1_{p,4} = 1_{q,3} - 1_{r,3} \qquad [5]$$

where $(p,q,r) = (x,y,z)$ or cyclic permatations. Using these operators and taking the adventages of the simplified model the rotating frame quadrupole Hamiltonian at each time t may be written as

$$H(t) = \begin{cases} -\tfrac{1}{2}Y = -\omega_Q 1_{y,4}/3 & (0 < t \leq 2T) \\ -1_x + Z = -2\omega_1 1_{x,1} + 2\omega_Q 1_{z,4}/3 & (2T < t \leq 2T+t_w) \\ Z = 2\omega_Q 1_{z,4}/3 & (t > 2T+t_w) \end{cases} \quad [4]$$

The density matrix written in terms of these operator is

$$\rho(t) = a_0 \mathbb{1}_{op} + \sum_{pi} a_{p,i}(t) 1_{p,i} \quad [5]$$

Assuming the usual equilibrium initial condition for the spin system describable by a density matrix proportional to 1_z, we can evaluate the density matrix at time $t = 2T+t_w+t_2$

$$\rho(2T+t_w+t_2) = \sum_{i=1,2} (a_{y,i} 1_{y,i} + a_{z,i} 1_{z,i}) \quad [6]$$

where

$a_{y,1} = \sin\beta \sin(\omega_e t_w/2) \cos(\omega_Q(t_2-T+t_w/2))$, $\omega_e = (4\omega_1^2 + \omega_Q^2)^{1/2}$

$a_{y,2} = \sin\beta \sin(\omega_e t_w/2) \sin(\omega_Q(t_2-T+t_w/2))$, $\tan\beta = 2\omega_1/\omega_Q$

The echo maximum appeared at $t_2 = T - t_w/2$, and $a_{y,1} = a_{y,1}(0) = \sin\beta \sin(\omega_e t_w/2)$ $= D(\omega_Q)$, $a_{y,2} = a_{y,2}(0) = 0$, here we define:

$D(\omega_Q) = \sin(\theta K)/K$, where $K = (1+(\omega_Q/(2\omega_1))^2)^{1/2}$ and $\theta = \omega_1 t_w$.

$D(\omega_Q)$ is so called the distortion factor of θ pulse.

The observed signal when sampling at time $t > 2T+t_w+t_2$ is

$$\begin{aligned}\langle 1_{y,1}\rangle &= \mathrm{Tr}(1_{y,1}\rho(t)) \\ &= a_{y,1}(0)\cos(\omega_Q t) + a_{y,2}(0)\sin(\omega_Q t) \\ &= D(\omega_Q)\cos(\omega_Q t)\end{aligned} \quad [7]$$

if we use two θ single pulse to construct the quadrupole echo sequence the total distortion factor would be aggravated to $(D(\omega_Q))^3$. These distortions may be reduced by shortening the pulse length (2), however, it leads to a considerably lower signal-to noise ratio. In CPMES, the distortion factor is only $D(\omega_Q)$, so we can use a smaller flip angles to obtain more uniform excitation without considerably loss of sensitivities. For example, if we choose $\theta = 45°$, the signal loss is $2^{-1/2}$, we can double the sampling time for accumulation and the distortion due to θ_x pulse finite width effects may be reduced significantly.

III. Computer Simulations

The theoretical conclusions are verified by the computer simulations These were produced by an exact dynamical calculation of the evolution involving a numerical diagonalization of the Hamiltonian during the pulses. The calculating expectation value of a operator

$$-\langle I_{p,i} \rangle = -\text{Tr}(\rho I_{p,i})/\text{Tr}(I_{p,i}^2) \qquad [8]$$

at time t is shown as a function of parameter $\omega_Q/(2\omega_1)$.

The comparision of the performance of CPQES and CPMES are shown in Fig.4. Where Fig.4(A) shows the operators expectation value $-\langle I_{y,1}\rangle$ and $-\langle I_{y,2}\rangle$ at the maximum of the quadrupole echo produced by $R_x(90)$ composite pulse. Clearly, $-\langle I_{y,2}\rangle$ is strong dependent on $\omega_Q/(2\omega_1)$ showing large phase distortion in this pulse sequence (phase factor $\varphi = \tan^{-1}(\langle I_{y,2}(0)\rangle/\langle I_{y,1}(0)\rangle)$, see in Eq.[7]). However, this phase distortions may be eliminated in the magic echo sequence constructed by $R_x(90)$, not only $-\langle I_{y,1}\rangle$ but also $-\langle I_{y,2}\rangle$ is independent on $\omega_Q/(2\omega_1)$ in the range of $|\omega_Q/(2\omega_1)| \leq 1.2$, as shown in Fig.4(B).

Hence, it is confirmed by the computer simulations that the CPMES has the ability to excited the spin systems uniformly, thus the correct lineshape can be obtained.

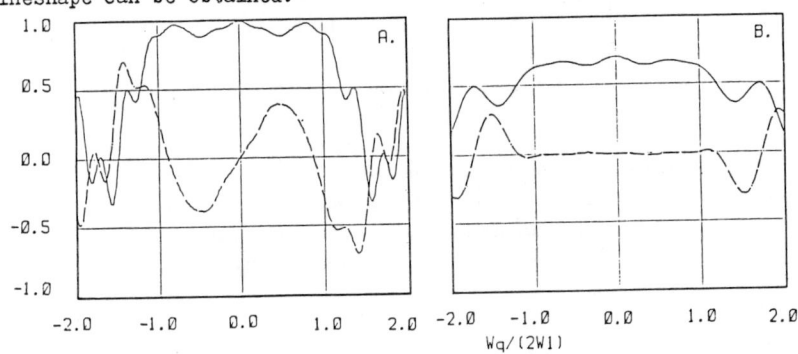

Fig.4 Computer simulations of the performance for A). quadrupole echo and B). magic echo sequence. The operator expectation value of $-\langle I_{y,1}\rangle$ and $-\langle I_{y,2}\rangle$ at the maximum of echo as a function of parameter $\omega_Q/(2\omega_1)$. Where in B). θ=45. The solid line for $-\langle I_{y,1}\rangle$ and the dash line for $-\langle I_{y,2}\rangle$.

IV. Experimental Verifications and Conclusions

Some experimental results are presented in Fig.5, these were quadrupole echo and magic echo ^{2}D-NMR spectra of polycristalline deuterated polymethylmethacrylate (PMMA) taken at 46.071 MHz on Brufer MSL-300 spectrometer. The spectrum is a superposition of two main powder patterns (1), one originating from the freely rotating methyl deuterium spins and one from the more rigid methine deuteriums. The former gives a strong pattern in the center of the spectrum, and the latter gives a broader pattern. The spectra were taken with a rf field strength of 31.25 KHz, corresponding 90 pulses of duration 8.0 us.

Spectrum 5(A) was taken with CPQES. The slopping wings of the methyl powder pattern are obvious and indicate the phase distortion effects due to the quadrupolar interaction acting during the rf pulses. This effects may be corrected by employing the CPMES as shown in Fig.5(B). Now the signals from the broader component become clear visible.

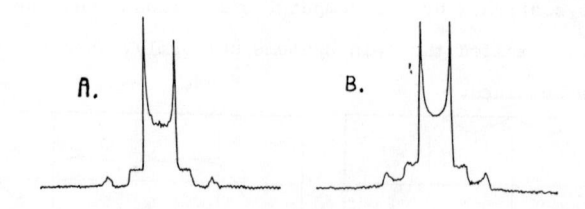

Fig.5 Experimental ^{2}D-NMR spectra of PMMA. A). spectrum resulted from CPQES. B). spectrum obtained with CPMES. The rf field strength $\omega_1/(2\pi)$ is 31.25 KHz. The spectrum in A). was averages of 200 scans, and in B). was averages of 400 scans, and θ=45, recycle time is 60 S.

We have shown that the CPMES are useful in the manipulation of three-level systems. The performance of the CPMES exceeds the CPQES, but only at the expense of increasing the sampling time. Deuterium NMR is widely used as a sensitive probe of motional processes(6) and the CPMES presented in this paper for eliminating the distortion of quadrupolar lineshape without increasing peak rf power and without considerably attenuating the signal will be wellcome.

REFERENCES

1. M.H.Levitt, D.Suter, and R.R.Ernst, J.Chem.Phys., 80(7),3064(1984).

2. M.Bloom, J.H.Davis, and M.I.Valic, Can.J.Phys., 58,1510(1980).

3. T.M.Barbara, J.Magn.Reson., 67,491(1986).

4. D.J.Siminovitch, D.P.Releigh, E.T.Olejniczak, and R.G.Griffin, J.Chem.Phys., 84(5),2556(1986).

5. R.C.Bowman, JR, and W.K.Rhim, J.Magn.Reson., 49,93(1982).

6. H.W.Spiess, Colloid Polym.Sci., 26,193(1983).

7. S.Vega, and A.Pines, J.Chem.Phys., 66,5624(1977).

LIST OF PARTICIPANTS

Peter Angelini
Oak Ridge National Laboratory
Oak Ridge, Tn 37830
USA

Tom Buck
AT & T Bell Laboratories
Room 1E-448, 600 Mountain Avenue
Murray Hill, NJ 07974
USA

Les Butler
Department of Chemistry
Louisiana State University
Baton Rouge, Louisiana 70803
USA

Chen Binjiang
Changchun Inst. of Optics and
Fine Mech., Academia Sinica
P O Box 1024, Changchun
CHINA

Chen Boliang
Shanghai Inst. of Tech. Physics
Academia Sinica
420 Zhong Shan Bei Yi Road, Shanghai
CHINA

J Callaway
Department of Physics
Louisiana State University
Baton Rouge, LA 70803
USA

B V R Chowdari
Department of Physics
National University of Singapore
Lower Kent Ridge Road
SINGAPORE 0511

J P Collins
JAPCO-WECCO Company
P O Box 14862
Baton Rouge, La. 70898-4862
USA

W E Collins
Department of Physics
Southern University
Baton Rouge, La 70813-0554
USA

Soraya Dehkordie
Computer Science Department
Southern University
Baton Rouge, LA
USA

Bill Dixon
Louisiana State University
Baton Rouge, La. 70803
USA

Ira Graham
M.E. Department
Southern University
Baton Rouge, La. 70813
USA

A M Karguppikar
Department of Physics
Dharwar University
Dharwar
INDIA

Sitka Kubatora
Institute of Physics
Czechoslovak Academy of Sciences
180 40 Praha 8-Liben, Na Slovance 2
Prague, CZECHOSLOVKIA

Devendra Kumar
Chemistry Department
Loiusiana State University
Baton Rouge, La. 70803
USA

Vikram Kumar
Department of Physics
Idian Institute of Science
Bangalore-560 012,
INDIA

K.V. Govindan Kutty
Scientific Officer
Radio Chemistry Programme
IGCAR, Kalpakkam 603 102,
INDIA

Lim Yung Kuo
Department of Physics
National University of Singapore
Lower Kent Ridge Road,
SINGAPORE 0511

S.V.J. Lakshman
Sri Venkateswara University
Tirupati,
INDIA 517 502

Velpuri Lakshminarayana
Andhra University
Visakhapatnam A.P. 530003
INDIA

Claude N. Lamb
Dept. of Chemistry/Henes Hall
North Carolina A&T State University
Greensboro, North Carolina 27411
USA

K.H. Liu
Physics Department
Southern University
Baton Rouge, La, 70813,
USA

Tian-Huey Lu
Department of Physics
National Tsing Hua University
855 Kuang Fu Road, Hsinchu, Taiwan
REPUBLIC OF CHINA

Steve McGuire
Department of Physics
Alabama A & M University
Normal, Huntsville, Alabama
USA

Ronald Mickens
Department of Physics
Atlanta, Georgia
USA

J.W. Mitchell
Analytical Chemistry Res. Dept.
AT&T Bell Laboratories
600 Mountain Ave, Murray Hill
NJ 07974-2070, USA

Rama Mohanty
Physics Department
Sounthern university
Baton Rouge, La. 70813
USA

Claude Mount
Material Evaluations Lab
Baton Rouge, La. 70813
USA

C.K. Mathews
IGCAR
Kalpakkam
Tamil Nadu 603-1102
INDIA

T. Nagarajan
University of Madras
Madras,
INDIA

Chand Patel
Materials Evaluation Lab
Baton Rouge, La. 70810
USA

D.C. Parashar
National Physical Laboratory
New Delhi 110060
INDIA

P. Sathya Sainath Prasad
Department of Physics
Indian Institute of Technology
Madras-600 036
INDIA

S. Radhakrishna
Dept. of Physics, IIT
Madras 600 036
INDIA

B. Rambabu
Physics Department
Southern University
Baton Rouge, La. 70813
USA

Sisla M. Rao
B.A.R.C.
Bombay 400094
INDIA

S. Bangar Raju
Department of Physics
Andhra Pradesh, Waltair
INDIA

Narasimha Reddy Katta
Osmania University
Hyderabad, 552612
INDIA

I. Ruffin
Physics Department
Southern University
Baton Rouge, La. 70813
USA

M. Salagram
Dept. of Physics
Nizam College, Osmania University
Hyderabad
INDIA

R.C. Sastri
Jadavpur University
Cacutta,
INDIA

Albert J Schultz
Ionwerks
Houston, Tx.
USA

Seow Pin Kwong
Analytical Chemistry Division
Rubber Research Institute
Kuala Lumpur 01-02,
MALAYSIA

Joseph Stewart, Jr.
Physics Department
Southern university
Baton Rouge, La. 70813
USA

U.V. Subba Rao
Department of Physics
Osmania University
Hyderabad,
INDIA 500 007

S. Pandyan
D.A.V.V.M. Sripushpam College
Poondi-613503
INDIA

A. Subrahmanyam
Department of Physics
IIT
Madras,
INDIA 600 036

S. Austin Suthanthiraraj
Department of Energy
Univ. of Madras, Guindy Campus
Madras,
INDIA-600025

Sun Ze
North China Res. Institute
of Electro-Optics
P.O.Box 8511, Beijing,
CHINA

John Tabony
Scientific Testing Laboratories, Inc
Baton Rouge, La.
USA

Steve Thomas
Physics Department
Atlanta University
Atlanta, Georgia
USA

Wang Dongsheng
Department of Physics
East China Normal University
Shanghai,
CHINA

T.H. Wang
Physics Department
Southern University
Baton Rouge, La. 70813
USA

Z. Wang
Southern University
Baton Rouge, La.
USA

C.H. Yang
Physics Department
Southern University
Baton Rouge, La. 70813
USA

Yang Huan-Wen
North China Res. Institute
of Electro-Optics
P.O. Box 8511, Beijing,
CHINA

Yu Mingren
Surface Physics Lab.
Fudan University
Shanghai
CHINA

Yu Wen-hai
Dept. of Physics
Univ. of Sci. & Tech. of China
Hefei, Anhui
CHINA